# GLOBAL
# ENVIRONMENTAL
# ISSUES

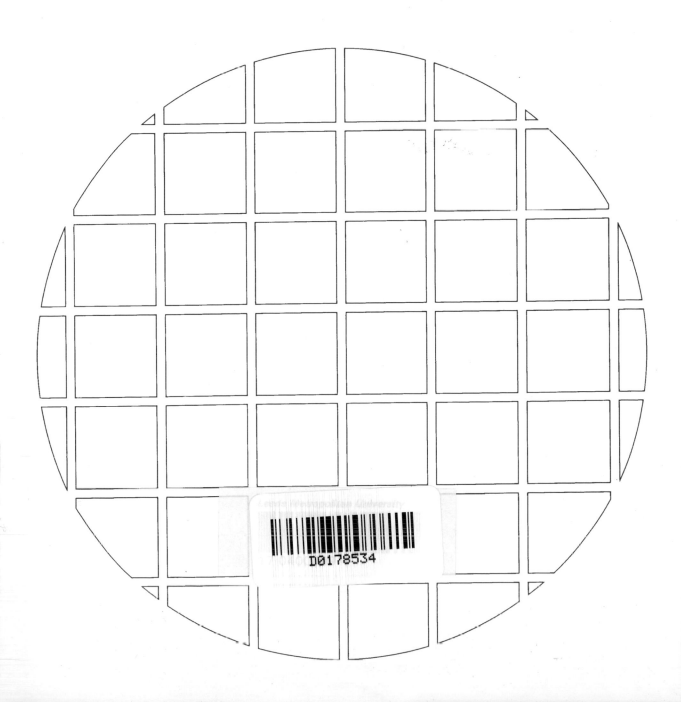

D0178534

This book is the fourth in a series published by Hodder and Stoughton in association with The Open University.

*Environment and Society*
edited by Philip Sarre and Alan Reddish

*Environment, Population and Development*
edited by Philip Sarre and John Blunden

*Energy, Resources and Environment*
edited by John Blunden and Alan Reddish

*Global Environmental Issues*
edited by Roger Blackmore and Alan Reddish

The final form of the text is the joint responsibility of chapter authors, book editors and course team commentators.

# GLOBAL ENVIRONMENTAL ISSUES

*EDITED BY*
*ROGER BLACKMORE AND ALAN REDDISH*
*FOR AN OPEN UNIVERSITY COURSE TEAM*

Hodder & Stoughton

A MEMBER OF THE HODDER HEADLINE GROUP

IN ASSOCIATION WITH

The Open
University

This book has been printed on paper made from pulp which was bleached without chlorine; other, environmentally-friendly, oxidation agents were used. The paper mill concerned emits very low levels of effluent, and uses raw material from sources renewable through reforestation.

*British Library Cataloguing in Publication Data*

A catalogue record for this title is available from The British Library.

ISBN 0–340–66356–1

First published in the United Kingdom 1991. Second edition 1996.

Edited and designed by The Open University.

Index compiled by Sue Robertson.

Typeset by Wearset, Boldon, Tyne & Wear.

Printed in the United Kingdom for Hodder & Stoughton Educational, a division of Hodder Headline Plc., 338 Euston Road, London NW1 3BH, by Butler & Tanner Ltd, Frome.

This text forms part of an Open University second level course, U206 *Environment*. If you would like a copy of *Studying with the Open University*, please write to the Central Enquiry Service, PO Box 200, The Open University, Walton Hall, Milton Keynes, MK7 6YZ.

# Contents

*Note: Chapter 2 is a revised version of the chapter authored by Kiki Warr which appeared in the first edition of this book.*

# Introduction

This is the fourth and last volume of a series taking a broad interdisciplinary view of the 'Environment', as it is perceived at the end of the twentieth century. Science, technology, social institutions and attitudes, local behaviour and global consequences, all interact in complex ways which the series seeks to illuminate. It aims to widen readers' *awareness* of environmental issues, to help in their *analysis* and to contribute to the *evaluation* of policies to influence them.

The first volume looked in a general way at the evolution of human society and its attitudes in relation to the physical conditions and life on Earth of which it is a part. The second and third volumes provided more detail on the complex structures which have supported an ever-increasing human population, first in settled agriculture, trade and the growth of cities, and then in modern industry, with its dependence on energy and minerals, and growing burdens of waste disposal and pollution. All of these developments have produced local environmental effects of varying severity throughout history but, with the growing scale of human society, increasingly 'globalised' by trade and communications, some environmental effects have been correspondingly widened to the Earth as a whole. The local institutions, from tribal to national, that were set up to monitor and control local change are not able to act on this global scale, and there has been a growing recognition in recent decades of the need to develop new institutions for international co-operation on such planetary questions.

The underlying issue is the way in which the natural world, both inanimate and living, is perceived by humans as providing 'resources' to satisfy our needs, as well as the scale of these demands. The 'tragedy of the commons', introduced in Book One and returned to in Chapter 6 of this book, describes the way in which free access to resources held in common by growing population inexorably leads to their exhaustion unless some social controls are exercised. Historically these have been provided on the local scale by the choice between moving on to new territory, private ownership or traditions and laws. But these are impossible, unacceptable or non-existent for major parts of the planetary system, like the oceans, the atmosphere or non-human life as a whole. Recent international negotiations have begun to address such questions, though it is a continuing process which has some way to go if it is to overcome deep-seated national rivalries and conflicts. The chapters of this book give accounts of a number of such separate issues, and then consider the way in which these can be seen as parts of a more general process of 'international regime' formation, and their implications for future lifestyles.

The first chapter, on the oceans, gives an introduction to modern views of their physical structure, properties and the life within them, the 'resources' these have been seen as providing for human use, and the growing international regulation of access to them. Negotiations culminated in the United Nations Convention on the Law of the Sea (UNCLOS) in 1982, but only ratified in amended form in 1994 after prolonged conflict over some of its provisions. This exhaustive agreement is an example of what is known as an 'umbrella' convention because it seeks to cover everything in agreed detail; as might be anticipated, such conventions are generally difficult to negotiate.

The next two chapters are concerned with damaging effects of human activities on the atmosphere. Chapter 2 deals with the urgent, but essentially circumscribed, problem of damage to the ozone layer in the stratosphere by the chlorofluorocarbons, manufactured chemicals not known in nature. Their use as refrigerants, spray-can propellants and so on grew from the 1940s. Concern about their possible effects began to be voiced in the 1970s, leading to the Vienna Convention to limit their use in 1985. In this year the dramatic discovery of the 'ozone hole' over the Antarctic sharpened demands for more drastic measures, which were progressively agreed in Montreal in 1987, London in 1990 and subsequently, requiring a complete abandonment of their production and use. This is an example of a more flexible 'framework' convention, which agrees general principles, to be defined in more detail later as national commitments and understanding are clarified, and is accordingly more readily negotiated.

A similar process is now under way on the much more intransigent subject of Chapter 3, the likelihood of more general climate change – global warming and sea-level rise – from the 'enhanced greenhouse effect' of human emissions of carbon dioxide, methane and other gases. These are much more deeply entrenched consequences of normal energy, agricultural and industrial practice, and the nature of the threat, and possible responses to it, continue to be argued. But a Framework Convention on Climate Change was agreed at Rio in 1992, recognising the need for *some* action; the details of what that should be are under review – in Berlin in 1995, Japan in 1997 – and will continue to be so. Both the ozone depletion and the climate change problems involve complex atmospheric science, only partially capable of making reliable predictions. So the interaction between scientific advice and political response, embodying the 'precautionary principle' of action before damage is certain but perhaps irrevocable, is a pronounced characteristic of these processes.

As, in a different way, is the subject of Chapter 4 – the preservation of biodiversity, which was also agreed in a Framework Convention at Rio in 1992. Here there is no doubt that human activities eliminate some other living species; the questions are more about the scale of this loss, the effect of it on the functioning of the global ecosystem – and the conflicts between aesthetic or moral outrage and economics 'rights' in the use of such 'resources', either directly or via biotechnology.

Chapter 5 discusses the broader, and less well-defined, topic of 'sustainable development', embracing all of the above and much more. Introduced in the Brundtland Report of 1987, this was intended as the main subject of the United Nations Conference on Environment and Development in 1992, but its far-reaching implications for the international economic order precluded firm commitments. The best that could be achieved was the lengthy and diffuse 'Agenda 21', and little change in the financial arrangements that were thought to be crucial. Nevertheless, a process was started there which is still going on, and developments since Rio in the international, national and local institutions created there are outlined.

In the last two chapters a broader view of these global developments is taken. Chapter 6 compares the process of negotiation involved in the formation of various international regimes, and draws some general conclusions about the procedures that are likely to succeed, and their developing influence on the way in which the world is governed.

As an epilogue to the book – and the series as a whole – Chapter 7 takes a more speculative look at the ways in which human society and its relationship to the environment might change over the rather longer term, over the next century – though with a clear recognition that the dramatic changes of the present century would have been hard to foresee in any detail in the 1890s. Thus, the series ends with a question to which the answer is by no means clear – but one which will be affected by all our local expectations and behaviour.

# Chapter 1    The oceans

## 1    Introduction

As you read this chapter, look out for answers to the following questions:

- How does the marine environment differ from the land environment?
- What natural resources are available in the oceans?
- What political and legislative initiatives on management of the ocean environment have been taken recently?
- How are, and how should, ocean resources be managed?

So far this series* has concentrated on the environment most familiar to us, the land environment. However, much more of the Earth is covered by ocean than by land – about 71% – so, in terms of its size at least, the ocean is an important environment on Earth. Up to about a hundred years ago we knew little about the ocean. Coastal-dwellers knew about the tides, and sailors something about currents, as in the days of fairly slow sailing-ships it took much longer to sail against a current. Those who fished also had some knowledge of what fish could be caught and where. But knowledge was generally limited to the surface layers of the ocean. What happened below the surface, and even how deep the ocean was, were unknown. In spite of the difficulties of researching the ocean, there are many urgent environmental issues which make it necessary to gain this knowledge, for example:

- It has been proposed that obsolete nuclear submarines could be disposed of by sinking them in deep water rather than dismantling them and dividing the materials between high-, intermediate- and low-level radioactive waste stores on land.

- Some communities in Iceland and Norway have proposed that there are enough minke whales to allow them to resume whaling.

Such proposals raise a host of questions: are there places in the ocean depths where no one will ever want to go? Can we be sure that ocean circulation will not bring radioactive materials back to the surface? How can we be sure that catching minke whales will not drive them to the verge of extinction as has happened with other species? Who decides about issues like these and who should decide?

To answer such questions, we need a basic knowledge of the main features of the ocean and the processes which occur within it. The features include the size and depth of its different regions and the salinity and chemical content of sea-water. The processes are both physical – the circulation of water both horizontally and vertically – and biological – from photosynthesis by phytoplankton to complex food webs supporting predatory fish and mammals. These are introduced in Section 2.

There are many other uses of the ocean apart from for food and dumping of waste: Section 3 looks at these uses and at how some have developed into abuses. It can be difficult to discern use from abuse, as this requires more than knowledge; it also depends on values. Competition for ocean resources has led to examination of these values and attempts to establish international laws for the ocean, an issue studied in Section 4.

So, finally, the important questions about the ocean depend on knowledge, values and the law of the sea. These are brought together in Section 5 which looks at the management of key issues, such as fishing stocks, pollution and whaling, and attempts to answer the questions outlined above.

*This book is the fourth in a series: the other titles are printed at the front of the book and on the back cover.

# 2   The marine environment

## 2.1   The oceans and features of the ocean floor

The Earth has three major oceans – the Pacific, the Atlantic and the Indian Oceans (shown in Figure 1.1). The Pacific, which is the largest, is 170 million square kilometres in area, about the same size as the Atlantic and the Indian Oceans combined. It covers more than a third of the surface of the Earth and stretches almost half-way around it from east to west. It is so large that all the dry land of the Earth's surface could fit within its borders. The Arctic Ocean is much smaller than the major three and is covered almost entirely with ice. The smaller water-covered areas on the Earth are calles *seas*, such as the Mediterranean and the Caribbean. Seas may be part of an ocean – the Caribbean, for example, is part of the Atlantic – or they may be nearly separated, like the Mediterranean.

This division of the ocean into named parts is not as precise as the division of the land into continents. Whereas, for example, it is easy to see the difference between the Australian continent and the South American continent, it is not so obvious where the Pacific Ocean ends and the Atlantic Ocean starts. This is because the continents are separated (or mostly separated), whereas the oceans are all interconnected, and water flows between them: there is really only one world ocean.

Q   What is the implication of this for pollutants dumped in the ocean?

A   It may be that a pollutant dumped in one ocean may be found anywhere in the other oceans.

In practice this does not always happen, and it is possible to estimate where a pollutant can travel from knowledge of water movement. For example,

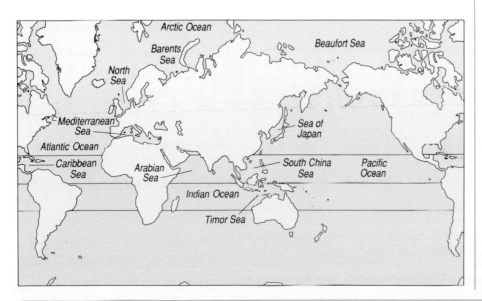

◄   *Figure 1.1*
*The oceans and some of the principal seas of the Earth.*

floating pollutants, such as plastics, in the Mediterranean Sea will remain there, as there is no flow of surface sea-water to take them out of the Mediterranean, whereas a pollutant that dissolves and sinks can flow into the Atlantic and the world ocean with a current that travels out of the Mediterranean below the surface at the Straits of Gibraltar.

How deep is the ocean? Does it get gradually deeper the further from land with its greatest depth in the middle of the oceans? If we think of the land, it does not necessarily get higher the further from the ocean, so perhaps we should not expect the ocean to get deeper further from the shore. The discovery that the ocean floor is as varied and rugged as the land – and not a flat, empty plain – is one of the most striking that oceanographers have made in the last hundred years: see Figure 1.2. The ocean floor has a number of major features, called continental margins, the deep ocean floor, ocean ridges and ocean trenches. Their position in the ocean in most cases falls into a simpler pattern than the position of features on land.

**Continental margins** border the land and have three distinct parts. The part closest to land and shallowest, the **continental shelf**, is generally only about 100 metres under the water and is relatively flat. At the outer edge of the continental shelf, called the *shelf break*, the steeper *continental slope* begins. The deepest part of the margin is the *continental rise*.

The continental shelf is the best known and most used part of the ocean. In width, the continental shelf ranges from almost nothing to 1500 kilometres, with an average of about 70 kilometres. It is widest where it borders low-lying land that was once covered by glaciers, or where large rivers empty into the sea (see Figure 1.3). The coast of Britain, and land around the Arctic Ocean, for example, have wide shelves. Where the land bordering the ocean is mountainous, as along the west coasts of North and South America, the continental shelf is narrow.

Although the continental shelf is covered by water, it is actually part of the coastal land. The edge of the ocean has moved back and forth over it as the sea-level has risen and fallen over geological time. At the maximum of the last glacial period when the sea-level was 130 metres lower than it is now, the shoreline must have been at the shelf break with the whole continental shelf above water.

Beyond the continental margin are the deep ocean basins. Because they are deep and flat, they are often called **abyssal** (from abyss) **plains**. They tend to lie at a depth of between four and six kilometres below sea-level. The flatness of abyssal plains is caused by a cover of sediments.

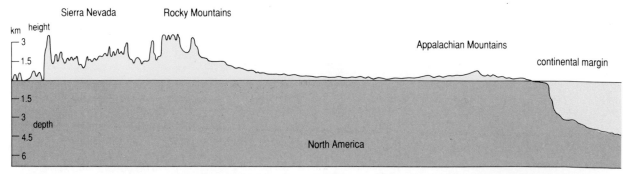

▲ *Figure 1.2   A comparison of the features of the land and the ocean floor on a line running across the USA and*

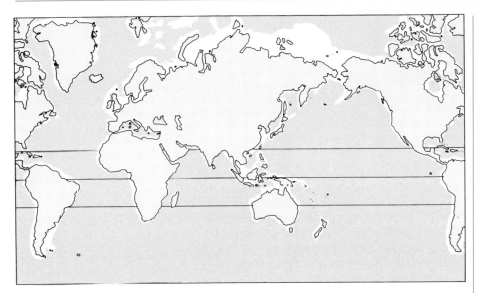

◀ *Figure 1.3*
*The continental shelves*
*of the world.*

Rising from the abyssal plains are enormous undersea mountain chains called **ocean ridges**. The tops of most are between two and three kilometres below sea-level. A few rise above sea-level: Iceland, for example, is the top of an ocean ridge. Ocean ridges have steep, rough sides and are similar in size and shape to mountain chains on land (look again at Figure 1.2), but they are much longer. The ridge in the Atlantic Ocean, for example, is longer than the Rocky Mountains, the Andes or the Himalayas. It runs the length of the Atlantic Ocean, and joins a ridge in the Indian Ocean which continues into the Pacific. In some oceans, including the Atlantic and the Indian Oceans, the ridges are central or mid-ocean, dividing the ocean basin roughly into two equal halves. In the Pacific, however, the ridge is closer to the eastern side of the ocean.

Dissecting the abyssal plains are **ocean trenches**, the deepest parts of the ocean, usually between seven and nine kilometres deep but descending to over 11 kilometres in some places. They are long, narrow and usually curved: they are about 100 kilometres wide and often thousands of kilometres long. There are no major trenches in the Atlantic Ocean. In the Pacific they run along the west, north and east sides, occurring just off the coast of western South America. There is a trench on the north-east side of the Indian Ocean.

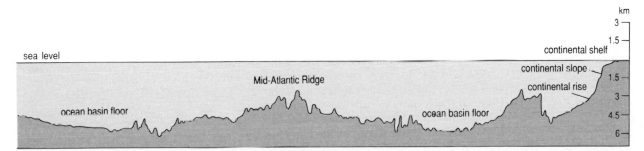

*the Atlantic Ocean to Africa. There is a vertical exaggeration of 100 times.*

*Activity 1*

How deep is the North Sea?

How deep is the middle of the Atlantic Ocean?

Is the greatest depth in the middle of the oceans?

## 2.2 Sea-water

In Book One of this series (*Sarre and Reddish*\* (eds) 1996) we learnt that the Earth's surface was covered with about 1380 million cubic kilometres of water, most of which – around 98% – is in the oceans. Sea-water has gases and solid substances dissolved in it. Although almost all the elements are present (in ion or compound form), the two most abundant are sodium and chlorine, the ingredients of common salt. These and four other elements – magnesium, sulphur, calcium and potassium – make up more than 90% in weight of the chemical elements dissolved in sea-water: see Table 1.1. The other elements are present only in extremely small quantities.

*Table 1.1 The concentration of chemical elements in sea-water. The other elements are present in less than 1 part per million*

| Element | Parts per million by mass |
|---------|---------------------------|
| chlorine | 19 500 |
| sodium | 10 770 |
| magnesium | 1 290 |
| sulphur | 905 |
| calcium | 412 |
| potassium | 380 |
| bromine | 67 |
| carbon | 28 |
| nitrogen | 11 |
| strontium | 8 |
| oxygen | 6 |
| boron | 4 |
| silicon | 2 |
| fluorine | 1 |

The **salinity** of sea-water is the amount of these chemical elements dissolved in the sea-water. In the open ocean, the salinity at the surface ranges from 3.3 to 3.7% (that is, 3.3 to 3.7 kg of elements in 100 kg of sea-water), with an average of 3.5%. Higher percentages occur where evaporation removes much of the pure water, thus concentrating the remaining sea-water, and where water is enclosed and cannot mix with the open ocean. The Red Sea, for example, which is partly enclosed in a hot, dry area, has higher than average salinity. Lower than average salinity occurs in coastal areas where large rivers of freshwater empty into the sea, and in

\*Authors' names in italics indicates that this is another book, or a chapter in another book, in the series.

polar regions where melting ice dilutes the sea-water. At depth in the oceans there is little variation in salinity, and it remains constant at about 3.5%.

Although salinity varies from one area of open ocean to another, the relative amounts of the six most abundant elements, shown in Table 1.1, and bromine, strontium and boron remain the same. By contrast the relative amounts of some of the less abundant elements in sea-water vary because marine plants use them and can deplete the supply. The most important of these elements are carbon and dissolved oxygen, but plants also need small amounts of other chemical substances, called **nutrients**, such as nitrogen and phosphorus, and although these are only present in almost undetectable amounts they are essential for life in the ocean.

Because of the dissolved substances in it, sea-water is colder than pure water when it freezes. While pure water freezes at 0°C, the temperature of sea-water must drop to about $-2$°C to freeze. (This principle is used when antifreeze is added to the water in a car radiator: the antifreeze lowers the freezing-point of the water.) When ice begins to form on sea-water, the dissolved substances in it remain in solution. Sea ice, therefore, is made of nearly pure water. Icebergs are not made of sea ice, but are huge chunks of glaciers on the Antarctic continent or the land around the Arctic Ocean which have broken off at the point where these glaciers meet the ocean. Winds and currents carry them out to sea. Icebergs can be dangerous to shipping and oil rigs: the most famous incident took place in 1912 when the luxury liner *Titanic* sank after it hit a relatively small iceberg in the North Atlantic.

*Activity 2*

Q1 The average concentration of gold in sea-water is $4 \times 10^{-6}$ parts per million. Taking the density of sea-water as around $10^3$ kg m$^{-3}$ how much gold is there in a cubic kilometre of sea-water?

Q2 The price of gold is around £10 000 a kilogram. What is the value of gold in a cubic kilometre of sea-water?

Given the value of the gold in only one cubic kilometre of sea-water, and considering how many cubic kilometres of sea-water there are in the oceans might justifiably make you wonder why we do not extract gold from sea-water. The answer is the cost of extraction: at the moment it is not economic as the cost of separating the gold, present in a relatively minute proportion by comparison with the other elements, would be higher than its value. Sea-water is an economic resource only for common salt, magnesium and bromine, all of which are present in much higher concentrations than gold.

## 2.3   Water movement

Sea-water is rarely still: usually it is in motion with waves, tides or currents.

### Waves

Waves range in size from ripples to towering masses of water over fifteen metres high. They are described by their *wavelength*, which is the distance between two crests, and their *wave height*, which is the distance between the bottom of the wave trough and the top of the crest. The highest open sea wave ever measured had a wave height of 34 metres.

The main cause of waves is wind blowing across the surface of the ocean. The speed of the wind, the length of time it has been blowing, and the distance it has travelled across the open ocean influence the height of the wave. Waves are the result of the water surface changing shape, not of water flowing with the wind. The water involved in this movement is only surface water: at depths greater than half the wavelength, the water moves very little, and there is no wave motion.

## Tides

Tides are caused by the gravitational pull of the Moon and the Sun on the Earth and modified by the rotation of the Earth. The pull is greatest on the side of the Earth facing the Moon, causing a high tide. On the side away from the Moon, where its pull is weakest, the water bulges away from the Moon, causing a corresponding high tide (Figure 1.4).

The Earth rotates once every twenty-four hours, and the Moon orbits around the Earth once in every twenty-eight days, so it takes longer than twenty-four hours (twenty-four hours and fifty minutes) for the Earth to rotate in relation to the Moon. During this time a place on Earth faces the Moon once and faces away from the Moon once and therefore most places have two high tides. Although most parts of the Earth have two high and two low tides every twenty-four hours and fifty minutes, a few areas have only one high and one low and some have a mixture of the two, with one high tide being much higher than the other. These differences are due to latitude, the effect of different depths of the ocean, the shape of a coastline, and the angles of the Moon and Sun relative to the Equator.

The Sun is much further away so its effect on the ocean is less than half that of the Moon. When both the Sun and Moon face the same side of the Earth at the same time, their pull is then combined and very high tides, called *spring tides*, result (Figure 1.4a). These occur every 14 days. When the Sun and Moon form a right-angle with the Earth, the gravitational pulls are in different directions, causing weaker tides, called *neap tides* (Figure 1.4b). Like the spring tides, neap tides also occur every fourteen days.

The difference between the high water level and the low water level of a tide is known as the *tidal range*, which varies from time to time and from place to place. The greatest tidal range in the world is at the head of the Bay of Fundy on the east coast of Canada, where it is up to 18 metres. The

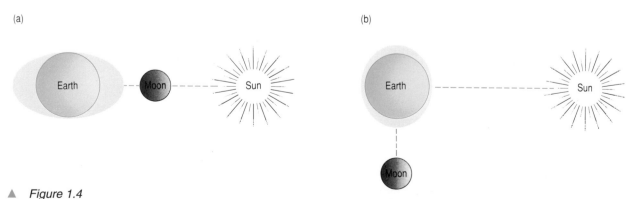

(a)                                                                          (b)

▲  *Figure 1.4*
*Tides are caused by the gravitational pull of the Moon and Sun on the Earth. (a) Spring tides. (b) Neap tides.*

Mediterranean, by contrast, has a very small tidal range. In the English Channel, the range in most places is about two metres, while in the Bristol Channel it can be 12 metres.

## Surface currents

Q   What is the main cause of the surface currents in the oceans? (This was introduced in Chapter 5 of Book One (*Silvertown*, 1996a).)

A   The winds.

Winds drive surface currents in the oceans, as well as creating waves. Winds create waves by pushing the sea surface up and down, and currents by dragging the water along with the wind. Currents do not flow directly downwind, but are deflected to the right or the left by the rotation of the Earth, depending on whether the currents are in the northern or southern hemisphere. The prevailing winds and the Earth's rotation, combined with the land masses which act as giant breakwaters on the currents, produce a pattern of rings of currents circulating in the open oceans called **gyres**: see Figure 1.5. These usually flow clockwise in the northern hemisphere and anticlockwise in the southern hemisphere.

   The currents in each gyre that carry water from the Equator towards the poles, such as the Gulf Stream and the Kuroshio in the northern hemisphere, are warm. They gradually lose their heat in higher latitudes.

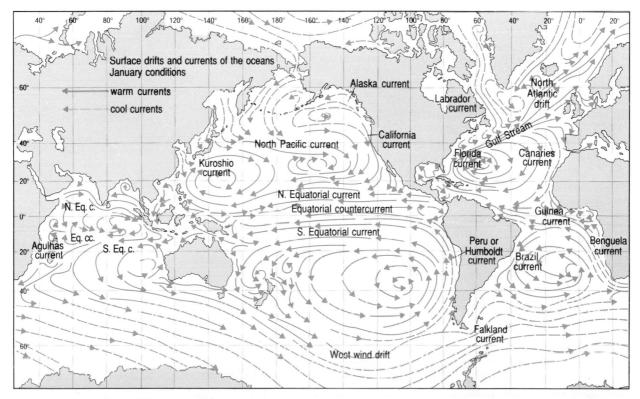

▲   *Figure 1.5   The surface currents in the ocean.*

The currents that return the water to the Equator are cooler. Because of this the surface currents moderate our climate, warming cool regions and cooling hot ones.

Currents are fairly weak over most of the ocean surface. Currents on the eastern side of the oceans may reach a speed of 2 kilometres per hour, while those on the western side, such as the Gulf Stream, are stronger and may reach as much as 8 km per hour. While most wind-driven currents are no more than 100 metres deep, the major currents transport enormous volumes of water. At any given moment the Gulf Stream alone is transporting more than 100 times as much water as all the rivers on Earth combined.

There is one major difference in the current systems in the northern and southern hemispheres: that is the presence of a current in the southern hemisphere which does not travel in a gyre but circulates around the world – the Antarctic Circumpolar current or West wind drift. This occurs because there are no land masses at these latitudes to deflect the current, which is driven by westerly winds.

Q   The Arctic region is generally less cold than the Antarctic region. This is due mainly to the surface current patterns in the ocean. Examine Figure 1.5 and consider the reason for this.

A   A major warm ocean current, the North Atlantic Drift, penetrates into the Arctic Ocean, bringing warm water and warming the Arctic region. There is no similar warm current reaching the Antarctic, as the Antarctic Circumpolar current prevents any warm current reaching the land area of Antarctica.

## 2.4   Vertical circulation, upwelling and mixing

In the last section we looked at currents in the surface water of the oceans. In this section we will examine the motion of water in the deep ocean, and the transfer of water between the shallow and deep ocean.

Deep ocean currents cannot be driven by winds as are the surface currents. Instead they are driven by density differences in the water. Colder water and water with greater salinity are denser than warmer or less saline water, and tend to sink. However, this occurs in very few places in the oceans as usually the warmest water is at the surface (Figure 1.6). The top 100 metres or so of the oceans are heated by solar radiation and are fairly warm (apart from in the polar regions), at about 10–15°C in middle latitudes and over 20°C in the tropics. The temperature decreases sharply below this depth, in a temperature gradient called a **thermocline**. Below about 1000 metres the temperature is constant at a few degrees, and this is independent of latitude: the deep waters of the tropics are at about the same temperature as the deep waters of the polar regions (Figure 1.6). The only area in the oceans where the temperatures are similar at all depths, and where the surface water might sink, is the polar regions. Here wind blowing off the ice-caps cools the surface water. When the water freezes, forming ice, this leaves the unfrozen water with a greater salinity, and therefore greater density, so this surface water sinks and flows along the ocean bottom to other parts of the oceans, even to the tropics. This water sinking in parts of the polar regions carries with it a good supply of oxygen and nutrients.

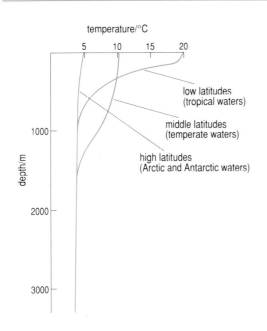

temperature/°C

low latitudes
(tropical waters)

middle latitudes
(temperate waters)

high latitudes
(Arctic and Antarctic waters)

◀ *Figure 1.6*
*The variation of temperature*
*with depth in the ocean.*

How does water in the deep oceans return to the surface? One way this occurs is by **upwelling** (*Silvertown*, 1996a). Surface water may be driven offshore by an offshore or along-shore wind (Figure 1.7). This water is replaced by cold, deep water. Main areas of upwelling are the western coasts of North and South America and the western coast of Africa.

Q  Why are upwelling areas important?

A  They support major fisheries. The upwelling water is rich in nutrients, which allows the growth of phytoplankton. These are fed upon by zooplankton, which are the food for fish.

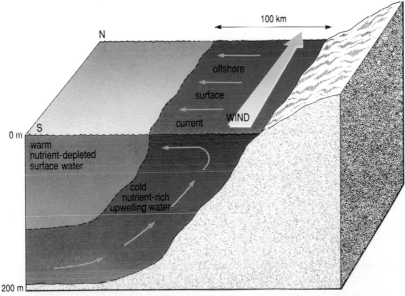

▲ *Figure 1.7  Coastal upwelling in the southern hemisphere.*

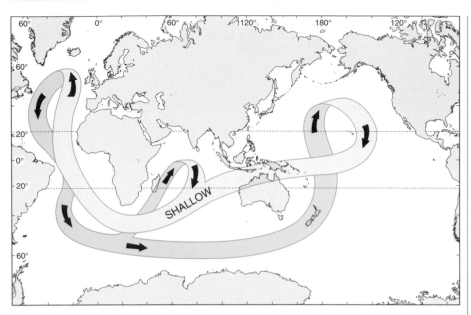

◀  *Figure 1.8*
*The North Atlantic conveyor-belt. The model represents the Atlantic as a radiator that imports warm water, releases the heat to the atmosphere, and exports the resulting cold water. Due to its influence, temperatures on land of coastal countries of western Europe are as much as 10°C warmer in winter than at comparable latitudes elsewhere.*

Upwelling can also occur in the open ocean, where wind patterns cause surface waters to move apart. This happens in parts of the Antarctic, the North Pacific and equatorial regions, and again produces areas of high productivity.

It has long been suspected that the Gulf Stream and North Atlantic Drift form part of some larger circulation in the oceans (see Figure 1.5). A 'conveyor-belt' model of the North Atlantic circulation has been suggested, based on observations that deep, cold water with identifiable North Atlantic characteristics can be traced to most other oceans: to balance this outflow, warm surface water must flow back into the North Atlantic across the equator. A simplified diagram of the 'conveyor-belt' is shown in Figure 1.8. Approximately 13 million cubic metres per second of warm water enter the north Atlantic across the equator (for comparison, the Amazon has an average flow of 0.02 million $m^3 s^{-1}$), returning as North Atlantic Deep Water formed in the Greenland, Norway and Labrador Seas. As the warm water flows north and cools to form the deep water, large quantities of heat are released to the atmosphere.

The ocean west of the British Isles is 4°C warmer than at a comparable latitude in the Pacific: the import of heat by the warm water of the conveyor-belt being equivalent to that reaching the ocean surface from solar radiation. This only happens to a significant effect in the North Atlantic, as Figure 1.9 demonstrates; the main beneficiary of the warming is Western Europe, particularly during winter months. Even though the scale and regional impact of the conveyor-belt are impressive, the processes that drive it are finely balanced. It is thought to have been interrupted, with dramatic results to the climate in the region, in the transition from glacial to inter-glacial climates about 10 000 years ago and may – again – be affected by climate changes caused by the release of greenhouse gases into the atmosphere. These aspects of the conveyor-belt are taken up again in Chapter 3.

But how fast does water circulate vertically in the oceans? Estimates vary but on average it is thought that water will spend between 200 and 1000 years in the deep sea before returning to the surface.

This has implications for the deep ocean disposal of radioactive waste.

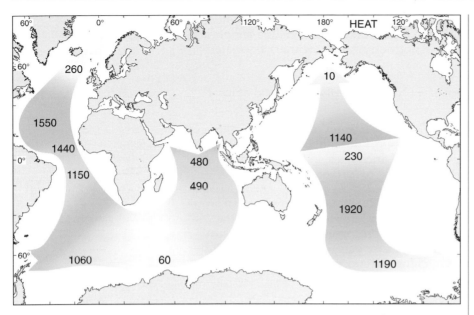

◀   Figure 1.9
*Estimates of heat transport by the oceans. Units are terawatts ($10^{12}$W or one million million watts). Note the linkages between the oceans, and the transport of significant heat northwards between the British Isles and Iceland.*

One of the questions asked in the introduction to this chapter was whether radioactive waste, which remains radioactive for thousands of years, could be safely dumped in the deep ocean. With a vertical circulation rate of a few hundred years, any leakage of waste from canisters in the deep sea would return to the surface ocean in a similar time – while it was still radioactive. Indeed one method of tracing ocean currents has been to follow the movements of radioactive isotopes such as Tritium from the time they were deposited on the ocean during atmospheric testing of nuclear weapons in the 1950s and '60s (see Section 3.3 below).

## 2.5   Oscillations in the ocean

While the oceans ultimately gain their energy from solar radiation, they also exchange large quantities of heat energy and moisture with the atmosphere, as we saw in Section 2.4. The heat removed from the ocean to the atmosphere (mostly in the form of latent heat as water is evaporated) is much larger than, for example, the energy imparted to surface currents by the winds. Because the oceans have a far greater heat capacity than the atmosphere, the way this heat is stored and released by the oceans has a major influence on the climate. In global terms the oceans act as a buffer, moderating daily and seasonal changes of temperature. But variations or rhythms also occur in the ocean currents, leading to anomalous patterns of sea surface temperatures which persist for months or longer. When sea surface temperatures are unusually high, the atmosphere is likely to receive large transfers of moisture and energy: a change of 1°C can have a significant impact on the regional climate.

The most celebrated example of this is an irregular oscillation in the sea surface temperatures of the tropical Pacific with the intriguing title of El Niño/Southern Oscillation (ENSO). 'El Niño' is a warm southerly current that sometimes appears off the coast of northern Peru at Christmas, hence the name given it by sailors – Spanish for 'the Christ Child'. Its appearance, at intervals of between 2 and 10 years, is associated with dramatic changes to the local climate: the drought conditions normally prevailing near the coast are replaced by heavy rains that bring a good harvest but also floods. It is also of great local significance to fishers because it signals a temporary disappearance of the normally cool northerly current with its associated upwelling (see Figure 1.7).

*Q*   How might the appearance of 'El Niño' affect the livelihood of local fishermen?

*A*   As Figure 1.7 shows, the cool, northerly current causes an upwelling of cold water, rich in nutrients that support phytoplankton. The latter provide food for fish (large shoals of anchovies, in fact). Disappearance of the upwelling signals an abrupt end to further fishing.

The name 'Southern Oscillation' was given by Sir Gilbert Walker to a large-scale, irregular, sea-saw of atmospheric pressure covering much of the tropical Pacific and Indian oceans. We now recognise that these pressure variations and the appearance of El Niño are part of the same event. Under 'normal' conditions, illustrated in Figure 1.10(a), air pressure is high over

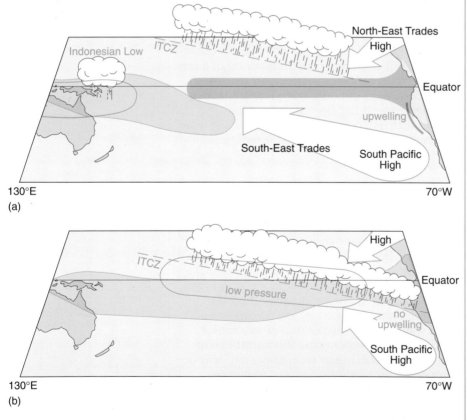

(a)

(b)

▲  *Figure 1.10*
*Schematic diagrams showing conditions in the Pacific in (a) a normal year and (b) during an El Niño event. Although no El Niño can be considered typical, the features shown in the diagram occur in most El Niños. The darker green tone indicates regions where the sea-surface temperature is higher than 28°C; the grey tone in (a) indicates regions that are normally dry. The band of cloud and rain shown on both diagrams is a migrating zone of wet weather, known as the intertropical convergence zone (ITCZ), that normally occurs near the equator.*

the eastern Pacific and low in the west over Indonesia, driving the prevailing south-easterly Trade Winds near the equator. This circulation pattern 'piles up' warm water in the western Pacific, increasing evaporation and rainfall there, while to the east the south-easterly trades maintain cooler surface waters near the equator through coastal upwelling. However, every few years an ENSO event occurs and the pressure patterns weaken, with two results: in the western Pacific the easterly winds weaken or reverse allowing currents to carry the warm surface-water eastward, while in the eastern Pacific the upwelling falters and is over-ridden by the warm El Niño current. The net result is a large area of anomalously warm water over the central and eastern Pacific, with associated low pressure and rainfall, while drought conditions prevail in the west (see Figure 1.10b). However, El Niño events rarely last for more than a year. The atmospheric pressure difference between Indonesia and the eastern Pacific is soon restored, the Trade Winds strengthen again and normal upwelling conditions prevail once more.

Nevertheless, the repercussions of this oscillation, with its large sea-surface temperature anomalies, are global: it is responsible for a range of extreme weather conditions, both droughts and floods, over much of the tropical region. There are also indications that the behaviour of ENSO may be linked to global warming. These themes are taken up in Section 4 of Chapter 3.

## 2.6   Marine life

Marine organisms can be divided into **pelagic organisms**, which live and feed in the body of the sea, and **benthic organisms**, those associated with the seabed.* Plants in the oceans have to live in the **photic zone**, the top hundred metres or so of the surface waters where there is light for photosynthesis. This means that benthic plants can only grow in the shallowest parts of the ocean, on the continental shelves. The deeper ocean has no benthic plants – here plants are all planktonic, unattached and floating in the surface layers as plankton.

Q   What are the major differences between terrestrial and marine plants?

A   Terrestrial plants are large, often trees and other long-lived plants. The main marine plants are microscopic, floating **phytoplankton**, which unlike terrestrial plants have brief lives, multiply very fast and decompose very quickly when they die.

Marine plants live in a very different environment from plants on land. They derive gases for photosynthesis and respiration, and their nutrient supply, directly from the water surrounding them rather than through special structures such as leaves and roots. Consequently, the ratio of surface area to volume of a marine plant must be large, so they must be very small (e.g. phytoplankton) or have many filaments or fronds (seaweeds). They must also have a mechanism for staying in the photic zone, either attached to a shallow sea bottom or able to control their buoyancy. Unicellular plants control their buoyancy by producing oils within their cells and floating seaweeds have small gas-filled bladders on their fronds.

Phytoplankton have a very high potential rate of population increase when conditions are right. They require light and a good nutrient supply and under these conditions have a high **productivity**.

*Life in the oceans was introduced in Chapters 3 and 4 of *Sarre and Reddish* (eds, 1996), where marine habitats and ecosystems were discussed.

*Q*   Which marine ecosystem has the highest and which the lowest **primary productivity** per unit area?

*A*   If you have read Table 7.1 in Chapter 7 of Book One (*Silvertown*, 1996c) you will know – but you should also have been able to work out – that the highest productivity occurs in algal beds and coral reefs, and the lowest productivity occurs in the open ocean.

Algal beds and coral reefs are very productive because nutrients are recycled within the community by animals and plants. Estuaries are also highly productive, because of the supply of nutrients from land. Continental shelves are fairly productive for the same reason. Upwelling zones are usually slightly more productive than continental shelves. (This can be seen in Plate 2.) The continental shelves (such as those around Britain) have high productivity, and so do coral reefs (look around the Bahamas) and upwelling areas (the coast of Peru). The high productivity in parts of the polar regions in spring is also visible.

Marine animals, like marine plants, are also adapted to a different environment from that of land animals. For example, an animal on land, if fixed in one place, would starve, as it would not be able to search for food. However, in the oceans there are animals that live fixed in position on the seabed, with a constant supply of food brought to them in the water, without any effort on their part, by currents, tides and organic debris falling from overlying waters. Animals that live in the open ocean may be very fragile, such as jellyfish, as they can rely on the sea for support.

The abundance and distribution of marine animals depend on that of the marine plants, which are their ultimate food source, so continental shelves and upwelling areas have large quantities of animals and the open ocean very few. In an upwelling area the phytoplankton are so dense that fish can feed directly on them (a two-stage **food chain**) and even some types of whales have only three stages in their food chain (phytoplankton – krill – whales). In the open ocean, where phytoplankton are widely scattered, food chains tend to be longer.

The deep ocean floor has relatively few animals because of the poor supply of food. The animals are mainly invertebrates, crustaceans and fish which feed on the remains of dead animals falling from the waters above, or on each other. The animals are small, and grow slowly.

## 2.7   *Summary*

The major oceans, although they have different names, are interconnected. The ocean floor is as varied as the land surface. Its main features are continental margins, abyssal plains, ocean ridges and trenches.

About 3.5% of sea-water is dissolved gases and solid substances. The most abundant of these substances are sodium and chlorine. Plant life in the oceans needs dissolved carbon, oxygen and nutrients such as nitrogen and phosphorus.

Sea-water moves in waves, tides and currents. Waves and surface currents are driven by winds. Tides are caused by the gravitational pull of the Moon and Sun. Deep currents are due to density differences in the water. Upwelling systems are driven by winds. It takes around 200–1000 years for water to circulate between the surface ocean and the deep ocean. Plants live in the surface waters of the ocean. The productivity of the oceans varies, the most productive regions being algal beds and coral reefs, estuaries and upwelling areas. Both marine plants and animals are very different to land organisms.

# 3   Uses and abuses of the ocean

## 3.1   Introduction

Today we use the ocean in many more ways than a century ago, when shipping and fishing were the only human activities associated with the oceans. Even these uses have now changed: ships have become larger and larger and fishing has increased to the extent that some species of fish are threatened by overfishing.

Many of our present uses of the ocean are harmless, or fairly harmless, but many are damaging. In considering the uses of the oceans, it is useful to distinguish between the uses which do not remove anything from the ocean (*non-extractive use*) and those which do (an *extractive use*, or resource).

The obvious non-extractive use of the ocean is as a separator or buffer between countries, a use that has been of great importance to the United Kingdom in particular. Another very important use is for transportation, both coastal and intercontinental. Transportation is, or can be, a relatively benign use of the ocean, except for the development of port facilities, frequently at the expense of estuaries and other important coastal habitats, and the pollution associated with transportation.

Other non-extractive uses of the ocean include seabed cables and oil pipelines, the former generally harmless but the latter potentially less so. The ocean is also used for a variety of recreational activities, most of which are relatively harmless. However, uses of the ocean that add material to it, such as building harbours and dumping of waste, are among the most detrimental uses.

The extractable resources of the ocean are of two main types, the living resources and the non-living energy and mineral resources. About 3 million tonnes of marine plants are harvested each year, mainly giant kelp and other large seaweed. This is used for fertiliser, food, animal feed and as a source of chemicals. About 70 million tonnes of marine fish are caught each year, and also shellfish and marine mammals such as whales and seals. In some countries the marine fishery provides a large proportion of the human protein intake, for example about 50% in Japan and about 20% in the former Soviet Union, although only about 3% in the United States.

After food, petroleum is the most valuable resource extracted from the ocean. About 27% of crude oil now comes from offshore production platforms, and about 20% of natural gas. The UK is one of the world's largest producers of offshore oil at this time: the only other nations with comparable offshore production are Saudi Arabia and Norway. Extraction of this wealth from the ocean comes at a price, as accidents during production (oil-rig blow-outs) and transportation (tanker spills) can cause major disasters for coastal habitats, although on a relatively local scale.

The mineral resources of the ocean are also valuable, but overall only a small proportion of the minerals we use come from the sea (a few per cent). There are two main sources of minerals, those that are extracted from sea-water, such as salt and bromine, and those that are mined from sediments on the seabed, such as tin. Extraction from sea-water has relatively minor effects on the ocean environment, but seabed mining can be very disruptive and harmful to the local marine life.

## 3.2   *Shipping*

The main commercial uses of ships are to transport passengers and cargo, and for fishing. Intercontinental passenger traffic by sea has generally declined over the last fifty years as travel by air became faster and cheaper, but shorter distances are still often travelled by ship, for example across the English Channel.

The rise of air travel has had little effect on cargo transport, as the size and weight of most cargo make air transport prohibitively expensive. There are three broad categories of cargo: bulk liquids, bulk dry cargo and liner cargo. Bulk liquids consist mainly of petroleum, which accounts for just about half of the world's cargo movements. Bulk dry cargoes consist largely of iron ore, coal, bauxite (an aluminium ore), phosphate (for fertiliser) and grain. Liquid and dry bulk cargoes form about 75% of world cargoes. The rest is shipped in mixed cargoes of small consignments of various items, carried on liner vessels and container ships instead of bulk carriers. The relative importance of the different ships and cargoes can be interpreted from Table 1.2.

Table 1.2   *World commercial shipping tonnages, 1995*

| Vessel type | Tonnage (million gross registered tonnes) |
|---|---|
| Oil tankers | 144 |
| Liquefied gas carriers | 15 |
| Chemical carriers | 12 |
| Bulk/oil carriers | 14 |
| Ore and bulk carriers | 138 |
| General (including refrigerated) cargo | 64 |
| Container ships | 39 |
| Roll-on roll-off ships | 20 |
| Ferries and other passenger vehicles | 17 |
| Fishing factories, carriers and trawlers | 13 |
| All other vessels | 15 |

Source: *Lloyd's Register of Shipping Statistical Tables*

Oil tankers have the greatest tonnage of all shipping on the ocean. The largest individual ships are also oil tankers. The first ships specifically built as tankers were small, carrying around 5000 tonnes of oil, but the decision of the British Navy to switch to oil from coal in its ships in 1912, a policy rapidly followed by other navies, led to the building of larger tankers, up to 15 000 tonnes. A further growth in the use of oil in vehicles and in boilers between the two World Wars led to a large growth in tanker tonnage, and the building of supertankers. Much of the oil was transported between the Middle East oilfields and western Europe and the tanker size was largely limited by the size of the Suez Canal. Closure of the canal following the Anglo–French attack on Egypt in 1956 and after the Arab–Israeli war in 1967, and a growing demand for oil, led to the building of even larger tankers, known as **very large crude carriers (VLCC)** of around 250 000 tonnes, and now the even larger **ultra large crude carriers (ULCC)** of over

▲   *A very large crude carrier (VLCC), here unloading crude oil at the oilport at Teesside, north-east England.*

400 000 tonnes. The VLCCs and ULCCs have their own problems of draught, limitation to deep water, manoeuvrability and stopping distance.

Q   From Figure 1.11, what are the two main oil transport routes?

A   From the Middle East to western Europe by the South Atlantic, and from the Middle East to Japan.

The Suez Canal, although open and offering a much shorter route to western Europe, is used much less than the South Atlantic route because of tanker size.

How safe is shipping? Over a million tonnes of shipping is lost at sea each year, and a far greater amount is damaged, often leaking cargo into the sea. Loss and damage is due to insufficient charting of the seabed, bad

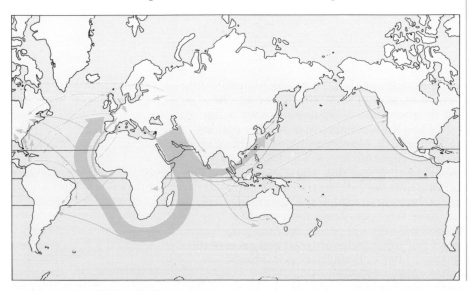

◄   Figure 1.11
*The main oil transport routes by sea in 1980. The arrow width represents the tonnage.*

weather, collisions or human error. Insufficient charting may seem to be surprising, but many areas are daily navigated by ships whose charts have not been updated since the days of steam. For example, the $25 million ore-carrier *Igara* was holed on an uncharted rock near Singapore in 1975: the most recent surveys of this area were done in 1907.

The greater size of present ships does not make them invulnerable to bad weather: in some cases their length makes them more vulnerable. Immense freak waves, for example, are known to occur at times in various areas of the oceans, particularly in the southern Indian Ocean off the east coast of Africa. Here the ships use the southerly Agulhas current (Figure 1.5) on passage to the Atlantic. When this current meets storm waves from Antarctic waters, very high waves can form, over 30 metres high. A small ship may ride over one wave, but two or more waves may hit a larger ship at the same time, raising each end of the ship but not supporting the middle section, causing the ship to break in two. Many ships have vanished totally in the Agulhas current but the route continues to be used, as it is the fastest and therefore most economic route around Africa.

Many ship casualties arise from human error of the 'dangerous driving' kind. The busiest shipping lane in the world, the English Channel, has had many of these. An example is the *Texas Caribbean*, a tanker which ran aground on Varne Bank in 1971. The same day, another ship, the *Brandenburg*, collided with the grounded tanker and sank, taking the further holed *Texas Caribbean* with her. A few days later a third ship, the *Niki*, hit the uncharted wrecks.

It is the loss and damage of ships that produces the greatest damage to the marine environment by shipping, through the loss of their cargo, and coastal states now usually have emergency action plans for coping with large oil spills, but they do not always work. The relatively small spill of 37 000 tonnes of oil from the VLCC *Exxon Valdez* in Alaska in 1989 was an environmental disaster, despite an emergency clean-up operation costing $1000 million. Twice the English Channel has had greater spills: 100 000 tonnes in 1967 from the supertanker *Torrey Canyon* (at this time there was no emergency action plan in existence) and 230 000 tonnes in 1978 from the VLCC *Amoco Cadiz*. These spills have highlighted the question of whether VLCCs and now ULCCs should be allowed passage in busy shipping lanes or in environmentally important areas and, if not, how to legislate to prevent their passage.

## 3.3   *Military use*

The military use of the ocean is generally to support a nation's land interests, but is also increasingly in support of its marine interests. The traditional purpose of a navy is the protection of the nation's shores, the safeguarding of its merchant fleet and the intimidation of smaller nations by a naval presence ('gunboat diplomacy'). More recently navies have been used to protect national interests in marine resources. The 1970s 'Cod War' between Britain and Iceland was an example of this, and less obviously the war between Britain and Argentina over the Falklands in 1982, since sovereignty of the islands confers the right to exploit the ocean resources around them and this area may have large petroleum reserves. The strategic use of navies has also become very important. Nuclear-powered submarines with intercontinental nuclear missiles are essential parts of the major navies, together with the associated anti-submarine warfare capabilities, which involves other nuclear submarines.

▲   *The underwater atom bomb explosion in Bikini lagoon, 27 July 1946. Atmospheric fall-out from the explosion released radioactive isotopes into the ocean over a large area of the Pacific.*

Military activities have similar pollution effects as civil shipping, and in addition are the main cause of radioactivity entering the ocean. Up to 1963, when the Partial Test Ban Treaty came into force, nuclear weapon tests were carried out in the ocean, which had lethal effects locally but also released radioactive isotopes into the ocean, which have had much wider effects. Radioactivity still enters the ocean from atmospheric nuclear test fall-out. Accidents involving nuclear weapons can also lead to contamination: for example, about 390 grams of plutonium were released into the Atlantic Ocean off Greenland in 1968 as a result of an accident involving four nuclear bombs, killing or contaminating marine organisms over a considerable area. Nuclear-powered submarines have also been lost: two at least from the USA, the *Thresher* in the western Atlantic in 1963 and the *Scorpion* in the central Atlantic in 1968, and at least six from the former Soviet Union. Leakage of radioactive isotopes from the break-up of the submarines' reactor power plants, as well as from nuclear missiles, is of concern.

Military activities also have the occasional local beneficial effect on the ocean. The prevention of fishing in large areas of the ocean owing to wartime activities permits the building of fish populations. Fish catches on the Atlantic continental shelf of Europe were remarkably better when fishing recommenced after the Second World War than before it. As a fish conservation plan, however, wars are a drastic solution.

## 3.4   Food resources

In Section 3.1 the living resources of the oceans were introduced as the most valuable resource we extract from the ocean, most of it as fish, with smaller amounts of other marine animals and plants.

*Q*  What is the general difference between the main type of food obtained from the ocean and that from the land?

*A*  The main foods produced on land are plants, cereals and root crops (*Grigg*, 1996), which provides mainly carbohydrates. The main food from the ocean is fish, which provides protein.

Protein is an essential part of the human diet, but is supplied only in small quantities by cereals and root crops, so that many people in developing countries suffer from a protein shortage. Later in this section we will consider whether the ocean could supply more protein than at present, to alleviate this protein deficiency.

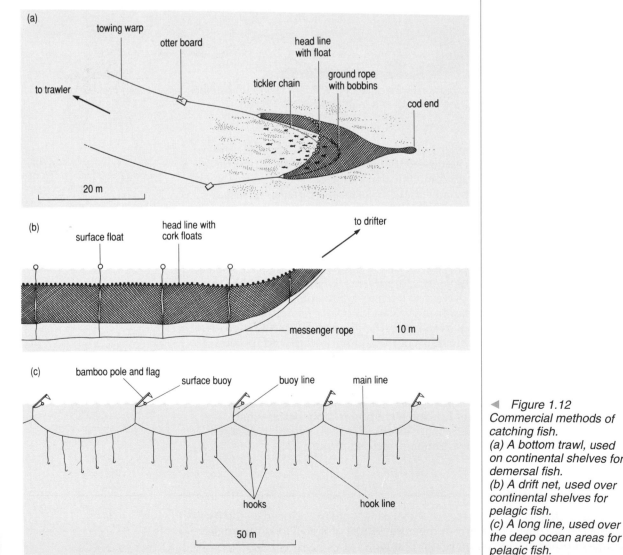

◀  *Figure 1.12 Commercial methods of catching fish.*
*(a) A bottom trawl, used on continental shelves for demersal fish.*
*(b) A drift net, used over continental shelves for pelagic fish.*
*(c) A long line, used over the deep ocean areas for pelagic fish.*

*Activity 3*

This activity relates to the distribution of fish and other animals in the ocean. Use earlier sections of this chapter (and *Silvertown*, 1996c) to list the factors that will influence the distribution of animals in the ocean, in terms of both location and depth.

Thus the food resources of the ocean are not distributed evenly throughout the ocean, but are concentrated in the surface layers of the ocean near to land (continental shelf areas) and upwelling areas. This uneven distribution influences the way that we get food from the ocean, in comparison to how it is done on land. Food gathering on land is almost entirely by farming, by planting crops and keeping domestic animals. In the ocean, food is mainly gathered by hunting, and this is easier in the areas where the animal populations are densest.

There are three main methods of catching fish, with the choice of method depending on the behaviour of the fish being hunted. **Demersal** (bottom-living) **fish**, such as cod, haddock and plaice, feed on the benthic fauna of continental shelves. These are caught by a bottom trawl (Figure 1.12a). Pelagic fish such as herring, anchoveta and sardines, which feed on the good supply of planktonic organisms over the continental shelves are caught by a drift net (Figure 1.12b), or a mid-water trawl. Fish that live in the upper parts of the deep oceans are much less abundant, and do not form shoals. These include tuna, bonito and shark. They are caught by a number of methods, including long lines (Figure 1.12c), rod and line, and nets.

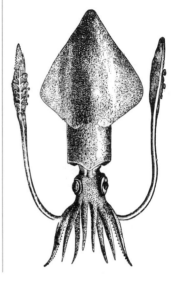

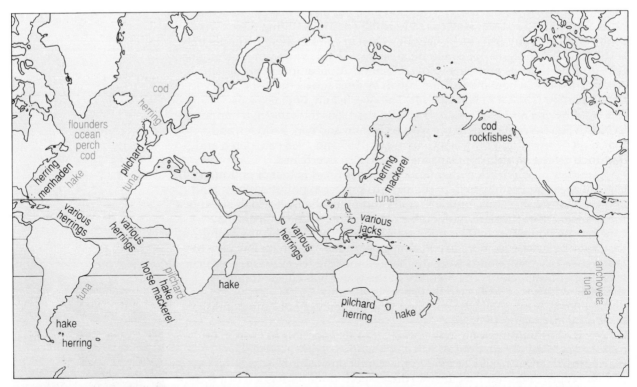

▲   *Figure 1.13   Major world fisheries. Those in green have been overfished, or are in danger of overfishing.*

The north Atlantic, the south-east Pacific (off Peru) and the north-west Pacific (near Japan) produce the highest catches of fish. The oldest fisheries are in the north Atlantic, catching herring, haddock, cod, plaice, mackerel and sardine mainly on and over the continental shelves of Europe and North America: see Figure 1.13. Nearer the Equator, pelagic fish such as sardine, menhaden and mackerel are fished, and demersal fish become unimportant. Many other pelagic species, such as hake and tuna, are fished in the tropical Atlantic. The south Atlantic produces smaller quantities of fish, mainly hake (often in the shallow waters around the Falklands), pilchards (near Africa) and horse mackerel.

In the Pacific, the Peru fishery catches just one fish, the anchoveta. The Japanese fishery catches anchovy, mackerel, saury, and also an animal which is not a fish – squid. Pollack, herring and cod are also important in the North Pacific, and salmon and shellfish off the west coast of North America.

The Peru anchoveta fishery is, or can be in some years, the largest fishery in the world ocean, catching over 12 million tonnes in 1970 and providing a third of Peru's export earnings. However, the catch is very variable: in 1972 it fell to 4.5 million tonnes and 1973, 1.8 million tonnes, with disastrous effects on the Peruvian economy. This was due to large-scale atmospheric and oceanographic variations. The anchoveta eat phytoplankton, which grow in large populations in the upwelling waters off Peru but they experience rapid population fluctuations with the irregular onset of El Niño conditions, described in Section 2.5.

Fish catches can also decline because of overfishing. This occurs when fewer fish are caught even though the fishing effort is increased by more boats or better fishing gear. Ideally a fishery should be regulated to produce a maximum catch with the minimum of fishing effort (fisheries management will be discussed in Section 5.2), but this is difficult to achieve, often for political reasons, such as individual countries setting their own limits on how many fish can be caught, when in combination the limits may result in overfishing, or for economic reasons: what is worse, overfishing or laid-up fishing boats and unemployed fishers? An example is the north Atlantic fishery, where overfishing began as long ago as the last century in the North Sea (shown in Figure 1.14). The overfishing of plaice was followed by cod and haddock by 1920. By the 1950s the same had happened to North Sea herring, hake to the west of Britain and cod, haddock and plaice off Norway and off Iceland. Between 1955 and 1966 cod, perch and haddock off the North American coast were also overfished.

Another problem affecting marine food resources is that of pollution. Coastal areas and estuaries in particular are subject to pollution from land, and the many fish, clams, mussels, crabs and other shellfish which breed in these areas are particularly vulnerable to pollution. These animals may concentrate pollutants in their bodies and even if they survive the pollution, they may be unsafe for human consumption. In the Minamata Bay area of Japan mercury compounds were discharged into the sea as waste products from a chemical factory. The mercury in the water was concentrated by shellfish, which were collected by fishers in the area. In 1963 it was discovered that the shellfish were the cause of mercury poisoning of the local people, causing 45 deaths.

As well as hunting food in the ocean, it is also possible to farm it, to harvest enclosed populations of fish and shellfish, a process called **mariculture**. The shortage of food in the developing countries and the value of some types of fish and shellfish in developed countries has led to a recent expansion in mariculture. In principle it is more efficient than conventional

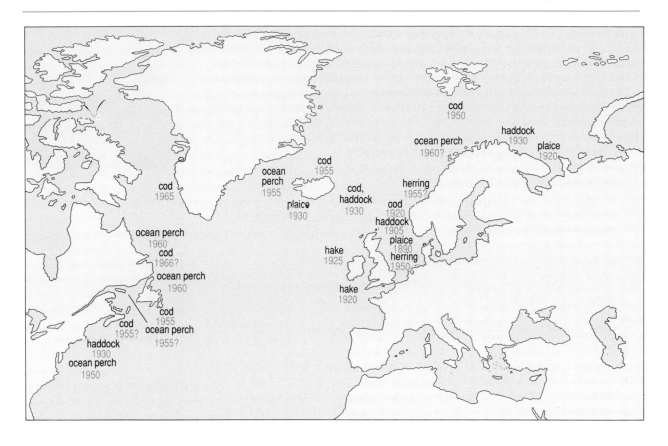

▲   Figure 1.14
The North Atlantic fishery,
with the dates of the
beginning of overfishing
for each fish population.

fishing, as the fish do not have to be hunted, and could provide continuous, not seasonal, supplies.

Shellfish (oysters, mussels and clams) are successfully cultivated on farms in Europe and America. They feed directly on phytoplankton carried to them by water currents, and so are most suited to coastal areas with strong tidal water movement. The farming of salmon, trout and lobsters is becoming important in Scottish sea lochs. Unlike the shellfish, these have to be supplied with food, often locally caught, poor-quality fish. Mariculture in developing countries is largely confined to South East Asia, farming milk fish (a herring-like fish) and mullet in coastal ponds, often in mangrove swamps. The fish feed on marine plants, mainly algae, which grow in the ponds. Although mariculture has many advantages of cost and supply over conventional fishing, it produces other problems, particularly to the local environment, such as the effect of uneaten fish food and fish wastes and chemical pesticides used for disease control in the fish, which can deoxygenate the local waters and upset the natural ecology of the area.

Can the supply of food from the oceans be increased by a large enough amount to make a significant contribution to world food shortages and protein deficiency? There is little hope of increasing the catch of the conventional fisheries in many regions of the world, but some areas are under-exploited (they could produce a larger catch), particularly the south Atlantic, west-central Pacific and the Indian Ocean, although none of these regions seems to have the productive capacity of the existing main fishing areas. There is also the possibility of greater use of non-fish species, which are underexploited in many areas. Squid, for example, have a large market only in Japán. Krill, a small shrimp-like animal, is abundant in Antarctic waters in summer and is the main food of some whales, but is not yet harvested in large quantities. Much of any new catch that is not directly attractive to human beings may be converted into fish-meal, a protein-rich powder that can be stored without refrigeration. This could provide a protein supplement in developing countries but at the moment is mainly used as animal feed (for most British chickens, for example). Most of the Peruvian anchoveta and other small fish are converted into fish-meal.

The marine catch by country is given in Table 1.3.

Q    Is the catch related to population size?

A    Partly: six of the countries in the top ten have populations greater than 100 million (Japan, the countries of the former Soviet Union, China, United States, Indonesia, India).

Q    Is the leading fishing nation a great consumer of fish?

A    Yes: marine protein is very important in the Japanese diet, about 50% of the protein intake (Section 3.1).

Q    Why do the low-population countries Peru, Chile and Iceland have large marine catches?

A    For Peru and Chile this is because of the large anchoveta fishery in their coastal waters. Iceland also has a high catch, partly because of availability of fish, but mainly because fishing is the traditional and main industry of the country. (Most of the catch from each of these countries is exported.)

## 3.5    Energy resources

The ocean can be used as a supply of two main types of energy: renewable energy from motion of the sea-water in waves and tides and fossil-fuel energy from oil and natural gas, stored in the rocks beneath the ocean floor.

The energy in waves comes from winds, and so ultimately from solar power. The energy in a wave depends on the height and speed of the wave, so potential wave-power sites need a reasonably constant supply of high and fast waves.

Q    What factors determine wave height?

A    Wind speed, length of time it has been blowing, and distance travelled across the open ocean (Section 2.3).

Some of the highest global wave energies occur in the North Pacific (a power $10^5$ W per metre) and to the north-west and north-east of Scotland ($5$–$7 \times 10^4$ W per metre). About $5 \times 10^9$ W of wave power could be supplied by the ocean around Britain, about 20% of Britain's requirements. Waves could also supply the same power for Norway (about 50% of Norway's requirements) and $2 \times 10^{10}$ W for the USA (about 7% of the United States' requirements).

The energy in tides is gravitational, from the Moon and Sun. A high tide is trapped behind a dam in an estuary, and the difference in water-level used to turn a turbine to generate electricity. The power generated

*Table 1.3    The marine catch in 1991, for countries with catches over 1 million tonnes*

| Country | Catch (million tonnes) |
|---|---|
| Japan | 9.10 |
| Former Soviet Union | 8.19 |
| China | 7.61 |
| Peru | 6.91 |
| Chile | 6.00 |
| United States | 5.20 |
| Thailand | 2.80 |
| South Korea | 2.48 |
| Indonesia | 2.38 |
| India | 2.34 |
| Norway | 2.10 |
| Denmark | 1.76 |
| Philippines | 1.70 |
| North Korea | 1.60 |
| Canada | 1.48 |
| Spain | 1.32 |
| Mexico | 1.26 |
| Iceland | 1.05 |

Notes: The UK catch was 0.83 million tonnes. Other countries with catches between half and one million tonnes were Argentina, Brazil, France, Malaysia, Morocco, Myanmar (Burma), New Zealand and Vietnam.

Source: FAO (1991) *FAO Yearbook*, Vol. 72.

depends on the difference in water-level on opposite sides of the dam (the tidal range) and the volume of water trapped behind the dam, so a large volume estuary (but with a narrow neck to minimise the dam construction cost) is necessary for a tidal power site. There are many estuaries with large tidal ranges that are suitable, particularly in the English Channel, Siberia, Alaska and Canada. A proposed tidal power station in the Severn Estuary could supply about $1.5 \times 10^9$ W, about 6% of Britain's requirements.

As well as being able to supply a significant proportion of the world's energy requirements, wave and tidal energy have the advantage that they are renewable energy sources. They are also non-polluting, with no acid rain, greenhouse gas generation or radioactive waste disposal problems. However, despite these advantages there is only one wave or tidal power station of significant size in operation, the $2.4 \times 10^8$ W tidal power station on the Rance estuary in Brittany.

There are two main reasons for this, the most influential one being the cost. Electricity generated by tides or waves is more expensive than that from conventional power stations, because of extremely high capital costs: the Severn barrage has been estimated at about £8500 million, though the long projected life of the power station, about 100 years, would make it competitive in the longer term. The Rance power station has been in operation since 1966 and will continue whereas conventional power stations of equivalent age have been closed down or are reaching the end of their life.

The second reason is the effect on the environment. Environmental effects involve changes to the estuary behind a dam, or to the ocean with offshore wave power barrages involving current and sediment movement and the build-up of pollutants. Locally, these effects will be substantial, but because they are local it is difficult to compare them with the national or global environmental side-effects of conventional power stations. Which are the most environmentally damaging? Would you prefer to live next to a tidal power station or a coal or nuclear one? If your answer is 'tidal', would you also be willing to pay a higher price for your electricity?

◀  The tidal barrage and power station across the Rance estuary uses the rise and fall of the tides to generate electricity. This has been in operation since 1966, but is still the only large tidal power station in existence.

Petroleum can occur beneath the ocean floor of the continental margins, the parts of the oceans which are geologically similar to the land. While petroleum was easily available and plentiful from sources on land there was little incentive to extract it from rocks under the ocean, as this requires working in a more difficult environment with more complex and expensive technology, leading to a higher petroleum price. However, increasing demand and the OPEC price rises in 1973 led to the development of offshore petroleum fields. The major offshore petroleum-producing countries are Saudi Arabia, the UK and Norway, but Abu Dhabi, the former Soviet Union, the USA and Venezuela are also important producers. About 27% of crude oil and 20% of natural gas now comes from offshore extraction.

Offshore exploration and production of petroleum produces environmental effects that can be more damaging than from land-based exploration and production, because of the ease with which any oil spill can spread over the ocean surface and be moved over large distances by tides and currents. In Section 5.4 we will examine oil pollution in more detail. Estimates of the contribution to oil released in the ocean by offshore production are around 0.1 million tonnes a year, which although large is only a small proportion, about a few per cent, of the total amount of oil that reaches the ocean.

## 3.6   Mineral resources

The mineral resources of the ocean can be divided into four main categories:

- the non-metallic aggregates (sand and gravel, calcium carbonate)
- minerals obtained from sea-water
- placers (shallow water metallic minerals)
- deep ocean metallic minerals.

*Aggregates* are the most important mineral resource extracted from the ocean. Sand and gravel are found on beaches, from cliff erosion, and on parts of continental shelves where they have been carried by rivers or are ancient beach sediments from times of lower sea-level. They are extracted by dredging. The cost of recovery is generally more expensive than on land and increases with water depth, so most aggregates are extracted from shallow water, less than 30 m deep. However, the final cost of marine sand and gravel may be cheaper than from land sources if it is used in coastal areas and shipped. The USA, Japan and the UK use large quantities of marine sand and gravel (the Channel Tunnel alone used 2.25 million tonnes) but on a global scale the marine supply is only a few per cent of the total.

Calcium carbonate is the other main aggregate mineral, and is used to make cement. In countries without limestone rocks on land, particularly in tropical countries, calcium carbonate offshore sands, formed from the shells of marine organisms, are mined.

Although sea-water is a gigantic store of many minerals, most are present in very low concentrations (Section 2.2) and only common salt, magnesium and bromine are extracted in significant quantities. Salt is produced in the coastal areas of hot countries by solar evaporation of water in shallow evaporating ponds. Magnesium and bromine compounds are precipitated directly out of sea-water by treating sea-water with other substances.

*Placers* are metallic minerals in beach and continental shelf sediments, mainly minerals of tin, iron and titanium. They have been eroded from ore bodies on land and carried to the ocean by rivers, like sand and gravel, and are separated by the movement of sea-water from the less dense sand and gravel. They are extracted by dredging, generally from shallow water, less than 50 m deep. The main mining areas are Indonesia and Malaysia for tin, Japan for iron and Florida, Sri Lanka and Brazil for titanium. The marine production of these metals is, however, only a small proportion of the world production: 5–10% for tin and titanium and less than 1% for iron.

There are also *metallic minerals* in the deep ocean, polymetallic (or manganese) nodules on the abyssal plains, and metalliferous sediments and crusts on ocean ridges. The main metals in the nodules and sediments are manganese and iron, but their main value lies in the smaller quantities of nickel, copper and cobalt. They exist in vast quantities, but are not being commercially exploited for two main reasons. The first is that the cost of mining in the deep ocean, involving raising the minerals 3–5 km from the ocean floor, is higher than mining on land. The second is the lack of agreement between nations on the rights and profits from mining in an international area: this problem is considered in detail in Section 4.

Extraction of minerals from the ocean can have considerable environmental effects. Dredging is damaging and often lethal to benthic organisms and also affects pelagic organisms by stirring up large quantities of sediment into the sea-water. Nearshore dredging can often lead to beach and coastal erosion as the dredged material may be part of a seasonal beach sediment cycle. Deep ocean mining would bring deep sea-water to the surface which would be colder, have a different salinity, and probably a higher nutrient content.

Comparing these environmental effects with the effects of land mining is difficult, as was the comparison between the effects of land and marine power stations. If done with care, marine mining is probably less environmentally destructive than that on land, but less is known about the long-term effects on the marine environment, especially for the deep ocean, so marine mining could have unexpected side-effects.

## 3.7   Waste disposal

Human society produces large amounts of waste. Disposal of this waste can be on land, to the atmosphere (mostly by incineration) or in the ocean.

There are two opposite and extreme views on waste disposal in the ocean. The earlier was that the ocean is so vast that any waste we add to it will be made more or less harmless by dilution and have no effect on the ocean as a whole. The opposite view is that the ocean is damaged by any introduction of materials from land alien to the environment of marine plants and animals. The reality lies somewhere between these extreme views: the oceans can accommodate a substantial, but not unlimited, part of our waste production and in doing so can protect valuable resources on land, such as space and the quality of the water we drink. Waste disposal to the ocean is also often the cheapest option.

The main questions of ocean disposal are what *kinds* and how *much* waste the ocean can take. The oceans are not naturally 'pure': they already contain large amounts of most substances, although at lower concentrations, and can tolerate more, but how much more can be tolerated without damage? Most ocean waste disposal takes place in coastal or continental shelf waters, with some local areas having heavy and continuous use, and where these are partially isolated from the rest of the ocean and not dispersed the waste can have undesirable effects. These effects include covering up of benthic habitats, stimulating the growth and death of plants (which uses up the dissolved oxygen necessary for other marine organisms) and toxicity.

The main wastes dumped in the ocean are:

- dredged sediments from rivers and harbours
- rock from land excavations
- sewage in raw form or as treated effluent or sludge
- various industrial wastes and heated water.

The waste may be discharged from land by a pipeline (sewage effluent and heated water) or loaded onto a ship and taken further offshore for dumping (dredged sediments and sewage sludge).

*Dredged sediments* are a natural substance of the marine environment, but by being moved elsewhere cause turbidity in the water (reducing photosynthesis) and can bury benthic communities. Unless the dumping is continuous these effects are short-lived. *Sewage* contains a number of substances, with different effects on the ocean. The effects of raw sewage on beaches and of disease organisms in sewage are obviously damaging. Sewage contains the plant nutrients nitrogen and phosphorus, and these may be beneficial or damaging: generally the ocean is poor in nutrients and their addition to the oceans in sewage disposal may benefit plant and animal communities in the same way as do nutrients in upwelling waters. The fish catch in many east coastal areas of England is greater than that in the rest of the North Sea, probably because of sewage disposal into the sea. However, it is possible to overwhelm a local area of the ocean with nutrients, producing excessive algal growths or phytoplankton blooms that poison marine animals. Sewage also contains organic matter which uses

◄ *Industrial effluent being discharged into the North Sea at Hartlepool, Cleveland.*

dissolved oxygen as it decomposes, and in large quantities may deplete an area of the oxygen necessary for marine plants and animals.

Many types of *industrial waste* are dumped in the ocean. Up until 1973 wastes from the china clay industry were dumped into the ocean off Cornwall. Waste from coal mining and slag from iron smelting have been dumped on beaches in northern England. These wastes have severely disrupted the ecology of the areas, reducing and changing the numbers and types of organisms, and are examples of disposal of waste in the ocean with little thought for the consequences.

Substances of potentially greater damage to the oceans are compounds of the heavy metals, such as mercury, cadmium and copper, which are waste-products from chemical, mining and other industries. These are toxic, and disposal on land is inadvisable as they can be dissolved in groundwaters and affect water supplies. Because of this, heavy metal waste has often been dumped in the ocean. Some marine plants and animals can build up high concentrations of these metals from sea-water with no apparent harm, but they are poisonous to other animals and humans that eat them: the problems with mercury in Minamata Bay (Section 3.4) were an example. This does not necessarily mean that no heavy metals should be dumped in the ocean, but does mean that we still need to investigate the effects and pathways of heavy metals in the ocean and that any dumping must be strictly controlled. For example, some heavy metals are essential in small quantities for life in the ocean, but how much more heavy metals can we dump in the ocean without causing problems?

The ocean may also be used for the disposal of *radioactive waste*. For example, nuclear fuel reprocessing at Sellafield discharges radioactive materials into the Irish Sea (*Sarre*, 1996). These include caesium-137 and plutonium which are discharged in liquid form through a pipeline. Radioactive materials may adversely affect all organisms. Although emissions from Sellafield and plants in other countries are limited by the International Commission on Radiological Protection (ICRP), the Sellafield experience has shown that radioactive disposal in shallow coastal waters can lead to human exposure, through consumption of marine shellfish and by accumulation of radioactive particles in beach sediments. The deep ocean floor has also been considered for the disposal of high-level radioactive waste in containers, the assumption being that containers could be built that would not leak, at least during the active lifetime of the waste, or if leakage occurred, that the deep oceans are sufficiently isolated for the leaked waste not to return to the surface ocean. We have seen that this last assumption is false (Section 2.4) and many countries, including the UK, have now abandoned plans for deep ocean radioactive waste disposal.

The ocean is also used as a repository for a completely different type of waste – *thermal waste*, hot water resulting from the use of sea-water for cooling in power stations or industrial processes. These plants are situated near a source of water, sometimes rivers and lakes, but often the ocean. In Britain, for example, all but one of the nuclear power stations are on the coast in order to use sea-water for cooling. The water is returned to the ocean about 10°C warmer than its original temperature. This temperature rise can have both beneficial and detrimental effects on the marine environment. The temperature difference is within the range of natural temperature variations in the oceans (except in tropical and polar regions) and is not harmful to many adult marine animals. An increase in temperature increases their metabolic rate and can be used in mariculture to increase growth rates. However, a temperature rise also decreases the solubility of oxygen in the sea-water and this, combined with increased metabolic rates, can lead to oxygen deprivation. Plankton and the small eggs and larvae of marine animals can be sucked into the power plant with the cooling water, and the sudden rise in temperature can be fatal.

Overall, our disposal of waste in the ocean has had little effect on it. Locally, however, some estuaries, coastal regions and semi-isolated seas have not fared very well. Some hazardous materials should perhaps not be discharged into the ocean at all, and there are some international agreements on this, such as the London Convention for the Prevention of Marine Pollution by Dumping of Wastes and Other Matter in 1972 (the **London Dumping Convention**). In all cases, waste disposal in the oceans should be compared with disposal on land and incineration to the atmosphere on environmental, technical and economic criteria. After all, we have to do something with our waste: we cannot just ignore it, but neither is the old idea of 'out of sight, out of mind – dump it in the ocean' necessarily valid.

*Activity 4*

List the advantages and disadvantages of disposal (a) on land and (b) in the North Sea of excavated material from the site of the National Gallery extension in London. The excavated material is a mixture of soil, rock and building rubble.

## 3.8   A need for legislation?

Many of our uses of the ocean are relatively harmless and exploit the vast resources of the ocean to the benefit of humankind. However, in some cases, overexploitation, such as overfishing, or inappropriate use, as in the disposal of certain wastes, has led to damage to the marine environment. There have also been disputes between states over which state has the right to use which bit of ocean: should the USA dump radioactive waste, for example, in the ocean at all, and if so, in its coastal waters, where they are a potential problem just for the USA, or in the deep ocean, where they become a lesser problem, but shared by other states? Issues such as these have led to attempts to establish an international legal regime for the ocean, which we will examine in the next section.

## 3.9   Summary

The traditional uses of the ocean were shipping and fishing, but we now use the ocean in many more ways. Some of these are non-extractive uses, and some are extractive uses, or resources.

The main non-extractive use is shipping. Extractive resources are of two main types, the living resources and the non-living energy and mineral resources, and as a repository for waste disposal.

Our more extensive uses of the ocean have led to abuses of this environment, by over-use or inappropriate use. For this reason, and because of disagreements about resource exploitation between states, it is necessary to have an international agreement on the legal regime of the oceans.

# 4   The new Law of the Sea

## 4.1   Introduction

Before 1945 the regime governing the use of the oceans was based on the principle of 'freedom of the seas'. Coastal states' territorial waters, where they had full sovereign power, generally extended three miles* from the coast – approximately the range of the first coastal artillery. Beyond that distance, the 'high seas' belonged to no-one. In these areas, according to common law, shipping and fishing could proceed unhindered. This customary law was enforced by the great naval powers, particularly Britain. These powers had a national interest in maximising their freedom of operations at sea, although in practice they did not shrink from interfering with other states' ships during international conflicts.

After 1945 this regime began to erode. Since the late 1960s the fundamental principles of the old law of the sea have been widely challenged. The demise of the old regime has been so rapid that three United Nations Conferences have been held to attempt to codify new rules. The first of these was held in 1958, closely followed by a second in 1960. The Third United Nations Conference on the Law of the Sea met from 1973 to 1982, and culminated in the **UN Convention on the Law of the Sea (UNCLOS)** which was signed by 119 states on 10 December 1982. This gigantic treaty – 320 articles and 9 annexes – has been called a constitution for the oceans, and must rank as one of the most important pieces of written international law of this century.

It provides for a massive extension of coastal state jurisdiction over adjacent ocean space. Beyond that, its clear message is that the resources of the high seas must be managed internationally, with the notion that the oceans are a 'Common Heritage of Mankind' joining that of 'Freedom of the High Seas' as the founding principles of international law.

The new Law of the Sea needed to be ratified by 60 states before it came into force (that, is, the legislatures of at least 60 states needed to confirm their adherence to the Convention in a constitutionally binding way). It took a long time before this threshold was crossed. The Convention did not come into force until 16 November 1994, and then almost entirely on the basis of ratifications by developing countries. The United States and other industrialised states strongly objected to part of the Convention. Until their objections were addressed, they not only refused to ratify the Convention themselves but also used their influence to delay others from joining. It was only in 1994, after the end of the Cold War and with wide acceptance amongst developing countries of market-based principles, that supplementary negotiations to meet western countries' concerns could be successfully completed.

By 1995 the prospects for wide membership of the (revised) Convention by all sections of the international community seemed to be good. Even before it came into force, its rules had already had a profound impact on the international management of ocean resources. However, many issues and disputes remain to be tackled and resolved. This section examines the politics of the development of the new Law of the Sea, and outlines its main

*It is traditional to use (nautical) miles, rather than kilometres, when referring to maritime limits and regulations, and we follow this convention in this book.

provisions. This provides a basis for the discussion of ocean management in Section 5. It also provides an important illustration of the issues and problems, as introduced in the Introduction to this book, relating to the establishment of new international regimes to manage global resources.

## 4.2   *The erosion of the old regime*

To its later regret, it was the United States that made the first direct challenge to the 'old' law of the sea after World War II. The Truman declarations of 1945 claimed US jurisdiction over certain high seas fisheries within 200 miles of the North American coastline, and over the resources of the continental shelf to a depth of 200 metres. The declarations were deliberately formulated ambiguously, in an attempt to avoid damaging the overall regime – which it was in the United States' interest to preserve.

However, the process quickly spread to Latin America. In 1947 Chile claimed rights over both fishing and the continental shelf along its ocean border, followed by Ecuador and then by Peru in 1953, which defined its economic jurisdiction at 200 nautical miles from its coast – the estimated breadth of the Peru current. By 1970 a total of 11 countries had claimed jurisdiction over about 1 900 000 square nautical miles of ocean. At the same time, there was a more widespread trend to extend territorial waters from 3 to 6 or 12 miles.

There were a number of powerful underlying processes eroding the old regime. After 1945 the oceans ceased to be regarded as an inexhaustible resource. Instead they became an area of intense competition for the extraction of potentially scarce goods. This was reinforced by advances in technology, which greatly increased the scale and efficiency of fishing and made it more feasible economically to extract oil and mineral resources from the seabed. Meanwhile, non-extractive uses of the ocean intensified. The density of shipping traffic in many areas made navigation complex and dangerous, and oil tankers and other vessels threatened to pollute coastal states. This generated pressures to redefine **innocent passage** (whereby vessels were allowed to pass through a state's **territorial waters** provided that they intended no harm to that state), and to limit freedom of the seas.

At the same time, the dissolution of the old European empires led to a rapid increase in the number of states: from about 40 to some 150 between 1948 and 1976. Many of these new states have sea coasts, and nearly all are developing countries. These typically had no stake in the old law of the sea, dominated as it was by the developed maritime powers. At first the more established ex-colonial states in Latin America took the lead in challenging the domination of the developed 'North' (as they saw it), but these were joined by newer states in the 1960s. Thus ocean politics became enmeshed in the highly-charged debates between the developed and developing states.

Differences of interest also existed within each of these groups. Developing states broadly divided between inland countries and those that had sea coasts. In the United States, United Kingdom and other maritime developed states, divisions developed between groups such as fishing communities and oil companies, most concerned with the exploitation of coastal resources, and those that wanted to preserve the freedom of the seas for global shipping and naval operations, and for fishing in distant areas.

In this context, revision of the old law of the sea became inevitable. Unregulated claims to ocean resources, and widespread attempts to impose unilateral restrictions on rights of passage or mineral extraction or fishing would lead to chaos and conflict. However, the increasing complexity of the political and economic issues and alignments ensured that the reform process would be difficult and contentious.

The First UN Conference on the Law of the Sea was held in Geneva in 1958, and was attended by 86 states. The agenda was set by the major maritime powers, who aimed to limit the erosion of the old regime. Four conventions were signed, relating to freedom of the high seas, conservation of fisheries, limits on territorial seas and fisheries jurisdiction, and economic rights over the continental shelf. In the first two of these, the principles of the old regime were bolstered fairly successfully and a framework for conservation policy and dispute settlement was agreed. The third and fourth issues proved more difficult, and the wording of the conventions had to be fudged pending the second conference that was reconvened in Geneva two years later.

The second conference came within a whisker of success. A compromise proposal to agree to 6-mile territorial limits and a 12-mile area of jurisdiction over fishing only just failed to obtain the necessary two-thirds majority. On the question of economic jurisdiction over the continental shelf, states had to be satisfied with an agreement to 'sovereign rights' (a fudge between full sovereignty and vague 'rights') out to either 200 metres depth or to 'exploitable' limits (a similarly vague formulation).

Although the North–South issues were already apparent in these conferences, the dominant political divisions ran along East–West lines. In retrospect, the 1960 conference represented a high point in international agreement. Between 1963 and 1966, the four conventions came into force. But as the 1960s progressed, the fragmentary process of unilateral claims and bilateral or regional agreements continued, and the exploitation of ocean resources intensified. The two conferences could not halt the demise of the old regime.

## 4.3   Setting the agenda for the Third UN Conference

In the mid-1960s, as East–West relations improved, the USA and USSR agreed between themselves that a Third UN Conference on the Law of the Sea should be called to update and further codify the customary laws of the sea. They were concerned that the extensions of territorial waters and areas of economic jurisdiction were eroding their rights of passage, and thus the areas of operation for their naval forces. Unless these customary rights were reaffirmed and defined in international law, the rights of innocent passage for their naval and commercial vessels might be limited. This was particularly worrying in relation to international straits – without clear transit rights, the extension of territorial seas to 12 miles could close some 125 straits of potential importance (such as the Bering, Gibraltar and Malacca Straits, and passage through the seas around Indonesia and the Philippines: see Figure 1.15).

In 1966 the USA and USSR worked out a skeletal Law of the Sea treaty comprising only three articles. On the first two they were in complete agreement: a maximum 12-mile territorial limit and free passage through international straits. The USSR had reservations about the third article, providing for preferential fishing rights in adjacent seas. The USA proposed this in order to win support from coastal states. To the United States it seemed worthwhile to accept extended jurisdiction over fishing and resource exploitation in coastal waters if this could ensure the freedom of the seas for naval forces.

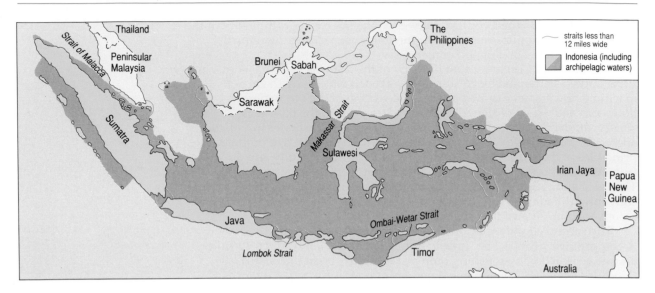

▲ *Figure 1.15   The seas around South East Asia: these waters include important international navigation routes, several of which have been threatened by an extension of territorial waters to 12 miles unless rights of innocent passage have been reaffirmed (as illustrated on the map).*

However, the agenda of the conference broadened in a way that was unwelcome to the superpowers. In November 1967 a Maltese minister, Arvid Pardo, made a historic speech at the UN General Assembly. He noted that advances in technology were opening up the possibility of the extraction of resources on or below the deep seabed. The potential wealth of these resources was staggering. In addition to sunken treasures, many analysts at the time estimated that 'manganese nodules' (containing manganese, nickel, copper and cobalt) lying on the ocean floor could be worth trillions of dollars. The possibility of deep-sea oil-wells, and mining of polymetallic sulphides and cobalt-, manganese- and nickel-bearing crusts under the seabed further emphasised the great value of the resources under the high seas.

Arvid Pardo argued that the wealth of the deep seabed should be declared to be part of the 'Common Heritage of Mankind', and that the Law of the Sea should be revised to ensure that it was shared by all members of the international community. To this end, he proposed that an International Seabed Authority (ISA) should be established to manage deep seabed mining. This position was soon incorporated in resolutions of the UN General Assembly. Third world states, co-ordinated through the **'Group of 77'** (a group of developing countries, originally 77, that operates mainly in the United Nations), adopted this proposal as an element in their campaign for a New International Economic Order and promoted it consistently thereafter.

*Q*   Consider the interests of the developed maritime states on this matter. How were they likely to regard such a proposal?

*A*   The USA, the UK and other developed states have an interest in a minimal framework to assure orderly seabed mining, but not in a supranational management authority and a scheme to redistribute wealth to developing nations. After all, it is the developed states, and

You call it the common heritage of mankind—what about us fish?

the large mining consortia based in them, that actually possess the capital and technology to mine the deep seabed. However, the notion of an area to be administered as the 'common heritage of mankind' was at least better than a national enclosure movement, where claims to exclusive rights over the seabed and fisheries extended even beyond 200 miles. Moreover, the maritime powers wanted to get the negotiations under way, in order to establish rights of passage. So the proposal for an International Seabed Authority was not rejected out of hand.

Pardo and others (such as Alva Myrdal of Sweden) also broadened the conference agenda in other ways. They foresaw that nuclear missiles or anti-ballistic missile systems might be based on the seabed, and argued for a ban on the use of the deep ocean floor for military purpose. Furthermore, they drew attention to inadequate controls over the dumping of radioactive and toxic wastes in the high seas, and over other sources of pollution such as oil spills. These issues were also taken up in General Assembly resolutions. Thus the major naval powers lost control of the conference agenda.

## 4.4   The Third UN Conference on the Law of the Sea

In formal terms, the Third UN Conference on the Law of the Sea began in 1973. In fact, however, the discussions really began in December 1967, when the General Assembly voted to establish the Ad Hoc Committee to study the peaceful uses of the seabed. After much debate, this resulted in a treaty forbidding the emplacement of weapons of mass destruction on or under the seabed in the high seas. This implicitly involved an agreement that the limits of territorial waters would be set at 12 miles from the coast. The treaty was signed in 1971 and, having received sufficient ratifications, came into force in 1972. For the superpowers, which in any case had no plans to place their nuclear weapons on or under the deep seabed, it seemed to be a small price to pay if it would clear the way to negotiations for a treaty to recognise naval rights of passage.

The negotiations – which began in earnest in 1974 and concluded in 1982 when the Convention was adopted – were often bafflingly complex, involving thousands of delegates and experts. Aware that the major powers were really most interested in establishing rights of passage, the majority of states agreed that the new regime would be negotiated as a 'package'. It would be regulated according to one single convention, which states would have to accept or reject as a whole: they could not sign up only to the parts that they liked.

*Q*   Consider the consequences of this approach for the negotiating process.

*A*   One consequence of this was that the negotiation process became very complex and prone to delay, as trade-offs were made. For example, inland states with no direct interest in regulations over fishing jurisdiction could obstruct the negotiations on this issue unless their own concerns about an entirely different article of the treaty were taken into account.

As the negotiating agenda became clear, states rushed to establish their claims. In the early 1970s, the notion of an **Exclusive Economic Zone (EEZ)** gained popularity, as an increasing number of states extended their claims to jurisdiction over sea and seabed resources within 200 nautical miles of their coast (see Figure 1.16). Between 1970 and 1977 the ocean area claimed by coastal states increased from 1.9 million to almost 20 million square nautical miles. Part V of the 1982 UN Convention on the Law of the Sea (UNCLOS) recognised such claims, implying that a total of about 28.5 million square nautical miles of the Earth's global commons were transferred to the jurisdiction of coastal states – a figure that exceeds the combined land area of Australia, Europe, Africa and North and South America (see Figure 1.17). For many island states, the areas of their EEZs greatly exceeded that of their land territory (shown in Figure 1.18).

This transfer was achieved more or less peacefully. One of the great successes of the 1982 Convention is that it lays out procedures for settling disputes, which became binding once it came into force. Provision is made for an International Tribunal for the Law of the Sea, as well as a number of

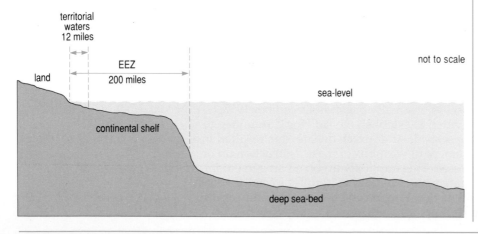

◀ *Figure 1.16*
*Simplified seabed profile, illustrating areas of coastal states' jurisdiction.*

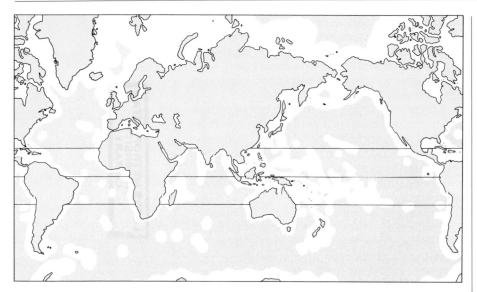

◀   *Figure 1.17*
*Map indicating the areas of*
*Exclusive Economic Zones.*
*If all coastal states claimed*
*200-nautical-mile EEZs,*
*they would cover 36% of*
*the area of the oceans.*

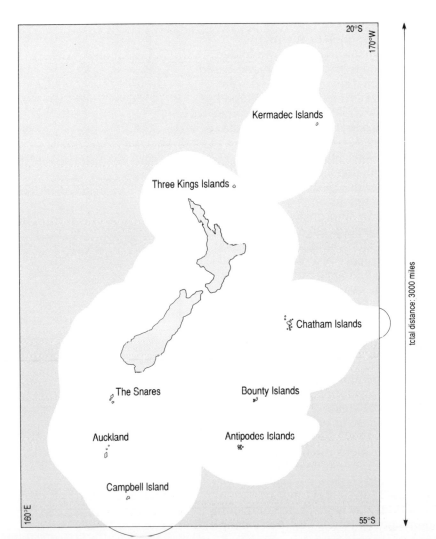

◀   *Figure 1.18*
*In many cases the area of a*
*state's EEZ is far greater*
*than its land area: the*
*example of New Zealand.*

other conflict resolution and adjudication bodies. Governments, companies and individuals all have standing in these dispute settlement procedures.

In its Exclusive Economic Zone a coastal state has sovereign rights over the resources of the zone, to exploit the minerals and to conserve and manage the fisheries. They also have jurisdiction over marine research and any installations such as oil rigs, and over the protection of the environment. Further, they have a duty to protect the marine environment in this area, although their right to impose more stringent regulations than the 'generally accepted international rules and standards' is very limited unless they can persuade the **International Maritime Organisation** of their case. An exception to this is in the Arctic Ocean, where Canada and other states bordering the Arctic Ocean won greater rights to regulate potentially polluting activities in ice-covered areas.

Parts II and III of the 1982 Convention establish the right of innocent passage of all naval and commercial vessels through the waters of an EEZ, and also of overflight by aircraft. Further, they provide for the right of innocent passage through the territorial waters of another state. The Convention also establishes the right of all ships (and submarines) to 'transit passage' through international straits, except in the case of 'historical straits' such as the Dardenelles where a separate regime in international law has long been established.

The 46 articles of Part XII of the 1982 Convention deal with the 'protection and preservation of the marine environment'. They avoided getting to grips with the problems of pollution from rivers and land-based sources: these raise the difficult issue of national sovereignty, and were opposed by many developing states as well as groups in the developed world. Developing countries argued that they could not afford to spend money on environmental protection. Indeed, a Brazilian delegate even implied that pollution was a symbol of economic progress in the developing world. By the end of the negotiations, in 1980–82, these attitudes were beginning to change, but negotiations on this part of the treaty had been completed and few wanted to risk reopening the issue. So the Convention limited itself to general exhortations on pollution from sources on land, and focused on pollution from ships instead.

In this area, a substantial number of agreements already existed such as the MARPOL Convention (1973) and the London Dumping Convention (1972), as described in Section 3.7. At the insistence of a number of states, such as Canada, Australia, Kenya, Indonesia and Spain, UNCLOS was conceived as an 'umbrella' convention which sought to bring these existing conventions, and the declarations of the 1972 Stockholm Conference on the Human Environment (discussed in Section 5 below) within one legal framework. Thus, various articles restrict 'harmful' dumping of waste or the release of oily waste from ships on the high seas. Further, there is an obligation on states to 'cooperate, to the extent possible, in eliminating all the effects of pollution and preventing and minimising the damage', and to 'jointly develop and promote contingency plans for responding to pollution incidents' on the high seas as well as in coastal areas.

In the main, the further development of specific environmental protection regimes beyond the conventions of 1972 and 1973 mentioned above was left to existing 'competent organisations' such as the International Maritime Organisation and the International Atomic Energy Authority. However, the 1982 Convention did create a new category of state with rights and duties: the 'port state'. Such states can start proceedings against a ship in one of its ports that is suspected of making illegal discharges on the high seas or in another state's EEZ. For example, if a ship unloading at Rotterdam was suspected of discharging pollutants illegally in the North Sea, the Dutch government would be entitled to begin proceedings against its owners.

Much the most contentious aspect of the negotiations related to Part XI of the treaty, which dealt with arrangements to control and manage seabed resources under the high seas. This is not surprising, in view of the differences of interest between most developing and developed states on this issue, as discussed above.

The third world governments, organised in the 'Group of 77' (but consisting of 116 members by 1980), argued strongly for an **International Seabed Authority (ISA)** that would control and manage the exploitation of deep seabed resources and the distribution of the profits to the governments of the whole international community. The prospect of receiving vast amounts of capital to contribute to economic development meant that the Group of 77 was unwilling to compromise on the principle that the deep seabed resources were part of the 'common heritage of mankind'. For inland states, which would not benefit from the establishment of EEZs, Part XI of the Convention was the only part that promised direct benefits, and they were prepared to obstruct negotiations on other parts of the treaty in order to obtain them.

Not all third world states had an unambiguous interest in the mining of minerals on the seabed. Several countries, such as Zambia, Zaire, Papua New Guinea and Botswana, depended greatly on earning foreign currency by mining copper and other minerals in their territory (Figure 1.19). Seabed minerals threatened to reduce world prices and thus foreign currency earnings. These states therefore joined with developed states in a similar position, such as Canada, to restrict and regulate the rate at which seabed resources could be mined. Interestingly, however, the political solidarity of the Group of 77 was sufficient for them to present a united front against the United States, United Kingdom and West Germany and other developed states that were strongly opposed to the proposed ISA.

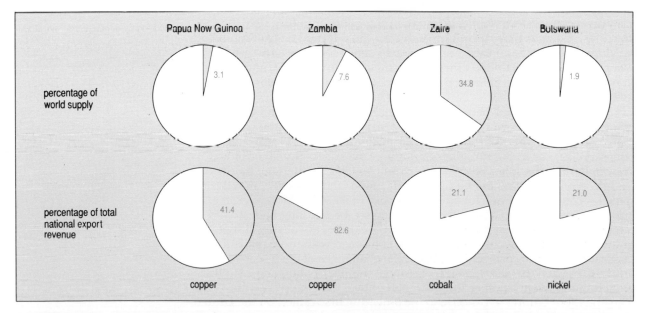

▲  Figure 1.19   Some land-based mineral producers that would be vulnerable to competition from seabed mining.

Most developed states were basically opposed to international control and management of these seabed resources. In opposing the Group of 77 proposal, they argued that the key to seabed mining was to provide assured access, security of tenure, and assurance of a reasonable profits that would make it feasible to risk the billion dollar investments that would be necessary even to begin exploiting these resources.

In the end, a compromise was formulated for the 1982 Convention. The ISA and mining consortia would mine 'in parallel'. Adjacent parcels of seabed would be exploited: one by a private company and the other managed by the ISA. Any company or state wishing to extract mineral nodules or mine the deep seabed would apply to the ISA for a licence, and would pay tax on its earnings. The ISA would have first choice of which parcel to manage, and its own mining operations would initially be underwritten by the West. It would have access to western technology 'at fair commercial rates'. To accommodate mineral producers such as Zambia, seabed mining would be limited to supply only a proportion of growth in world demand.

## 4.5   Bringing it into force

The UNCLOS negotiations finished in 1982, when the new Convention was signed at Montego Bay, Jamaica, by 119 states. Another 38 states signed the agreement in the following two years. Since it was not possible to agree only to parts of the treaty, the governments of the USA, UK and West Germany refused to sign, declaring opposition to Part XI – the international regime for deep seabed mining. Twelve other states also decided against signing, for a mixture of reasons.

Legally, the 1982 Convention could only come into force after it had been ratified by 60 states. By spring 1991 only 45 states had ratified the treaty, and further progress seemed likely to be slow. Moreover, although there was enough support amongst developing countries to ensure that it would ultimately come into force, it was widely recognised that the new regime would be gravely weakened if the USA and most other developed countries continued to refuse to join it. In this context, in July 1990 the UN Secretary General took the initiative to convene informal consultations to explore whether the concerns of developed states could be addressed in a way that was acceptable to other UN members. Between 1990 and 1994 fifteen meetings were held on possible amendments to the disputed Part XI of the Convention.

Much had changed between 1982 and the early 1990s. With the end of the Cold War, there was a new spirit of co-operation within the UN between the members of the old 'Western' and 'Eastern' blocs and their allies. At least as important, the developing countries' campaign for a 'new international economic order' had run out of steam. Faith in the merits of central economic planning had greatly declined, and market-oriented policies commanded much wider acceptance or support. Moreover, the earlier expectations of a bonanza from deep sea-bed mining had proved illusory. The decline in mineral prices and the discovery of substantial mineral deposits on continental shelves, combined with more realistic assessments of the costs and technological challenges of deep seabed mining, pushed the anticipated commercial mining of manganese nodules well into the twenty-first century. In this context, developing countries became more willing to concede to the amendments wanted by the United States and other industrialised countries.

Initially the discussions were held amongst a relatively small group of concerned developed and developing states. After 1992 they were opened to all UN members, with a growing sense of urgency as it became clear that the Convention would soon come into force. On 28 July 1994 the UN General Assembly adopted a resolution – by a vote of 121 in favour, none against, with seven abstentions (Colombia, Nicaragua, Panama, Peru, Russia, Thailand and Venezuela) – to adopt an 'Agreement relating to the Implementation of Part XI of the United Nations Convention on the Law of the Sea of 10 December 1982' and to urge states to adhere to it and to the Convention. More than fifty states signed the new Agreement almost immediately, including the United States and most other developed countries.

The 1994 Agreement is to be interpreted and applied together with Part XI of the Convention as a single legal instrument. Where these are inconsistent, the Agreement will prevail. Essentially, the Agreement accommodates all the earlier objections of the United States and the other developed states to the deep seabed mining provisions in the 1982 Convention. It embraces market-oriented policies, and eliminates provisions such as mandatory transfer of technology and ceilings on mining production. It streamlines the ISA and curtails its regulatory discretion, substantially increases the influence of developed countries in its decision-making, and delays and sharply confines the role of the ISA's own mining operation.

With this agreement in place, the prospects for establishing the new Law of the Sea as a global and effective convention greatly increased. Many issues and disputes relating to the international management of ocean resources are not resolved by UNCLOS – for example the management of fisheries, whaling, pollution and semi-enclosed seas, as discussed in the next section. States such as Indonesia remain dissatisfied with rules that could be interpreted to give intrusive rights of passage throughout its archipelagic waters, and there are continuing disputes about the boundaries and rules of EEZs in some regions. However, the amended Convention provides an important framework for tackling these issues and problems, and a major example of efforts to construct environmental regimes.

## 4.6   Summary

The UN Convention on the Law of the Sea (UNCLOS), signed in 1982, provided for a new international regime for the management and exploitation of the oceans. This section has examined the development of this new regime, and discussed some of its main provisions.

*Activity 5*

Review this section and prepare notes on:

(a)  the main reasons why a new law of the sea was needed;

(b)  the main interest groups involved in the UNCLOS negotiations; and

(c)  the extent to which these interest groups have achieved their objectives.

# 5   Ocean management

## 5.1   Introduction

This section examines the management of the ocean and its resources. UNCLOS has, in some situations, helped to clarify rights and responsibilities, but does not solve all management problems. Only a few issues can be discussed here, so these have been selected as representative of the different aspects of ocean management.

Fisheries management and pollution control are large-scale ocean management topics in this section, involving all parts of the ocean. A study of the whaling industry is used as an example of fisheries management in the high seas area that involves international co-operation and agreement.

The management of the North Sea is used as a comparison. This is an area which is also under the control of more than one state, but is within the states' EEZs, with national legal rights and responsibilities.

## 5.2   Management of fisheries

It may seem initially that UNCLOS has solved one of the major problems of fisheries, that of international disagreement over fishing rights and quotas. It establishes the right of each coastal state (and places a responsibility on each coastal state) to implement policies which will achieve and maintain an optimum yield from the living resources within its 200-mile wide EEZ and to decide on allocations of fishing rights to fishing vessels of other countries. We have already seen that coastal regions have the most abundant and important food resources (Section 3.4) so the national control of fisheries in these regions clarifies the responsibility of management for most of the world's fisheries. However, fisheries management still has problems. These arise from the mobility of fish (they do not recognise national boundaries! ), interaction between various fish and other species, and overfishing. To understand these problems, it is necessary to look at what controls the population in a fishery.

The total biomass (weight) of a fish species (the **stock**) is increased by the growth of adults in the population ($G$) and from young fish (recruits) joining the population ($R$): see Figure 1.20. ('Stock' is a term that may also be used to mean number of fish.) The stock is depleted by death from disease, old age and predation (natural mortality $M$) and capture by fishing (fishing mortality $F$). The stock, gains and losses are related by the equation:

$$\text{end stock} = \text{beginning stock} + (G + R) - (M + F)$$

In an equilibrium fishery, the stock does not change, so that:

$$\text{end stock} = \text{beginning stock}$$

and:

$$(G + R) = (M + F)$$

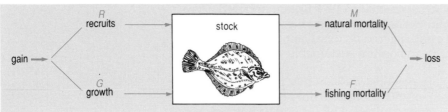

◄ Figure 1.20
The controls on stock size
in a commercial fishery.

The efficient management of a fishery should give the greatest catch year after year (the **maximum sustainable yield**) while maintaining an equilibrium stock. It is unfortunate that the maximum sustainable yield can often only be estimated once a fishery has been overfished. The following discussion illustrates this.

A stock of fish is generally greatest when it is not being exploited. In this state it includes a high proportion of larger and older fish. When fishing begins, the large stock yields large catches to each fishing boat, but because there are few boats the total catch is small. Increased fishing reduces the stock, so the loss due to natural mortality decreases, and is less than the gains from growth and recruitment ($G + R$ greater than $M$). If the catch is less than the difference between natural gains and losses the stock will tend to increase ($G + R$ greater than $M + F$); if the catch is more, the stock will decrease ($G + R$ less than $M + F$): this is **overfishing**. When the stock neither increases nor decreases, it is in equilibrium, making a sustainable yield possible. This sustainable yield is a maximum when the stock is somewhere between two-thirds and one-third of the unfished stock. In this state the average size of the individuals will be smaller and the age will be younger than the unfished stock, and individual growth will be highest.

The maximum sustainable yield can also be related to fishing effort. Initially the total catch increases as the fishing effort increases (more or bigger boats or better fishing gear). This tends to encourage an even greater fishing effort, until the total catch begins to fall due to overfishing (shown in Figure 1.21). The maximum sustainable yield is possible when there is a moderate amount of fishing. This can either be obtained by fishing the older fish heavily and leaving the younger ones alone, or by catching moderate amounts of fish of all sizes.

There are two types of overfishing – growth and recruitment. **Growth overfishing** occurs when too many small fish are caught, so that their growth potential is not exploited. **Recruitment overfishing** occurs when too many adults are caught, leaving not enough fish to spawn to produce new recruits. This occurs rarely, as fish lay so many eggs that usually recruitment does not depend on the adult stock but on other factors, although if it does occur it is more serious than growth overfishing as it can threaten the survival of the whole stock.

The herring fishery in the North Sea is an example of overfishing. The fishing effort increased gradually between 1950 and 1962 but the catch remained fairly stable at 0.85 million tonnes. There was a big increase in effort from 1963 onwards, which combined with good natural recruitment in 1961 produced a peak catch of 1.4 million tonnes in 1965. The catch decreased to 0.7 million tons in 1970 and continued to decrease throughout the 1970s. This decline was due to first growth overfishing, then recruitment overfishing: by 1969 herring larvae numbers had declined to only about 20% of their numbers at the beginning of the 1960s, severely depleting the recruitment to the adult stock and endangering the

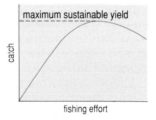

▲ Figure 1.21
The relationship between
fishing effort and maximum
sustainable yield.

population. The situation was so severe that in 1977 Britain imposed a ban on North Sea herring catches. Stocks began slowly to increase again, but by the mid-1990s, were once more endangered by overfishing. If this fishery could be managed effectively, the maximum sustainable yield is estimated at 0.75 million tonnes a year.

The North Sea herring fishery demonstrates another problem of multinational fisheries management that UNCLOS does not solve, that of the migration of fish between waters controlled by different countries. There are three groups of North Sea herring, which lay eggs on the seabed in different spawning areas, in waters controlled by Britain and France (see Figure 1.22). The eggs hatch to produce larvae which then drift to the German and Danish coasts. The larvae develop into young fish (whitebait) and swim offshore to the nursery area in Dutch, German and Danish controlled waters. When they are two to three years old the young herring join the adult population (recruitment) in the adult feeding grounds in British and Norwegian controlled waters. The action of one country to control the herring fishery is therefore ineffective without controls by other countries. For example, the British ban on herring catches in 1977 would need not only an equivalent ban by Norway, but agreement between the Netherlands, Germany and Denmark to stop or reduce fishing for whitebait in the nursery grounds (which threatens recruitment) and agreement between Britain and France to protect the spawning areas.

This discussion of the North Sea herring fishery has considered herring in isolation from any other species (**single species management**) but a more realistic approach is to also consider the effects and changes in other species on the herring (**multi-species management**). Adult herring, for example, feed mainly on planktonic animals and are in competition with other pelagic fish, and the herring are also a food source for cod and other benthic fish, so a change in the stock of another species by fishing may also affect the herring stock.

Unfortunately multi-species fisheries management is far from straightforward, as often the detailed effects of one species on another is not known, and the importance of each species may be difficult to balance. The reduction of pelagic fish species in the North Sea has coincided with an increase in some demersal species, but there is no direct link. The North Atlantic cod catch was 3–4 million tonnes a year throughout the 1960s, but has fallen since then as a consequence of overfishing (though a contributory factor may have been growth of the fishery for capelin, one of the important food items for cod). The North Pacific fur seal has declined in numbers

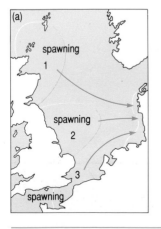

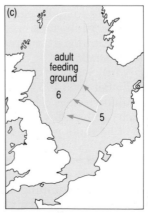

◄   *Figure 1.22*
*The movement of herring in the North Sea.*
*(a) Herring lay eggs mainly in three different sprawning areas (1, 2, and 3). The planktonic larvae drift to shallow water off Germany and Denmark (4).*
*(b) The young herring swim offshore to the nursery ground (5).*
*(c) Young herring join the adult population in their feeding grounds (6).*

since the growth of the Alaska pollack fishery, an important food for the seal: which is more important, the commercially more valuable pollack, or the seals, essential to small populations of Inuit?

Fisheries can also affect each other when boats hunting for one species of fish catch – without intending to – many species, some of which may even be wasted. The US shrimp boats catch a wide variety of animals in their trawls, but usually only the shrimp are retained, and the rest of the animals, often three or four times the quantity of the shrimp, are discarded at sea, because a boat's limited hold space is more profitable to use for the higher-value shrimp. Prawn fishing in the Irish Sea has a similar effect and kills (among other things) many young cod – a particularly damaging form of growth overfishing. The tuna fishery that uses nets in the Pacific catches many dolphins, which – being air-breathing – drown when caught in the nets. The dolphins are not used: they are thrown overboard as they are commercially less valuable than tuna. This need not happen, as tuna can be caught by long-line methods which will kill far fewer dolphins, but this is often a commercially less efficient way of catching tuna. Do you prefer to have cheaper tuna even if it means killing dolphins that are not then used for food? Or would you be prepared to pay more to avoid this?

There are various controls that can be used to regulate a fishery, in an attempt to obtain a maximum sustainable yield.

1   *Net size/landing size*   A minimum size can be set for the mesh of a net, which lets the younger fish go free, and/or a minimum size of fish can be landed. This may be able to control growth overfishing. Regulations of this type were applied partially successfully to the North Sea herring in the 1950s and 1960s. It provided some protection for the adult herring, but recruitment still declined due to a whitebait fishery with much smaller net sizes in the herring nursery ground.

2   *Annual catch limit and fishing vessel limit*   The total catch can be limited by quota, whereby fishing stops when a specified tonnage has been caught,

and/or by controlling the size and number of vessels. Properly implemented, such measures can be effective, but the social and economic costs can be severe: whole fishing communities can be put out of work, their vessels laid up or de-commissioned. The biggest problem with quotas is that of multi-species fishing. Reaching the quota for a target species commonly involves exceeding the quota for one or more others. In the case of the North Sea and north-east Atlantic fisheries, for example (and probably elsewhere too), exceeding quotas carries severe legal and financial penalties, so great tonnages of dead and dying fish are routinely dumped at sea (of the order of 25 million tonnes annually worldwide in the mid-1990s, compared with the total annual global commercial catch of around 80–90 million tonnes per year).

3    *Restricted fishing seasons and areas*    This can be used to encourage recruitment, by protecting spawning areas and times and nursery grounds. This is more draconian than either of the others, and its effect on fisheries is correspondingly greater. It is also more controversial, because evidence of drastically declining fish stocks is rarely unequivocal. It is very difficult to demonstrate that an abrupt decline in stocks results from overfishing because stocks can fluctuate from one year to the next by an order of magnitude or more, through *natural* causes (such as unfavourable conditions at the spawning grounds, excessive predation of eggs and larvae en route to or from the nursery grounds: see Figure 1.22).

In the mid-1990s it was estimated that two-thirds of the world's commercial fisheries were in serious decline, and it was becoming clear that the alternative to sensible implementation of (preferably) all of the measures listed above would be collapsed fisheries and all fishing communities out of work.

Implementation of these controls in the past has often been accompanied by international disagreements over restrictions and quotas. The Law of the Sea has clarified the position in some cases but will not prevent other disagreements. For example, the Anglo-Icelandic 'Cod Wars' of the mid-1970s, in which Iceland wished to limit fishing within 50 miles of its coast to protect stocks, would now be a clear case of Iceland having the right to do so. There have been many multinational regional fisheries commissions, often supported by the United Nations Food and Agriculture Organisation (FAO), but states often disagree on what action to take. The North-East Atlantic Fishery Commission was unable to reach agreement on action after the 1955 collapse of part of the North Sea herring fishery: states acted individually and therefore ultimately ineffectively.

## 5.3    *Whaling*

Whales are not fish, but air-breathing mammals. There are two kinds of whale, the toothed whales and the baleen whales. The toothed whales include the sperm whale, which is the largest toothed whale (up to 20 metres long), the killer whale and dolphins. They eat fish and squid, and the sperm whale can dive to depths of around 1000 metres in search of food. The baleen whales do not have teeth, but instead horny baleen plates ('whalebone') that hang from the roof of the mouth, forming sieves which filter their food, usually planktonic crustaceans, such as krill, from sea-water. Baleen whales include the blue whale, which can be up to 27 metres

◀   *Whaling in 1835.
Whaleboats (large rowing
boats) were used to get
close enough for a thrown
harpoon to reach the
whale. The whales were
usually larger than the
whale boats and could
overturn them, making
whaling a very dangerous
occupation.*

long and is the largest animal that has ever lived on the Earth, the sei at 18
metres long and the minke, up to 9 metres long.

Whales have very different birth rates from fish, which produce
thousands of eggs. Whales produce a maximum of one calf (young whale)
every two years. The baleen whales live mostly in polar waters, where there
is high productivity of their food in spring and summer. They migrate to
warmer tropical waters in winter where their calves are born. The toothed
whales are not restricted to polar waters, because their food occurs
throughout the oceans.

The history of whaling provides an extreme case of over-exploitation of
a marine living resource. Whaling began around a thousand years ago in
the North Atlantic, catching whales that swim close to shore. By the
seventeenth century, whales were caught in deeper water, by the British
and Dutch. These were bowhead whales, also called right whales as they
did not sink when they died and could be retrieved by small boats, so they
were the 'right' whales to hunt. The North Atlantic right whale stock
declined, so whalers began to hunt sperm whales in other oceans as well as
the North Atlantic. The development of larger and faster motor-boats and
of the explosive harpoon about a century ago led to an expansion of the
whaling industry and the ability to catch other whales, the fin and
sei whales.

This led to a decline in whale stocks in the North Atlantic, so British
and Norwegian whalers began hunting whales in the Antarctic at the
beginning of this century, establishing land-based processing factories. The
whaling industry expanded further when Japanese, German and Russian
ships also began whaling in the Antarctic, and by 1925 factory ships began
to be used to process the whales.

This expansion began to affect whale stocks in the Antarctic, with the
catch of the largest and therefore most profitable whale, the blue whale,
falling in the 1930s (Figure 1.23a). The catch per unit effort of blue whales
also fell (Figure 1.23b) indicating that more whales were being caught than

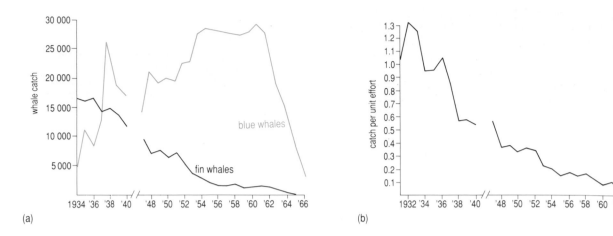

▲  *Figure 1.23   (a) The catch of blue and in fin whales in the Antarctic. (b) The catch per unit effort (catch per catcher–day's work) of blue whales in the Antarctic.*

the maximum sustainable yield. As blue whale numbers declined, whalers began to catch more fin whales (Figure 1.23a) and then the smaller sei and humpback whales.

The falling whale catches caused the whaling states to set up the **International Whaling Commission (IWC)** in 1946, to set guidelines on protected species, restricted areas, catch quotas, size requirements and the import of whale products. The obviously over-exploited blue and humpback whales gained complete protection from the 1960s and catch quotas were set for many others, such as the fin and sei whales.

The IWC tried to steer a course between what was good for the whale stock, and what was good for the whalers. The whaling industry was based on a large labour force and capital investment in ships and did not want to stop catching whales, but neither did it want to over-exploit them to the extent of destroying their own livelihood. The IWC had to conserve whale stocks, with a sensible balance between the current catching of whales and maintaining the stock for future use. Unfortunately this has proved difficult to achieve, for a number of reasons.

One reason is that the IWC jurisdiction extends only to those states who are members, and in addition not only are there no means of legal enforcement to members, but also an IWC member can avoid being bound by decisions by filing an objection, although diplomatic and economic pressure has sometimes been used successfully to prevent this. Past violators have included the USSR in 1974 and 1975, Chile, Peru and South Korea in 1978 and Norway in 1986. An objection filed by Japan in 1987 was withdrawn after US threats to cut the Japanese fishing quota in the US EEZ. UNCLOS has been of little use to the IWC as a legal means to enforce its jurisdiction, as most whaling occurred in the high seas areas.

Another problem for the IWC occurred when many states who were not involved in whaling joined in the early 1970s. These states joined with the aim of stopping whaling altogether: whales, along with pandas and seals, had become a major symbol of the environmental movement. This polarised the IWC into the whalers, who wanted to conserve the stock but still exploit it as any other living resource, and the non-whalers who believe killing whales is wrong, either because the whales are in danger of extinction or regardless of how many whales there are. There is no doubt that catching whales by harpooning is far more cruel than the slaughter of domestic land animals, but how endangered are whales?

Some are endangered to the point of extinction. The population of such widely scattered marine animals cannot be estimated accurately, and blue whales have such a low population that even their numbers have proved especially difficult to estimate: as many as 14 000 or as low as 200. If the lower estimate is correct the blue whale has almost no hope of survival. The North Atlantic right whale is estimated at about 1000: the stock has not increased since the 1920s and there is little hope of recovery. There are about 2000 Pacific bowhead, but these are in real danger, as 20–30 of these are still caught each year in the Arctic by the Inuit, a catch allowed by the IWC to support Inuit culture (whales are the traditional prey of the Inuit). Humpbacks number about 10 000 and grey whales 21 000 and both are considered as endangered species. On the other hand, sperm whales, at around 2 million and minke whales at 750 000 appear to be in no danger. The minkes may even have benefited from the depletion of the other baleen whale stocks by reducing competition for food.

Because of the protection of some whales, and catch quotas on others, the whalers began to catch mostly minkes in the 1970s. The maximum sustainable yield for whales is between 3–5% of the stock (this is a very different value from the maximum sustainable yield of a fish stock because of the difference in birth rates), so a minke whale stock of 750 000 would allow 22–38 000 minke a year to be caught whilst still conserving the stock. However, the IWC decreed a total ban on commercial whaling in 1985, a victory for the environmental non-whalers, even though some minkes and sperm whales could probably be exploited without endangering the species.

The ban is, in part, an understandable reaction – or over-reaction – to the endangering of so many of the whale species, but the lack of scientific grounds for banning whaling of all species caused some states to exploit a loophole in the ban, which permits the catching of some whales for scientific purposes. In the mid-1990s the only two nations still catching whales (mainly minkes and no more than a few hundred annually) were Japan and Norway, despite some international condemnation and harassment by ships of environmental groups.

## 5.4   Pollution

Pollution of the ocean occurs on every scale, from the local collection of bottles, floats and other plastic objects on beaches, to the international effects of major oil spills and the worldwide presence of the pesticide DDT in the oceans. These examples are unquestionably pollution, but how about ocean dumping: is the disposal of sewage in the oceans, which can have beneficial as well as detrimental effects, pollution? Is the natural seepage of oil from the ocean floor into sea-water that occurs in some areas pollution? It depends on how pollution is defined. The definition used here is that stated at the UN 1972 Conference on the Human Environment:

> Pollution means the introduction by human beings, directly or indirectly, of substances or energy into the oceans resulting in harm to living resources, hazards to human health, hindrance to marine activities including fishing, impairment of quality for use of sea-water and reduction of amenities.

Q  Are (a) sewage disposal, and (b) natural oil seepages, pollution as
   defined above?

A  (a)  Sewage may or may not be a pollutant, depending on whether it
   has harmful effects on living resources, human health, sea-water
   quality or amenities. (If dumped in limited quantities in appropriate
   places it will not usually be a pollutant.)
   (b)  Natural oil seepages, as they are not human-made, are not considered
   pollutants. (This still holds, even in areas where natural seepages are
   much higher than the introduction of oil to the oceans through human
   causes, or where natural seepages have damaging effects.)

Marine pollution can be continuous (a cooling-water outfall ) or episodic
(sewage dumping). It can be deliberate (radioactive waste dumping) or
accidental (oil spills). Pollutants vary in how long they remain in the ocean
before being broken down into other substances or removed into sediments
or to the atmosphere. Naturally occurring substances, such as domestic
sewage, usually persist for a short time in the ocean, but human-made
substances, such as pesticides, persist for a longer time (Figure 1.24). It is
the very persistent and permanent pollutants, such as pesticides and toxic
metals, that have continuous and long-term effects on the ocean and are of
most concern.

   Pollutants enter the ocean by diverse routes. The main source is directly
from land, through outfall pipes into the ocean or from rivers, which can
carry sewage, industrial waste, fertilisers and pesticides. Pollution also
comes from the ocean, both accidentally and deliberately from ships and by
mining of ocean resources. Some pollutants reach the ocean by fall-out or
wash-out from the atmosphere.

   Oil is often thought of as the major pollutant in the ocean, prompted
largely by the visible results – tarred beaches and oil slicks – with the
attendant damage and death to marine life. Estimates of the amount of oil
reaching the ocean from human-made sources range from 1.9 million
tonnes to 4.8 million tonnes annually. The oil industry is responsible for

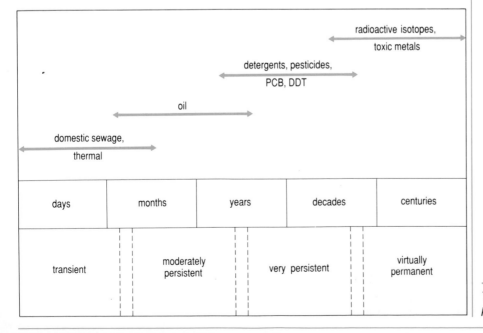

◀  *Figure 1.24*
*The persistence of*
*pollutants in the ocean.*

about one-third of this, from tanker transportation, offshore production and refineries; shipping accounts for about another quarter, the rest coming from land sources, in sewage and rivers. Much of the latter arises from careless or thoughtless discharges: in Britain about 20 000 tonnes of oil annually from do-it-yourself car oil changes is poured down the drain, most of which reaches the sea. Only about 5% of the oil reaching the ocean is accidental, due to oil-rig blow-outs or tanker accidents, although these are the most spectacular and often most damaging incidents.

Oil varies in its persistence in the oceans from days to years depending on the type of oil: see Figure 1.24. Persistent oils are crude oil, heavy fuel oil and lubricating oil. Non-persistent oils are gasoline and light fuel oil. In a crude oil spill at sea, about 25% of it, the lighter volatile fraction, will evaporate in a few days (athough this takes longer under cold conditions) (Figure 1.25). This fraction tends to be the most toxic. The oil vapour is finally broken down by reactions with oxygen in the atmosphere. Part of the remaining oil dissolves, and part is stirred by wind and waves into an emulsion with sea-water, most of which floats on the surface. It looks like, and is referred to as, 'chocolate mousse'. Micro-organisms can break down the oil, both in the water and in the ocean-floor sediments. Around 60% of the oil is degraded within a few months of the spill but, like evaporation, this process is much slower in the colder oceans. The remaining 15% or so of oil forms dense, persistent clumps called tar balls, which may float, wash ashore or sink to the bottom.

◀   Clean-up of a beach in France after an oil spillage from a tanker. Oil spills are most damaging in coastal areas.

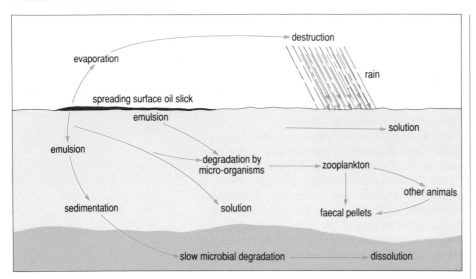

◄ *Figure 1.25*
*The dispersal and*
*degradation of an oil*
*spill in the ocean.*

Oil can have damaging and often lethal effects on marine life. These effects are usually not severe in the open ocean, where most of the organisms live below the surface, as most of the oil floats. The major casualties are seabirds, whose buoyancy and thermal insulation depend on feathers which repel water but attract oil. Fortunately, the population of seabirds in the open ocean is usually low. Oil has the most damaging effect in shallow water and coastal areas where the oil cannot disperse so easily and can affect benthic habitats and beaches. Here oil penetrates the fur of seals, otters and other marine mammals, which are affected in much the same way as birds. Fish are relatively immune to coating with oil, and so are some seaweeds, as they have a surface that repels oil, but most smaller organisms can do nothing to prevent being smothered, and are killed by lack of light, lack of food or asphyxiation. The lighter oil fractions are also often poisonous to marine organisms, but the effect varies with type of organism, the season and the stage in the life-cycle.

The oil spill off Alaska from the VLCC *Exxon Valdez* was one of the most studied, and environmentally disastrous, oil spills. The tanker collected 140 000 tonnes of crude oil from the oil terminal of Valdez, which is at the southern end of the Trans-Alaska oil pipeline. 37 000 tonnes of this oil were released into Prince William Sound on 24 March 1989 when the tanker foundered on a rock outcrop off Bligh Island. This is the largest oil spill to date (mid-1996) in US coastal waters. The response to the spill was very slow, which allowed the slick to disperse into the Sound, and high winds and rough seas made the dispersed slick difficult to recover or contain. The oil spread southward from Bligh Island, to the west of Montague Island, and further out of the Sound and on some of the shorelines in the area. In two weeks the slick had travelled up to 250 kilometres from the holed tanker, spreading over an area of 3–4000 square kilometres and washed up on 1100 kilometres of shoreline.

The Alaskan coast is an area with a rich diversity of wildlife and living resources (fishing is the main industry), but is also a very fragile environment, with many species living at the limit of their ecological ranges, so this oil spill had a catastrophic effect. Most of the sea otters in the area were killed, and many of the seals and vast numbers of seabirds,

estimated at 30 000. The main fisheries are for salmon, herring, cod, crabs and other crustaceans; almost the whole 1989 season's fishing was prevented and the effects extended over subsequent seasons, by the damage to spawning areas, killing of crustaceans and tainting of fish. The shoreline, despite an attempted clean-up operation that cost the Exxon oil company $1 million a day, remained polluted for years. Assessments of the situation highlighted the inadequacy of contingency plans for an oil spill of this size in such bad weather in a polar environment. The slick spread over too large an area, and was too broken up by bad weather for methods of recovery or containment – skimmer boats, booms, ignition and chemical dispersants – to be effective. Once the oil hit the shoreline there seemed to be no effective method of removing it.

Most of the environmental problems caused by the *Exxon Valdez* oil spill were caused by its occurrence in an area that was coastal, polar and fragile. This is illustrated by the very different impact of another spill the same year. The VLCC *Kharg 5* was in the Atlantic on passage from the Middle East around Africa to Europe with 284 000 tonnes of crude oil when it was holed by an explosion on 19 December 1989. 70 000 tonnes of oil leaked into the ocean, but this has had little environmental effect. The oil had a high proportion of the lighter oil fractions, and this more toxic part rapidly evaporated from the warm waters. The spill occurred steadily over a period of about two weeks, and most was emulsified and dispersed by wave action. There was concern that the slick might be washed up on the Moroccan coast and ruin beaches and hence the tourist industry, but although it came within 30 kilometres of the coast the slick never reached the shore. A few months later the slick seemed to have largely disappeared due to dispersal and micro-organism degradation. World War II was a time of many oil spills into the ocean from sunken shipping, but the ocean was able to recover from such temporarily devastating pollution.

*Activity 6*

Q1 What was the maximum speed of movement of the oil slick from the *Exxon Valdez* up to 7 April 1989, in km hr$^{-1}$?

Q2 What area could the slick cover at this speed one day after its release, if it spread out in all directions? (The area of a circle is $\pi r^2$ where $\pi$ is approximately 3 and $r$ is the radius.) What implications does this have for recovery of the oil?

Another group of pollutants of great concern in the ocean are some synthetic compounds containing carbon, hydrogen and chlorine. These include dichlorodiphenyltrichloroethane (DDT), a pesticide, and polychlorinated biphenyls (PCBs) which are used in a number of products, such as plastics, paints and electrical equipment. DDT and PCBs enter the ocean mainly from the atmosphere, so they are most concentrated in surface waters, and they are ubiquitous throughout the world oceans. They are stable compounds that persist for a greater time in the ocean than does oil (Figure 1.24), and although they are present in low concentrations in the ocean, around 1 part per billion parts of sea-water, organisms tend to accumulate DDT and PCBs in their bodies at higher concentrations than those in the surrounding sea-water. Oysters, for example, living in sea-water with 0.1 parts per billion (ppb) DDT were found to have 7000 ppb DDT. This process is called **biological magnification**, and continues up the food chain, so it can ultimately affect human beings. DDT and PCBs are toxic, and can have adverse effects even in low concentrations: they can reduce the productivity of phytoplankton, cause abortion and foetal abnormalities in marine mammals, thin and break the eggshells of birds and cause skin diseases.

An intrinsic problem in the management of marine pollution is that a pollutant that enters the sea in one part of the ocean can move into other parts, so that pollution-control measures, to be effective, often have to be international. For example, in an attempt to control pollution of the Mediterranean Sea, seventeen Mediterranean coastal states agreed on a Mediterranean Action Plan in 1976. The North Sea is an area where there are similar international agreements. The Law of the Sea has had some success in controlling pollution from dumping in the high seas, such as banning the discharge of oily water from tank washing. This has reduced the pollution, but not stopped it: the problem of detection and legal enforcement remains.

The ocean, because of its immensity, has been able to dilute, disperse and so mitigate the worst effects of many of the pollutants, but pollution occurs to some extent in all parts of the ocean. Some pollutants are a problem everywhere because of their toxicity and persistence, and some are only a problem in local areas. Some seas which are partially isolated or have a limited water exchange with the open ocean are very polluted, for example, by sewage at the mouth of the Ganges and off New York, by oil in the Persian Gulf, sewage and industrial waste off Buenos Aires and by mining waste off British Columbia. In contrast to these areas, the centres of the oceans are relatively much less polluted: however, the author of this chapter was very saddened to recover tar balls and plastic bottles when towing a net to collect plankton in the centre of the Atlantic – it seems that no part of the ocean, however remote, is completely immune to pollution.

## 5.5   *The North Sea*

The North Sea is a partially isolated part of the ocean that is relatively shallow and mostly less than 100 metres deep (Figure 1.26). The southern part is less than 50 metres deep, but it gets deeper to the north. The North Sea is bounded by eight coastal states – Belgium, Britain, Denmark, France, the Netherlands, Norway, Sweden and Germany. It is about 500 kilometres wide between Britain and Norway in the north, but in the south the distance between Britain and the other European states is much less, so that the EEZs of each state extend not to 200 miles, as they would overlap, but to a median line at lesser distances from the shoreline. The relatively small size of the North Sea, its natural resources and proximity to areas of high population and industrialisation have made the North Sea the most intensively exploited part of the ocean, with a management problem that has to involve eight states.

The North Sea, particularly the southern part, has always been one of the busiest shipping areas in the ocean, particularly into the ports of Europort/Rotterdam, Zeebrugge/Ostend and Harwich. The ship traffic density on the approach to these ports is such that ships must keep to traffic lanes (i.e. drive on the right) in the major shipping channels, like on roads

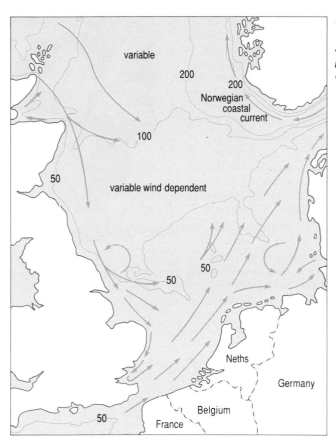

◀   *Figure 1.26*
*The depth (in metres)*
*and surface current*
*pattern (green arrows)*
*in the North Sea.*

on land. There is also a considerable amount of ferry traffic between Britain and the rest of Europe which cuts across the main shipping lanes. The northern part of the North Sea has had an increase in tanker traffic in the last twenty years because of the development of oil and gas fields. The effects of this traffic density on the marine environment include the construction of docks and dredging of estuaries, deliberate discharge of waste, accidental loss of cargo overboard, collisions and sinking. However, no port-of-refuge has been designated around the North Sea for a damaged oil tanker: no state wants to accept a dirty and dangerous stricken tanker.

Fishing is of major importance in the North Sea. The fish catch doubled from 1948 (1.75 million tonnes) to 1974 (3.5 million tonnes) but reduced to around 2.4 million tonnes from 1978.

*Q*   What proportion of the world fish catch is provided by the North Sea?

*A*   The world fish catch is around 70 million tonnes (Section 3.1), so the North Sea provides (2.4 ÷ 70) × 100%, about 3.5% of this.

This is a very significant catch for such a small area of ocean. In addition to fish, 0.17 million tonnes of molluscs (oysters, mussels and other shellfish) and 0.04 million tonnes of crustaceans (crabs and lobsters) are caught.

The major pelagic fisheries of the North Sea are for herring and mackerel, which made up a large proportion of the catch in the 1950s and 1960s. The main demersal fisheries in the North Sea are for plaice, sole, cod, haddock and whiting. The plaice catch has risen from 0.07 million tonnes in 1957 to 0.15 million tonnes in 1984. Sole catches have been stable at around 0.02 million tonnes for the last ten years. However, the cod and haddock fisheries are in trouble. Cod catches rose to 0.34 million tonnes in 1972 and have since declined to 0.2 million tonnes. Haddock catches have fluctuated between 0.05 million tonnes and 0.55 million tonnes in the last twenty years.

Another type of fishery has developed in the North Sea from the 1980s, which takes smaller species – sand-eels, sprats, pout and blue whiting – for conversion to fish-meal. The catch is large (around 0.35 million tonnes) and has implications for other fisheries as these smaller species are often the prey for other fish, as well as for other animals such as sea birds.

We have seen that overfishing of herring led to a collapse of the stock in the 1950s and a total ban on fishing in 1977. The stock had recovered sufficiently for the ban to be lifted and limited fishing allowed by 1982. The herring has not been the only example of overfishing: in a 1977 report on North Sea fish stock, cod, haddock, whiting and plaice were identified as subject to growth overfishing, mackerel to recruitment overfishing and sole to both. The situation with cod and haddock developed into recruitment overfishing, and by 1988 both were subject to strict control of catch and net sizes.

This overfishing of many of the stocks in the North Sea was partly caused by the past absence of any fisheries agreements or legal enforcements for stocks other than those close to the shore. In the late 1960s fishery resources of the North Sea were managed by the 1966 European Fisheries Convention, by which the coastal state has sole rights to fish less than six miles from shore, and to share rights between six and twelve miles with those states that had fished in those waters between 1953 and 1962. This left the rest of the North Sea (most of it) as an unmanaged free-for-all fishery. After the UK, Denmark and Ireland joined the EC, the **Common Fisheries Policy (CFP)** changed management rights and responsibility from the states to the Community, with equal access to all member-states. Norway, which has an economy that is more dependent on fishing than the economies of other states, rejected membership of the EC largely because of this CFP. Following Iceland's claim to a 200-mile fisheries limit in 1976, in 1977 the EC states also claimed jurisdiction up to 200 miles, which put most of the North Sea under CFP control. The main effect of this was to stop fishing by Eastern bloc countries, Iceland, the Faeroes, Sweden and Spain and it did, at least in theory, provide the power for the EC to manage fish stocks effectively. UNCLOS III confirmed these 200-mile fisheries limits. Britain, although a non-signatory, takes full advantage of the limits.

CFP management of the North Sea has, however, been subject to much disagreement between members on individual quotas and multi-species management objectives. The UK, for example, argues that as a large share of the fish stocks lie within its water it is entitled to a large proportion of the total catch. There are also different consumer preferences in each country, leading to disagreements over which stocks have priority: the English want to eat and maintain the cod stocks, the Germans redfish and saithe, the Dutch plaice, the French whiting, the Scots haddock. The result is that CFP management of the North Sea fisheries is not as effective as it could be.

The other two main resources that are extracted from the North Sea are aggregates and petroleum. Aggregates (sand and gravel ) are used in the construction industry and although they are available from land sources, from the 1960s onwards increasing amounts have been extracted from the North Sea, to about 6 million tonnes annually from the UK North Sea

waters, about 15% of the total UK supply of aggregate. Extraction of marine deposits were considered as often more economic for coastal areas and less environmentally disruptive than extraction on land, leaving no large holes in the ground. The damage to the marine environment has often seemed less important (out of sight, out of mind). Marine aggregate extraction in UK waters is controlled by the Crown Estate Commission, which consults with the NCC, DoE, MAFF and the Hydraulics Research Station over environmental implications. In the early 1990s there were six principal extraction areas, mainly between 7 and 20 miles offshore, as well as extraction from shoreline areas. Environmental problems arising from marine aggregate extraction include the destruction of benthic communities, including fish spawning and nursery grounds and mollusc fisheries, damage to bottom trawls by an uneven seafloor and an increase in turbidity of the sea-water.

Production of gas from the North Sea began in 1968 and has been rising steadily: 70 000 million cubic metres were extracted from UK fields in 1994. Oil production started in 1971 and, including natural gas liquids, exceeded 250 million tonnes in 1994, with roughly equal contributions from Norwegian and UK fields. This has resulted in the location of many exploration rigs and fixed production platforms in the North Sea. These can have beneficial effects – they become rapidly colonised by benthic organisms which attract fish – but also increase the risk of oil pollution from blow-outs or collisions between a ship and platform. The largest oil spill in the North Sea so far has been the blow-out on the Ekofisk Bravo platform in 1977. This was in the central North Sea and was less of an environmental disaster than it could have been, as the wind kept the slick away from the shorelines, allowing it to be broken down by wave action and by chemical dispersants.

The use of the North Sea that has caused most international disagreement and most environmental concern is for waste disposal. The North Sea receives waste from rivers, coastal discharge, direct dumping and atmospheric incineration of waste at sea.

The southern part receives most of the river-borne waste, from major rivers that flow through highly populated and industrialised parts of Europe, such as the Rhine, Weser and Elbe. Much of this river water is highly polluted with sewage and industrial waste (Figure 1.27). What happens to these pollutants when reaching the North Sea depends on the

◀   A fireboat continues safety spraying of sea-water onto the Ekofisk Bravo platform after the blow-out.

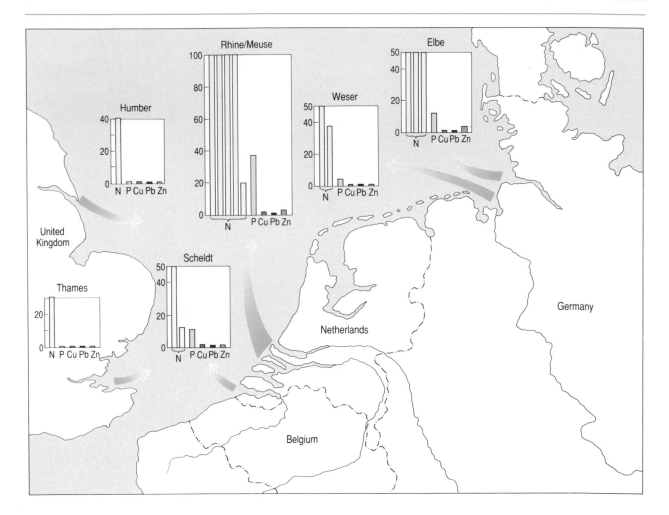

*Main sources of river pollution in the southern North Sea (thousand tonnes per year): nitrogen (N), phosphorus (P), copper (Cu), lead (Pb), zinc (Zn).*

circulation within the sea and water exchange with the Atlantic Ocean. Water generally circulates in an anticlockwise direction in the North Sea (Figure 1.26) which means that water discharged from UK crosses the North Sea towards German waters in the east. Pollutants from continental rivers also move east to Dutch and German waters, causing a build-up of pollution in the eastern coastal zone: Figure 1.28 shows the distribution of dissolved zinc as an example of this. This happens to be fortunate for Britain – its pollution becomes someone else's problem – but not much fun for the Netherlands, Germany and Denmark.

Whether pollutants will accumulate in the North Sea or be diluted by the large body of water in the ocean depends on the rate of exchange of water between the two. Table 1.4 gives the major inputs and outputs of water to the North Sea.

The volume of water in the North Sea is about 47 000 cubic kilometres, so with a water input or output of 56 550 cubic kilometres a year this gives a residence time for water in the North Sea of 47 000 ÷ 56 550 years – a bit less than a year. This is often referred to as the *flushing time* for the North Sea, that is on average water (and any associated pollution) will move out of the North Sea within a year. This prevents an extreme build-up of pollutants, but many problems remain. The increased amounts of nutrients

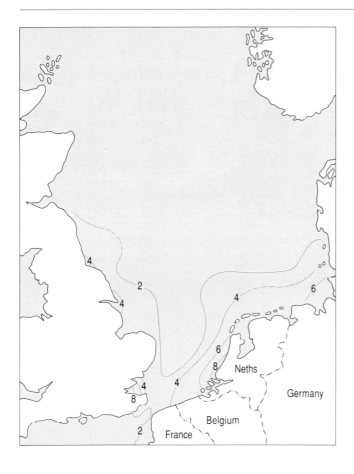

◀   Figure 1.28
The distribution of dissolved
zinc in the North Sea
($10^{-6}$ grams per litre).

Table 1.4    Annual water budget for the North Sea

|  | Volume/km$^3$ |
|---|---|
| *Input* | |
| Between Scotland and Norway | 50 450 |
| Through Straits of Dover | 4 900 |
| From Baltic Sea | 500 |
| Precipitation | 330 |
| From rivers | 370 |
| | |
| *Output* | |
| Norwegian coastal current | 56 300 |
| Evaporation | 250 |

(nitrogen and phosphorus) particularly in the coastal waters of the
Netherlands, Germany and Denmark is a major concern, and has caused
algal blooms that have contained enough toxins to kill animals which
consume them and also to deoxygenate the sea-water when they die and
decompose. Another concern is the build-up of persistent pollutants (toxic
metals, DDT and PCBs) in the water and sediments in the same areas. Some
of the diseases in fish have been linked to the high levels of these
contaminants and there are many local examples of their effect on
organisms. The epidemic in 1988 that killed about half of the common seals
in the North Sea by a distemper virus was suspected to have been
(although not proven to be) exacerbated by the effect of pollution on the
seals.

There have been many multinational agreements by the coastal states
on specific environmental problems in the North Sea, such as the 1974
Convention for the Prevention of Marine Pollution by Dumping from Ships
and Aircraft, but until 1984 no overall environmental management
agreement. In 1984 West Germany, prompted by the pollution problems in
West German waters and by the growing political influence of the Green
Party, organised a ministerial conference of North Sea states for protection
of the North Sea, to review the state of the North Sea and formulate
proposals to improve its quality. A second conference was held in London
in 1987 and agreed on declarations of intent for protective measures. These

included a ban on dumping rubbish overboard from ships, the reduction of riverborne pollutants, and the phasing out of waste-dumping or incineration at sea. This sounds as if it should solve all pollution problems, but this has proved not to be so, because of a loophole in the declaration, which is 'No material should be dumped in the North Sea unless there are no practical alternatives on land and it can be shown to the competent international organization that the material poses no risk to the marine environment'. Some states, although signatories of this declaration, are using this loophole to continue dumping, claiming no practicable alternatives on land or no risk to the North Sea. The UK, for example, continued to dump 5 million tonnes of sewage sludge a year in the North Sea, claiming no risk, and saying that it has reduced the concentrations of toxic substances in the sludge. The other states disagree and have stopped dumping sewage sludge. The UK also continued to dump colliery spoil and ash from power stations at sea off the north-east coast of England, claiming no practicable alternatives on land. However, at the third conference at The Hague in March 1990 the UK agreed to end dumping of sewage sludge by

▲  *The North Sea is heavily used, for shipping, fishing and waste disposal. Sludge dumping from a UK barge in the North Sea, a practice which is to be phased out over the 1990s (top left). Protests over the continued pollution of the North Sea by the dumping (top right). Tanker and petrochemical storage tanks and refinery at Europort in the Netherlands (below). Europort is the largest petrochemical port in Europe.*

1999 and industrial waste by 1993, mainly because of condemnation and pressure from the other states and environmental groups.

Despite disputes, the 1987 and 1990 Conferences on the Protection of the North Sea were the first attempt to manage all types of pollution in the North Sea on a whole North Sea basis. Management of the North Sea has to balance environmental integrity and exploitation. The problems have not yet been solved, but they have at least been recognised and plans agreed for improvement.

## 5.6   Summary

Fisheries management is complicated by the mobility of fish, as they ignore national boundaries, so fisheries policies often involve multinational agreements. Most fisheries occur within EEZs, which gives the coastal state the legal fishing rights, and also the responsibility for conservation of fish stocks. Ideally, fishery policies, including catch quotas and net sizes, should allow a maximum sustainable yield of a fish stock. To achieve a maximum sustainable yield the population dynamics of the fishery must be known. A falling catch despite a greater fishing effort is known as overfishing. Growth overfishing occurs when too many small fish are caught. Recruitment overfishing occurs when too many adults are caught and not enough fish are left to spawn and produce new recruits. Fisheries policies, to be effective, need to involve multi-species implications.

Whaling is a fishery that takes place mainly in the high seas area, so is not under the legal control of one or even a few nations. It has a history of over-exploitation and some species of whales are in danger of extinction. At present there is a ban on commercial whaling, although some whales could be exploited without endangering the species.

Pollution, like fishing, is a problem without national boundaries. Substances that may be harmless in some areas of the ocean or at low concentrations may become pollutants in other areas or at higher concentrations. The substances of greatest concern are the persistent pollutants that are ubiquitous and increasing in concentration in the world ocean. Sewage, although locally damaging, has a short-term effect. Oil spills are extremely damaging in coastal areas, but less so away from land. The ocean is able to dilute many pollutants and mitigate the worst effects, but pollution is present throughout the oceans and can be a major problem in coastal areas.

The North Sea is an area with unique management problems: it is bounded by eight coastal states with high population and industrialisation, and, although relatively small in size, is intensively exploited by shipping, fishing, aggregate extraction, petroleum extraction and waste disposal. It has a limited water exchange with the open ocean, so pollutants are not quickly diluted. Management of the North Sea has required multinational agreement and is now within the control of the coastal states as it is within the EEZs.

Ocean management involves a balance between conservation and exploitation. Unfortunately in many instances exploitation has gone too far, resulting in problems such as overfishing or oil spills. In contrast, in the present management of whaling, the total ban may be taking conservation further than necessary.

# 6   *Conclusion*

This chapter on the marine environment is an investigation of an area that we know less about than the land, even though it is larger in size. We have looked at the main features of the ocean – the different regions, the salinity, chemical composition and movement of sea-water and oceanic life – as a basis for understanding the marine environment. The chapter examined our uses of the ocean in the past, and looked at how the uses have developed to the present. The ocean has important natural resources, both food and physical resources, which are exploitable, but have often been over-exploited, damaging the environmental integrity of the ocean.

Competition for these resources was a major factor leading to the establishment of a Law of the Sea, which divided the ocean into areas under the control of the coastal states, out to 200 miles from the shoreline, and the high seas area, to be managed for the benefit of all. UNCLOS, as modified by the 1994 Agreement, is upheld by most nations.

Environmental problems in the ocean arise from inappropriate use or over-exploitation. It is, for example, an appropriate place to dump limited quantities of some substances, or to catch controlled amounts of fish, but it is not a universal waste disposal site or unlimited larder.

Management of the ocean environment is as critical as management of the land environment, and similarly has to balance conservation and exploitation. In the ocean these views can reach wider extremes than on land, from the ultra-conservationist view that any exploitation of marine environment is wrong, to the out of sight, out of mind view that we can do what we like to the ocean as it is far enough away not to bother us or big enough to look after itself. A more realistic view is between the two: the ocean is a very useful part of the Earth, but must be used sensibly.

## *References*

GRIGG, D. (1991) 'World agriculture: productivity and sustainability', Ch. 2 in Sarre, P. and Blunden, J. (eds) *Environment, Population and Development*, London, Hodder and Stoughton/The Open University (second edition) (Book Two in this series).

SARRE, P. (1996) 'Environmental issues in Cumbria', Ch. 1 in Sarre, P. and Reddish, A. (eds).

SARRE, P. and REDDISH, A. (eds) (1996) *Environment and Society*, London, Hodder and Stoughton/The Open University (second edition) (Book One in this series).

SILVERTOWN, J. (1996a) 'Earth as an environment for life', Ch. 5 in Sarre, P. and Reddish, A. (eds).

SILVERTOWN, J. (1996b) 'Inhabitants of the biosphere', Ch. 6 in Sarre, P. and Reddish, A. (eds).

SILVERTOWN, J. (1996c) 'Ecosystems and populations', Ch. 7 in Sarre, P. and Reddish, A. (eds).

## *Further reading*

First, two general books on oceanography, the former elementary, the latter more advanced:

SMITH, S. (1982) *Discovering the Sea*, Longman (UK), Time-Life (USA).

MEADOWS, P. S., and CAMPBELL, J. I. (1990) *An Introduction to Marine Science* (second edition), Glasgow, Blackie.

To follow up the issues in Section 4 see:

UNITED NATIONS (1983) *Official Text, UN Convention of the Law of the Sea*, (Sales No E.83.V.5).

The UN also published a useful set of background pamphlets on UNCLOS III, including *A Quiet Revolution* (E.83.V.7), and *Law of the Sea – the new UN Convention* (a set of nine papers).

SANGER, C. (1986) *Ordering the Oceans: the making of the law of the sea* (London, Zed Books) is a useful historical discussion of UNCLOS III.

CHURCHILL, R. and LOWE, A. (1988) *The Law of the Sea* (second edition) (Manchester, Manchester University Press) is a detailed description and discussion of the terms of the treaty.

See also 'Law of the Sea Forum: the 1994 Agreement on Implementation of the Seabed Convention on the Law of the Sea', *American Journal of International Law*, Vol. 88, No. 4, pp. 687–714 (1994).

A comprehensive report and assessment of current marine pollution and future threats is:

GESAMP (1991) *The State of the Marine Environment*, Oxford, Blackwell Scientific.

Finally, two journals which cover marine environmental topics are *Marine Policy* (Butterworth Scientific) and *Marine Pollution Bulletin* (Pergamon).

## Answers to Activities

### Activity 1

We can see from Figure 1.3 that the North Sea is mainly part of the continental shelf so should be around 100 metres deep.

Figure 1.2 shows that the Mid-Atlantic Ridge lies at the middle of the Atlantic, and the top of this should be between 2 and 3 kilometres deep.

The ocean trenches are the deepest parts of the oceans, and these are not usually in the middle of the oceans.

### Activity 2

*Q1*  1 km$^3$ of sea-water is $10^3 \times 10^3 \times 10^3$ m$^3$ = $10^9$ m$^3$.
If the density is $10^3$ kg m$^{-3}$, 1 km$^3$ has a mass of $10^3 \times 10^9$ kg = $10^{12}$ kg.
A gold concentration of $4 \times 10^{-6}$ p.p.m. is:

$4 \times 10^{-6} \times 10^{-6}$ kg in 1 kg sea-water.

that is $4 \times 10^{-12}$ kg gold in 1 kg sea-water.
The amount of gold in 1 km$^3$ = $10^{12} \times 4 \times 10^{-12}$ kg = 4 kg.

*Q2*  Value = amount $\times$ price = $4 \times$ £10 000 = £40 000.

*Activity 3*

- The main factor is the distribution of food for the marine animals, and this is dependent on the distribution of plants, the ultimate food source of all animals.
- Plant growth needs (a) light, which means they live in the surface layers of the oceans (the photic zone), and (b) nutrients.
- The nutrient supply is greatest where nutrients are supplied from land or by upwelling.

*Activity 4*

(a)  Land disposal

Advantages:
- Apart from building rubble, natural land material.
- Useful for land-fill.

Disadvantages:
- Transport costs to land-fill sites, at a distance from central London.
- Environmental impact of road transport.

(b)  North Sea disposal

Advantages:
- Material not toxic.
- Short transport distance by road to docks.

Disadvantages:
- Unless dumping strictly controlled, can damage benthic communities.

*Activity 5*

(a)  After 1945 the old regime governing the use of oceans, based on freedom of the high seas and three-mile territorial waters, began to erode. As the use of ocean resources intensified, the oceans came to be regarded as an area of competition for the extraction of potentially scarce goods rather than as the inexhaustible resource of old. The USA and Latin American states began to declare jurisdiction over fisheries and minerals up to 200 miles from their coasts, followed by other states. An increasing number of states had little stake in the old regime, which was widely seen to work in the interests of the developed maritime powers and their allies. In this context, agreement on a new Law of the Sea regime became a priority. Otherwise, unregulated claims to ocean resources and unilateral actions and restrictions threatened to lead to chaos, conflict and over-exploitation.

Earlier UN Conventions aimed to slow the erosion of the old regime. In this they were partially successful. But by the mid-1960s it became clear that a new UN Convention on the Law of the Sea (UNCLOS) was needed to update and further codify existing laws and create a new international regulatory framework for the use and exploitation of the oceans. Negotiations started in 1973, and continued for ten years. Not surprisingly, the task of agreeing to regulations governing most aspects of the use and exploitation of the oceans was extraordinarily complex and prone to delay and obstruction. However, to properly understand the difficulties experienced in the talks and the ultimate results, it is important to appreciate the major interest groups involved.

(b)  There was a common interest in establishing an agreed regulatory framework to avoid conflict and mismanagement of ocean resources, though the details would obviously be controversial. But the major maritime powers were primarily interested in establishing rights of passage and 'freedom of the seas'. Coastal states were interested in establishing rights to an Exclusive Economic Zone stretching up to 200 miles from their coasts. Developing countries wanted to ensure that the resources of the deep seabed were shared amongst all states, and not just by the developed states and large companies with the technology to extract the resources.

(c)  The final UNCLOS document of 1982 provided for all of these interests, and dealt with many other issues besides. It provided the basis for the international management and exploitation of the oceans, and is among the most important pieces of international law ever agreed. The USA, UK and other developed states wanted most parts of the Convention, but took the position that the provisions of Part XI of the convention dealing with deep seabed mining set a number of undesirable obligations, precedents and constraints on private mining consortia, and gave too much discretionary power to the ISA in which they had too little power. By linking agreement with Part XI to acceptance of the rest of the Convention, developing states were able to drive a hard bargain in the UNCLOS negotiations.

After 1982 most developed states sought to establish most of the parts of UNCLOS as customary law and international practice, while refusing to ratify the Convention itself without amendments to Part XI. This threatened the effectiveness and legitimacy of the whole regime, while making it clear that the seabed mining regime, as envisaged in the 1982 Convention, would not materialise. In the changed political and economic circumstances of the 1990s, and with much reduced expectations of the reliable wealth from deep seabed mining, many developing states reassessed their interests and bargaining power. They agreed to make concessions to developed states in order to salvage some elements of the deep seabed mining regime, and also in order to reinforce the other elements of the Convention, in which many developing countries had come to believe they had stronger interests than they had earlier recognised.

## Activity 6

*Q1*  The slick travelled up to 250 km in 14 days. 14 days is $14 \times 24$ hours = 236 hrs. This gives a maximum speed of about 1 km hr$^{-1}$.

*Q2*  Oil could have travelled up to 24 km from the tanker in 1 day, so could cover a circular area of radius 24 km. This is $3 \times 24^2$ km$^2$ = 1800 km$^2$. After just a day the oil has covered a considerable area of ocean, needing recovery methods capable of coping with a large area of oil.

# Chapter 2   Damage to the ozone layer

## 1   Introduction

As you read this chapter, look out for answers to the following key questions:

- What is the role of ozone in the atmosphere and how is it being changed?

- What are the problems of modelling the effects of enhanced concentrations of ozone-depleting substances?

- What recent changes have there been in ozone concentrations?

- What are the likely consequences of ozone depletion?

- What political negotiations have taken place and what changes are needed?

The Montreal Protocol, signed on 16 September 1987, was the first international agreement to restrict release to the atmosphere of substances deemed damaging to the global environment – specifically, the ozone layer. The substances controlled by the Protocol include a group of chemicals called **chlorofluorocarbons** – or **CFCs**, for short. These compounds first came on stream in the 1930s, as the 'ideal' – stable, non-flammable, non-toxic – substitutes for the decidedly noxious chemicals then used as refrigerants. They are also cheap to produce, and their uses soon expanded into the now familiar sectors – as propellants in aerosol spray-cans, as blowing agents (to put the bubbles into all manner of 'foamed' plastics), and as solvents and cleaning fluids in many speciality areas (notably, the electronics industry). Unless steps are taken to prevent it, all of these uses can lead ultimately to the CFCs being released into the atmosphere.

The signatories to the original Protocol pledged effectively to halve their use of CFCs by the end of the century. Just eighteen months later, international opinion was hardening behind the need for a complete phase-out: this aim was finally adopted in June 1990. A powerful driving-force here was undoubtedly the real – and growing – concern about the massive loss of stratospheric ozone over Antarctica that occurs with the return of sunlight each southern spring – the so-called **ozone hole** – first detected in 1984. By 1988 there was irrefutable evidence linking this seasonal loss with the presence in the stratosphere of chlorine atoms, largely derived from CFCs. It would appear that the international community acted with commendable speed in tackling this problem.

But did it? This action was actually the culmination of a long-running saga about CFCs, initiated in 1974 by F. Sherwood Rowland and Mario Molina, two scientists at the University of California, Irvine, in the United States. In fact, fears of a human threat to the ozone layer were first voiced even earlier (in 1970/71) through concern about the effects of supersonic aircraft (like Concorde and the American rival then on the drawing-board at Boeing) that fly within the stratosphere. That the projected fleet of 'super' supersonic aircraft never materialised owed more to simple economics than to their possible impact on the ozone layer, so we shall not dwell on the debate here, other than to note that the link between ozone loss and skin cancer was first raised in this context. The cancer scare made headlines, generating a degree of public concern and media attention that was to spill over into the new debate about CFCs (evident in Figure 2.1).

One aim of this chapter is to trace this debate through from its early focus on spray-cans in the United States, thence on to the international agenda, and finally culminating in the decision to phase out all CFC usage worldwide. In doing so, particular emphasis is given to the way in which an evolving scientific understanding of the problem – in terms of both model studies and direct observations of the atmosphere (for example, the ozone hole itself) – has helped to shape the scientific, public and political debate. To follow that analysis in a critical way requires some acquaintance with the science itself – the myriad of processes that control the amount and distribution of ozone in the stratosphere. So a further theme of the chapter is to present the underlying science – and it necessarily includes some chemistry. An effort has been made to keep the chemistry, and other

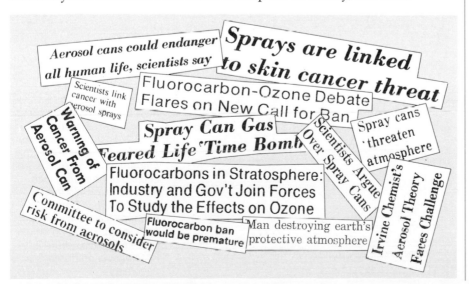

◀ Figure 2.1
Montage of 1974
newspaper headlines

technical terms used, to a minimum. If you do not have a science background, however, you may not wish to spend too long on the 'technical' questions in the Activities before turning to the Answers given at the end of the chapter; the other questions should help you to focus on the most important points.

With this scientific background in place, the chapter then looks briefly at the stepwise process that has led to the phase-out of the CFCs and related substances. The aim here is to highlight important interactions between scientific evidence, and the uncertainties associated with it, on the one hand, and the many other factors (political, economic and so on) that are involved in achieving truly international agreement to tackle a global environmental problem.

## 2   *Ozone in the atmosphere*

Ozone gas is perceptibly blue in colour and has a characteristic and pungent smell (*ozein* is Greek for 'to smell'). It occurs throughout the atmosphere, but only ever in small amounts, never exceeding one molecule in every 100 000 present. Indeed, if all the ozone contained in a vertical column of the atmosphere could be brought down and assembled at the Earth's surface, it would form a layer only some 3 mm thick. And yet all life depends on this minor by-product of its own existence (*Silvertown*, 1996) – most obviously through ozone's role as a filter of the Sun's **ultraviolet (UV) radiation**, but also in more subtle ways, as we shall see. Nearly all of the ozone in the atmosphere is contained in the lower two layers, with about 90% in the stratosphere. The typical profile shown in Figure 2.2(b) illustrates the way in which the amount of ozone changes with height.

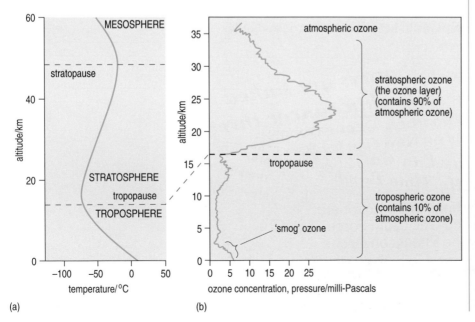

◀   *Figure 2.2*
*(a) The variation of temperature with altitude through the atmosphere. (b) A typical profile for ozone showing the variation in the amount of ozone (measured in units of pressure) with altitude. Ozone in the stratosphere is beneficial because it serves as the primary UV radiation shield, but in the troposphere it has toxic effects on humans and vegetation.*

Q   At sea-level the average atmospheric pressure is one hundred thousand
    Pascal*, or $10^5$ Pa. Use Figure 2.2(b) to estimate the concentration of
    ozone at sea-level in both parts per million (ppm) and parts per billion
    (ppb).

A   The figure shows ozone pressure at sea-level to be approximately
    5 *milli*-Pascal (or five thousandths of a Pascal). Its *concentration* in the
    atmosphere is the ratio of this pressure to atmospheric pressure, that is
    five parts in a thousand divided by a further hundred thousand, or five
    parts in a 100 million. (For the mathematicians: $5 \times 10^{-3} / 10^5 = 5 \times 10^{-8}$.)
    This is the same as 0.05 ppm or 50 ppb.

Ground-level concentrations much higher than this are a serious
problem, because ozone is highly toxic: even in very low concentrations, it
irritates the respiratory system and can cause severe damage to human
health. Plant growth may also be impaired. Indeed, ozone is now believed
to be responsible for some of the damage to forests in Europe and Canada
originally ascribed to acid rain.

Altogether, then, ozone is one of the more noxious pollutants of the
troposphere. The beneficial effects of ozone depend on the fact that most of
it *is* in the stratosphere, well away from direct contact with life. The region
of highest ozone concentrations occurs between altitudes of around
18–35 km** (Figure 2.2b) – the so-called **ozone layer**. The debate about
'damage to the ozone layer' essentially revolves around concern that
human activities are changing this characteristic profile, leading, in
particular, to a net loss of ozone from the stratosphere. To understand this
concern requires a closer acquaintance with the processes that determine
the ozone budget in the stratosphere.

Q   Using Figures 2.2a and b (and bearing in mind the different scales used
    for height), try to identify the major consequence of the formation of the
    ozone layer in the stratosphere.

A   The absorption of UV radiation in this region by oxygen and ozone
    molecules increases their 'thermal energy' and raises the temperature of
    the surrounding atmosphere. In a sense the ozone layer is the cause of
    the stratosphere.

Pressure falls with increasing height, and is lower in the stratosphere than
in the lower atmosphere (or **troposphere**). While this has an important
influence on the chemical reactions in the stratosphere, the details need not
concern us here.

There are three other differences between the environment of the
stratosphere and the troposphere which will help you to understand the
role of CFCs in ozone depletion. The first difference is the obvious one: that
the stratosphere contains most of the ozone (90%), and has higher
concentrations than in the troposphere. The second difference is in the
temperature profiles. In the troposphere, the temperature falls with height
giving rise to convective mixing of all gases in the atmosphere and the
weather patterns we are all familiar with. In the stratosphere, the opposite
is the case: it is a region where temperature increases with height, but in
general is considerably colder than at ground level. A consequence of this
**temperature inversion** is that the stratosphere has its own circulation
patterns and little mixing of air occurs with the troposphere except in the
lowest levels of the stratosphere. The final difference may be less obvious

* The Pascal is the SI unit of pressure
(force per unit area) and is equal to
one Newton per square metre.

** The ozone layer is found mostly
in the lower and middle regions of
the stratosphere. However, the
altitude of the tropopause – the
boundary between the lowest part
of the stratosphere and the
troposphere (see Figure 2.2b) varies
with latitude. It is higher at the
equator than the poles. The altitude
of the ozone layer likewise varies
with geographical location.

but is a key to understanding the effect of CFCs and related compounds. The stratosphere is rich in high-energy UV radiation which reacts photochemically to break down almost all organic (carbon-based) compounds. The troposphere receives little UV radiation and organic compounds released into the atmosphere are far less affected by photochemical attack.

## 2.1    *The ozone balance: a first look*

Ozone is a form of oxygen, but whereas molecules of ordinary oxygen each contain two atoms, the ozone molecule has three (see Figure 2.3). In industrial societies ozone is now being generated at ground-level by the action of sunlight on gaseous pollutants, but the Earth's *natural* ozone factory is the stratosphere. Here the raw materials are ordinary oxygen seeping up from the troposphere, and sunlight, but the processes involved depend on how *both* oxygen and ozone respond to that radiation. The curve in Figure 2.4 indicates that the incoming radiation from the Sun peaks – is strongest, or most intense – in the visible region of the spectrum. Nevertheless, there are significant amounts of radiation on either side of this region – both at longer wavelengths (in the infra-red) and at shorter wavelengths (in the ultraviolet). Whereas infra-red radiation merely conveys the warmth from the Sun, UV radiation is more energetic and can be harmful to unprotected humans if they receive too large a dose – otherwise known as sunburn!

There is a connection between the wavelength of radiation and the energy associated with it: stated explicitly, the shorter the wavelength, the

oxygen atom

oxygen molecule

ozone molecule

▲   *Figure 2.3*
*Experiments reveal just how the atoms in a molecule are 'stuck' together. A 'picture' of the molecules can then be built up by representing each atom as a tiny sphere. In this chapter, different colours or tones are used to represent atoms of different elements.*

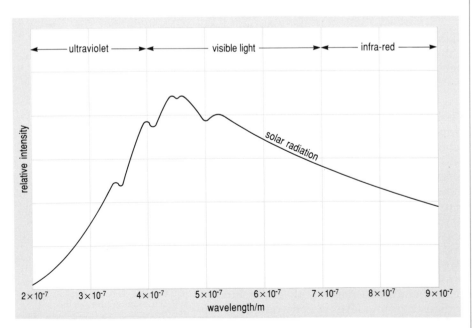

▲   *Figure 2.4*    *Solar radiation peaks in the visible region, but spans wavelengths from the ultraviolet through to the infra-red.*

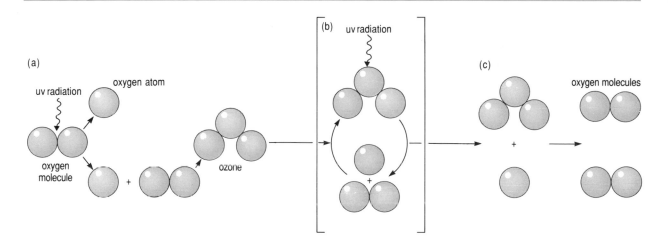

higher the energy. So absorbing UV radiation imparts more energy to a molecule than does the absorption of infra-red wavelengths – sufficient to break chemical bonds in fact, thus splitting the molecule into fragments.

Ordinary oxygen absorbs UV with wavelengths below about $2.4 \times 10^{-7}$ m: this provides the energy needed to split up – or **photodissociate** – the molecule into a pair of highly reactive oxygen atoms. Once released, an oxygen atom can combine with an intact oxygen molecule, forming ozone (Figure 2.5a). This process not only produces ozone: it also thereby filters out most of the incoming solar UV with wavelengths of less than $2.0 \times 10^{-7}$ m, and some of that in the $2.0 \times 10^{-7}$ to $2.4 \times 10^{-7}$ m region as well.

But up in the stratosphere, the odds are stacked against ozone. The bonds holding the molecule together are significantly weaker than the one in oxygen, so rather less energy is required to dismantle the molecule – implying that radiation of longer wavelengths would be required. In fact, ozone mainly absorbs in a region from $2.15 \times 10^{-7}$ to $2.95 \times 10^{-7}$ m, although wavelengths up to $3.20 \times 10^{-7}$ m are also absorbed to some extent. Further study of the curve in Figure 2.4 reveals that the Sun's radiation is richer in these longer wavelengths, so you might expect ozone to have a somewhat fleeting existence in the atmosphere. Concentrations build up because there is so much oxygen around, providing numerous opportunities for the interaction that makes ozone to take place. In effect, ozone is 'pushed' through a cycle of reactions: it absorbs UV, breaking into its constituent parts, only to be formed again (Figure 2.5b). After many trips around this cycle, the end comes when an ozone molecule encounters a free oxygen atom, and is converted back into ordinary oxygen (Figure 2.5c).

The reactions collected in Figure 2.5 comprise what is called a **chemical mechanism**, essentially a description of the individual steps believed to contribute to a particular chemical process – the chemistry that shapes the ozone layer in this case. It was first proposed as such in 1930. The enormous burst of research in recent years has revealed that matters are actually a good deal more complicated than this. Nevertheless, this simple scheme does capture the essence of the ozone budget, so it is worth dwelling on for a moment.

The crucial point is that ozone is constantly being created and destroyed in the stratosphere. But *given constant conditions* (and the force of this proviso will become apparent later), it settles into a **dynamic steady state**: its concentration stays the same because the *rate* at which it is formed

▲   *Figure 2.5*
*(a) Ozone is created when UV radiation breaks up an oxygen molecule, freeing its atoms to combine with other oxygen molecules.*
*(b) The ozone so formed is repeatedly broken up and reformed until*
*(c) it is destroyed by collision with an oxygen atom.*

is balanced by the rate of its removal. A useful analogy here is with a bucket of water: think of it being filled from a tap at a steady rate, but also drained through a hole in the bottom of the bucket at the *same* rate. The level of water in the bucket (equivalent to the concentration of ozone) does not change.

## 2.2   *The influence of trace gases: catalytic cycles*

If the scheme in Figure 2.5 is subjected to a detailed *quantitative* test, a major discrepancy is revealed: the computed ozone profile mimics the shape of the observed profile surprisingly well – but it predicts steady-state concentrations of ozone that are too large. This is because the scheme seriously underestimates the rate of ozone loss. Think of the bucket again. Suppose you have a steady state set up, and then the hole in the bucket is made larger. Now the water can get away faster than before, so the level in the bucket starts to fall. But as it falls, the rate at which water runs out of the hole slows down again – until it once more matches the filling rate. A new steady state is established, but now the water level is somewhat lower than before.

Returning to the ozone budget, the element missing from the picture so far is the presence in the stratosphere of a number of other *naturally occurring* gases in truly trace amounts – a few parts per billion (ppb) ($10^9$) or even parts per trillion (ppt) ($10^{12}$) will do. They exert their influence by engaging in a kind of 'atomic quadrille' of changing partners that effectively speeds up the destruction of ozone, while allowing them to re-emerge constantly for a further round. To see how this works, it helps to move away from the pictures of atoms and molecules in Figure 2.5 to a more symbolic representation – using the universal language of chemistry. Thus we now write the ozone-loss step (Figure 2.5c) as a chemical equation as follows:*

$$O_3 + O \longrightarrow O_2 + O_2 \qquad\qquad\qquad \textbf{1}$$

Among the first of the trace gases to be recognised as an ozone destroyer was nitric oxide (NO). Here the 'dance' or cycle can be written as follows:

$$O_3 + NO \longrightarrow O_2 + NO_2 \qquad\qquad\qquad \textbf{2}$$
$$O + NO_2 \longrightarrow O_2 + NO \qquad\qquad\qquad \textbf{3}$$

In the first step (equation 2), nitric oxide (NO) 'steals' an oxygen atom from ozone, itself being converted to nitrogen dioxide ($NO_2$). It can then 'hand' this on, as it were, to a free oxygen atom, which turns $NO_2$ back into NO (equation 3). Notice that the *overall* effect of this cycle is just the same as equation 1: ozone and atomic oxygen are converted into molecular oxygen – a net loss of ozone. The crucial point is that nitric oxide emerges unscathed at the end of the cycle – free to destroy many more molecules of ozone. A substance that acts in this way is called a **catalyst**, and the cycle it passes through (here equations 2 and 3) is a **catalytic cycle**. Analogous cycles exist for oxides of hydrogen (see Activity 1 at the end of this section) and, of course, for chlorine – the focus of concern about CFCs.

Q   Try to write a cycle (a pair of reactions) that shows how a chlorine *atom* (Cl) can destroy ozone catalytically – just like NO in equations 2 and 3.

A   The desired effect is achieved if a chlorine atom first attacks an ozone molecule, forming a molecule of **chlorine monoxide** (ClO), and then

* The crucial thing to check when you write an equation is that you haven't lost (or gained) any atoms. *Atoms are neither created nor destroyed during chemical reactions, although they can – and usually do – change partners.*

this collides with another oxygen atom – freeing the chlorine atom for further destruction. In symbols:

$$O_3 + Cl \longrightarrow O_2 + ClO \qquad\qquad\qquad\qquad 4$$

$$O + ClO \longrightarrow O_2 + Cl \qquad\qquad\qquad\qquad 5$$

The collection of atoms and molecules involved in each of these cycles have become known as *'families'* – the *nitrogen family*, the *hydrogen family* and the *chlorine family*. You may find many of these substances referred to elsewhere as *free radicals* or *radicals*, for short. Examples include atoms like O and Cl and molecules like ClO. Just what is meant by the term radical is beyond the scope of this book. Suffice it to say that they are highly reactive – eager, that is, to join up with other atoms or molecules. The ones involved here are avid 'scavengers' of ozone – thus ensuring that the indirect routes to ozone destruction via the catalytic cycles are faster than the direct one (equation 1). More telling is the fact that the cycles *are* catalytic: this is why trace constituents can have such a marked effect on the ozone budget. In particular, the chlorine cycle is especially efficient: current estimates suggest that every free atom of chlorine in the stratosphere can destroy *as many as 100 000* molecules of ozone before it is 'inactivated' or removed in some way.

This possibility of inactivation reveals another level of complexity in the chemistry of the stratosphere. The nitrogen, hydrogen and chlorine families are *not independent* of one another: rather, they are coupled together by reactions between the atoms and molecules of different families. Because of this, the rate of ozone destruction for one cycle of reactions may well depend on the concentration of a catalytic species that strictly 'belongs' to another family.

Particularly important here are couplings that produce so-called holding cycles, whereby active radicals become 'locked up' as more stable **reservoir molecules** – and hence unavailable (on a more or less permanent basis) for participation in ozone-destroying cycles. As far as the chlorine family is concerned, the important long-term reservoir is hydrogen chloride (HCl): indeed, at any given moment some 70% of stratospheric chlorine is thought to be present as HCl. A second, but more temporary reservoir, is a molecule known as chlorine nitrate – formed by coupling into the nitrogen family (Figure 2.6), and hence tying up both active chlorine and active nitrogen. Do bear in mind, however, that the chlorine (and nitrogen) in these reservoir molecules is not actually *removed* from the atmosphere: if a process can occur which releases it into its active form again, then it can set about destroying ozone.

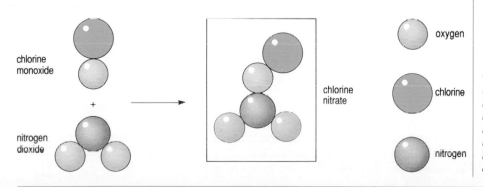

chlorine monoxide

+

nitrogen dioxide

chlorine nitrate

oxygen

chlorine

nitrogen

◄ *Figure 2.6 Formation of the reservoir molecule chlorine nitrate – by reaction between chlorine monoxide (ClO) and nitrogen dioxide (NO$_2$) – interferes with the catalytic cycles for ozone destruction.*

There seems to be a paradox here: first nitric oxide was described as an ozone destroyer, now it appears to interact with the chlorine monoxide, another ozone destroyer, to form reservoir molecules of chlorine nitrate. As always, when competing reactions occur, the overall balance depends on factors like temperature and the concentrations of the chemicals involved. The nitrogen family is thought to be mainly responsible for ozone depletion in the upper stratosphere. In the lower stratosphere where most of the ozone exists, chlorine is thought to be the major cause of its destruction. The amount of ozone depletion that actually occurs depends on how much of the catalyst, in this case the chlorine family, is available in the active forms and how much is locked up in reservoir molecules. You will see later that this is believed to be the key to the special chemistry of the 'ozone hole' over Antarctica.

If you are reeling somewhat from this sudden onslaught of molecules and chemical formulae, take heart! The main aim here is to give you a feel for the true complexity of ozone chemistry, but don't worry too much about the details. The crucial point to grasp is the notion of a catalytic cycle and the way in which this allows minute traces of certain gases to speed up the rate of ozone destruction. In fact, we have only scratched the surface. The most recent descriptions include some 100–200 interactions among the atoms and molecules of the families that control the ozone budget – far removed from the original mechanism in Figure 2.5.

It is important to be clear that most of these processes appear to be at work in the 'natural' atmosphere, unpolluted by human activities. The radicals active in destroying ozone are all derived from other gases that percolate up from the troposphere – the so-called **source gases** – and there have always been natural sources of these. This is what keeps the natural rate of ozone production in check and the ozone budget balanced at the 'normal' level. The problem is that in the contemporary atmosphere, the background concentrations of catalytic species – especially of chlorine – have already been supplemented by different and *growing* anthropogenic (human) sources. There is evidence that this rise *has* started to chisel away at the hole in the bucket – and not only over Antarctica.

*Activity 1*

Write a cycle for the hydrogen family, showing how an atom of hydrogen (H) can catalyse the destruction of ozone. Assume that the first step produces a hydroxyl radical, usually written OH.

## 3    *Source gases*

The source gases are trace constituents of the atmosphere, produced – or emitted – at ground-level. They carry the elements of the catalytic families (nitrogen, hydrogen and chlorine), but in the form of *stable* molecules, that are usually rather insoluble in water, the latter ensuring that they are not rapidly 'rained out'. These trace gases can thus linger in the troposphere for long enough to be moved around and well mixed by the sort of large-scale atmospheric motions that global circulation models of the climate describe (see Chapter 3, Section 3.6). As a result, molecules of these gases are not only distributed *around* the globe, they are also transported *upwards* – and some, at least, are eventually carried across the tropopause into the stratosphere. Mostly this happens over the tropics – a region of rapidly rising air currents, often driven by violent storms. Once in the stratosphere, the compounds eventually encounter a more hostile environment. Molecules that were inert in the lower atmosphere can now be broken down – often by sunlight (as are oxygen and ozone), but sometimes chemically – to produce the active radicals that speed up the destruction of ozone. Information on the main *natural* source gas for each family is provided in Table 2.1.

The point to bear in mind is that the sinks (factors which remove a substance) and sources of these gases are not well understood. And neither are they in balance – which is why the atmospheric burden of each gas is increasing. Nitrous oxide is effectively inert in the troposphere, hence the long lifetime recorded in Table 2.1. In the stratosphere, however, UV radiation promotes reactions which transform it into nitric oxide NO – and so feed it into the nitrogen family. Methane (Figure 2.7a) actually affects the ozone budget in a number of ways, but the most direct of these is its role as the main stratospheric source of radicals for the hydrogen family.

*Table 2.1  Tropospheric concentrations and lifetimes of natural source gases (1992), and global trends*

| Source gas | Concentration (ppb by volume) | Rate of increase 1986–92 (%/yr) | Lifetime (years) |
|---|---|---|---|
| nitrous oxide, $N_2O$ | 310 | 0.2–0.3 | 120 |
| methane, $CH_4$ | 1714 | 0.8–1.0 | 14.5 |
| chloromethane, $CH_3Cl$ | 0.6 | – | 1.5 |

Source: Houghton, J. T. *et al.* (1994) *Climate Change 1994: radiative forcing of climate change and an evaluation of the IPCC IS92 emission scenarios*, Cambridge, Cambridge University Press/Intergovernmental Panel on Climate Change

Turning to the chlorine family, here the main *natural* source gas is chloromethane (Figure 2.7b; think of it as methane, but with one hydrogen atom replaced by a chlorine atom). Mostly this is of biological origin – being generated during wood-rotting and natural forest fires – but some arises from slash-and-burn tropical agriculture. Notice the short lifetime recorded in Table 2.1. Because of this, rather little of the gas that enters the atmosphere (only some 10%) actually reaches the stratosphere, where the chlorine atom is stripped off and enters its cycle of ozone destruction (equations 4 and 5 in Section 2.2). This input is now dwarfed by that from the many other organic (that is, carbon-based) compounds containing chlorine that are accumulating in the troposphere from strictly *anthropogenic* sources: this is where the now infamous CFCs enter the scene.

Q  The report from which Table 2.1 was taken estimates that the total **'chlorine loading'** of the atmosphere – that is, the concentration of chlorine atoms effectively 'carried' in organic molecules – was around 3.6 ppb by volume in 1992. What proportion of this was due to human emissions?

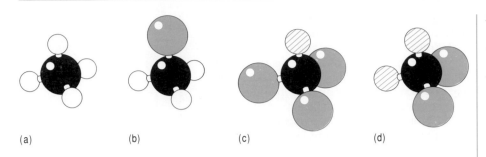

(a)                    (b)                    (c)                    (d)

◀ *Figure 2.7*
*Representations of*
*(a) methane (CH₄)*
*(b) chloromethane (CH₃Cl)*
*(c) CFC-11*
*(d) CFC-12.*

◯  hydrogen

◔  chlorine

▨  fluorine

●  carbon

*A*   Taking $CH_3Cl$ to be entirely natural, at least $(3.6 - 0.6)$ ppb $= 3.0$ ppb by volume, or $(3.0 \div 3.6) \times 100 = 83\%$ was anthropogenic. Emissions of CFCs have provided the major contribution to this source.

## 3.1   *Chlorofluorocarbons (CFCs)*

As a class of compounds, CFCs *all* contain atoms of carbon, fluorine and chlorine, and only these elements – but combined together in varying proportions. To the cognoscenti, the precise composition is reflected in the code name used to identify each compound. The details need not concern us. Suffice it to say that the two most widely-used compounds – labelled **CFC-11** and **CFC-12** (Figures 2.7c and d) – can each be thought of as derived from methane.

*Q*   Look at the representations in Figure 2.7, write formulae for these two CFCs. (The chemical symbol for fluorine is F.)

*A*   Methane is $CH_4$. From Figure 2.7(c), in CFC-11 one hydrogen is replaced by fluorine and three by chlorine, to give $CFCl_3$. From Figure 2.7(d), CFC-12 has two chlorines and two fluorines – giving $CF_2Cl_2$.

Information on the five CFCs included in the original Montreal Protocol is presented in Table 2.2. The striking feature is the very long lifetimes of these compounds. Ironically, this stems from the very characteristics that made CFCs seem so ideal: they are non-toxic, non-flammable – in short, effectively completely inert in the lower atmosphere. Despite prodigious research efforts, no evidence has emerged for *any* tropospheric sink for CFCs. On the contrary, the experimental programmes set up to monitor these (and other) trace gases at stations around the globe have only confirmed an inexorable build-up in the atmospheric burden of CFCs, although recent years have seen the *rate of increase* of concentrations in the atmosphere slow down significantly as countries respond to the controls adopted by the Montreal Protocol and its subsequent amendments; some typical data are included in Figure 2.8.
    The central thesis in the seminal work published by Rowland and Molina in 1974, in recognition of which work they were awarded in 1995 the Nobel Prize for Chemistry, was the tropospheric inertness of CFCs. Given this, they argued that the only plausible fate for these compounds is transport upwards until they reach altitudes in the stratosphere where the incoming UV radiation is sufficiently energetic to break them up – some 25–40 km for CFCs 11 and 12. Here, photodissociation strips off their chlorine atoms – the net effect being to add to the natural burden of

Table 2.2   *Tropospheric concentrations (1992) and atmospheric lifetimes of artificial chlorine and bromine-containing source gases, known collectively as halocarbons.*

| Compounds grouped by type | Chemical formula | Cl (or Br) atoms per molecule | Lifetime (yr) | Ozone depletion potential | Concentration (ppt by volume) (1992) |
|---|---|---|---|---|---|
| 1  Chlorofluorocarbons (CFCs) | | | | | |
| CFC-11 | $CCl_3F$ | 3 | $50 \pm 5$ | 1.0 | 268 |
| CFC-12 | $CCl_2F_2$ | 2 | 102 | 0.8–0.9 | 503 |
| CFC-113 | $CCl_2FCClF_2$ | 3 | 85 | 0.9 | 82 |
| CFC-114 | $CClF_2CClF_2$ | 2 | 300 | 0.85 | 20 |
| CFC-115 | $CClF_2CF_3$ | 1 | 1700 | 0.4 | <10 |
| 2  Miscellaneous groups | | | | | |
| Carbon tetrachloride | $CCl_4$ | 4 | 42 | 1.20 | 132 |
| Methyl chloroform | $CH_3CCl_3$ | 3 | $5.4 \pm 0.4$ | 0.12 | 160 |
| Chloromethane | $CH_3Cl$ | 1 | 1.5 | 0.02 | 600 |
| 3  Hydrofluorochlorocarbons (HCFCs) | | | | | |
| HCFC-22 | $CHF_2Cl$ | 1 | 13.3 | 0.04–0.05 | 102 |
| HCFC-141b | $CFCl_2CH_3$ | 2 | 9.4 | 0.1 | 0.3 |
| HCFC 142b | $CF_2ClCH_3$ | 1 | 19.5 | 0.05–0.07 | 3.5 |
| 4  Hydrofluorocarbons (HFCs) | | | | | |
| HFC-23 | $CHF_3$ | 0 | 250 | 0 | |
| HFC-125 | $CHF_2CF_3$ | 0 | 36 | 0 | |
| HFC-134a | $CH_2FCF_3$ | 0 | 17.7 | 0 | |
| 5  Bromine source gases (halons and methyl bromide) | | | | | |
| H-1211 | $CF_3Br$ | 1(Br) | 2.0 | 5 | 2.0 |
| H-1301 | $CF_2ClBr$ | 1 + 1(Br) | 65 | 12–13 | 2.5 |
| Methyl bromide | $CH_3Br$ | 1(Br) | 1.3 | 0.57–0.64 | 12 |

Source: Adapted from WMO (1994) *The Scientific Assessment of Ozone Depletion*, Tables 2.1 and 13.4

stratospheric chlorine. The threat thereby posed to the ozone layer is no longer in any doubt. But a more detailed examination of the size and consequences of that threat – and the effectiveness of measures being taken to mitigate it – is in order. There is one final point to be made before we embark on that.

So far, the CFCs have been lumped together. However, the differences between them are important – and not only because this influences their uses, and hence emission rates. It also determines the 'efficiency' with which a given compound can destroy ozone. In general, this is enhanced by a long lifetime (mainly because this ensures optimal transfer into the stratosphere) and a high number of chlorine atoms per molecule. This idea is given quantitative expression by calculating an **ozone depletion potential**, or **ODP**, for each compound, with the behaviour of CFC-11 being

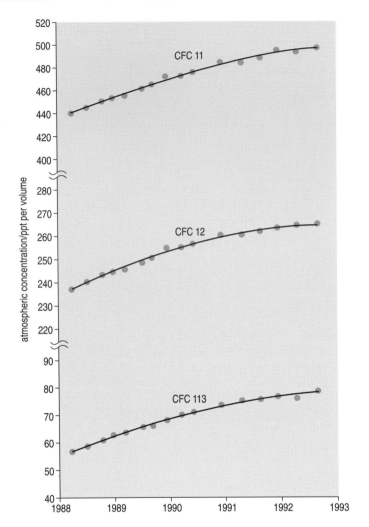

▲   Figure 2.8
*Recent trends in the globally averaged concentrations of the three most common CFCs: (a) CFC-11, (b) CFC-12 and (c) CFC-113.*

taken as a kind of bench-mark against which others are compared. Some recent estimates for CFCs and other ozone-depleting chemicals discussed in Section 4 are given in the penultimate column of Table 2.2: numbers *less* than one mean the compound is *less* destructive than CFC-11, and *more* than one that they are more destructive. Some of the other compounds are actual or potential CFC substitutes. Note the high ODPs of certain *bromine* compounds, as well as chlorine compounds.

The precise numbers need not concern us, but the ODP concept does raise an important general point. The extent to which the *potential* threat posed by a given compound is likely to be realised depends, of course, on the atmospheric burden of that gas – and this, in turn, depends on the amount of it being released into the environment. Thus the ODP effectively allows global emissions of different chlorine compounds to be 'translated' into 'contributions to ozone depletion'. An example is shown in Figure 2.9, based on figures from 1985 when worldwide emissions of CFCs were at their height.

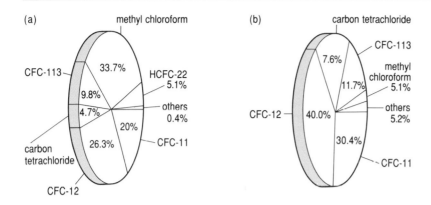

▲   *Figure 2.9*
*(a) Estimates of global emissions of anthropogenic chlorine compounds during 1985.*
*The tonnage of each compound is expressed as a percentage of the total tonnage.*
*'Others' include CFCs 114 and 115 and certain halons (Activity 2).*
*(b) These figures translated into percentage contributions to ozone depletion, using*
*the ODP of each compound.*

## 3.2   *Summary of Sections 2 and 3*

Ozone ($O_3$) concentrations in the atmosphere vary, peaking in the
stratosphere, at around 18–35 km.

Ozone concentrations result from a dynamic balance between creation
and destruction. Creation is initiated by the photodissociation of oxygen
($O_2$); ozone is destroyed by reaction with atomic oxygen (O):

$$O_3 + O \longrightarrow O_2 + O_2$$

Loss of ozone (by this reaction) can be catalysed by a number of reactive
atoms or molecules (radicals), that are constantly regenerated. To generalise
the cycles you have met:

$$O_3 + X \longrightarrow O_2 + XO$$

$$O + XO \longrightarrow O_2 + X$$

where **X** can be nitric oxide (NO), or hydrogen (H), or chlorine (Cl) (or
bromine, Br – see Activity 2, below). The catalytic families interact, forming
reservoir molecules (such as hydrogen chloride and chlorine nitrate) that
act as holding tanks for active radicals.

The active radicals for these cycles are produced from insoluble source
gases that are transported from the troposphere where they are relatively
stable, to the stratosphere where they are attacked photochemically by UV
radiation.

The principal source gases for the nitrogen family (nitrous oxide, $N_2O$)
and hydrogen family (methane, $CH_4$) have both natural and anthropogenic
origins.

The principal source gases for the chlorine family are now non-natural
– including the long-lived CFCs and carbon tetrachloride, together with
other less inert compounds like methyl chloroform and some CFC
substitutes. The ODP assigned to each compound reflects its efficiency at
destroying ozone.

*Activity 2*

The following questions may be used to revise some of the key concepts of Sections 2 and 3.

Q1  The concern about supersonic aircraft (e.g. Concorde) referred to in Section 1 centred on the fact that their engines produce nitric oxide (NO): any engine that draws in air and uses the oxygen in it to burn fuel at high temperatures inevitably produces some NO.

(a)  Why should this raise concerns about the ozone layer?

(b)  Can you suggest why the vast amounts of NO produced by ground-based vehicles do not raise such concerns?

The following questions are a bit more technical. Do not spend too long on them if you don't have a scientific background, but do read the comments at the end of the chapter.

Q2  The Montreal Protocol also controls emissions of compounds known as **halons**, widely used in fire extinguishers. These are organic compounds that carry atoms of bromine (Br) and sometimes Cl as well, into the stratosphere. Write a catalytic cycle showing ozone destruction by bromine.

Q3  One of the compounds mentioned in Q2 is labelled halon-1301: each molecule contains one atom of bromine and no chlorine. The estimated atmospheric lifetime is 65 years, and the compound is assigned an ODP of 12–13. What do these figures suggest about the relative efficiencies with which Cl atoms and Br atoms destroy ozone?

(Hint: Refer to Table 2.2, and compare the figures for halon-1301 (H-1301) with those for CFCs of comparable lifetimes.)

# 4  *Modelling global ozone change*

In this section we take a brief look at the use of models in predicting the effect on ozone of changing concentrations of the source gases – an essential backdrop for the material to come in Section 5, which looks at the evidence of damage to the ozone layer, and Section 7, which charts the international reaction which resulted in the Montreal Protocol and further steps to stem the causes of the damage. Here we outline some of the assumptions and approximations that have to be incorporated into computer models designed to represent what is actually happening; this is necessary in order to make possible the task of simulating such complex interactions as the distribution of ozone.

In the next chapter we shall return to the role of modelling in predicting the impact of enhanced levels of greenhouse gases on the climate.

Modelling global warming is similar to forecasting ozone depletion: they share many of the same difficulties – and the results that come out of computer model studies are subject to the same kinds of uncertainties. As we shall see, these fall into two broad categories: on the one hand, those linked to the scientific description 'built in' to the computer model; and, on the other, those that stem from the difficulty of projecting future emissions of different source gases – especially when these are subject to changing patterns of human activity.

We shall begin by delineating the problem.

## 4.1  Modelling the distribution of ozone: the problem

Central to any model of the ozone layer is a description of the chemistry outlined in Section 2 – a complex network of coupled, interacting and often competing chemical and photochemical reactions. The basic information comes from laboratory studies of the individual reactions – the rate of each process, and how this depends on the concentrations of the different species involved, and on other conditions, such as temperature*, or, where appropriate, solar radiation. The results of such studies are incorporated into the model as a set of equations that constitute a description of 'ozone chemistry'. But this captures only part of the problem. In the real atmosphere, the observed distribution of ozone is shaped by subtle, all-pervasive interactions between ozone chemistry and atmospheric dynamics.

* The rate of a reaction invariably depends on the temperature – usually speeding up as the temperature is raised. You may have noticed this yourself. For example, bicarbonate of soda (sodium bicarbonate) reacts with the acid in fruit to produce carbon dioxide – almost imperceptibly when cold, but with a vigorous fizzing when heated.

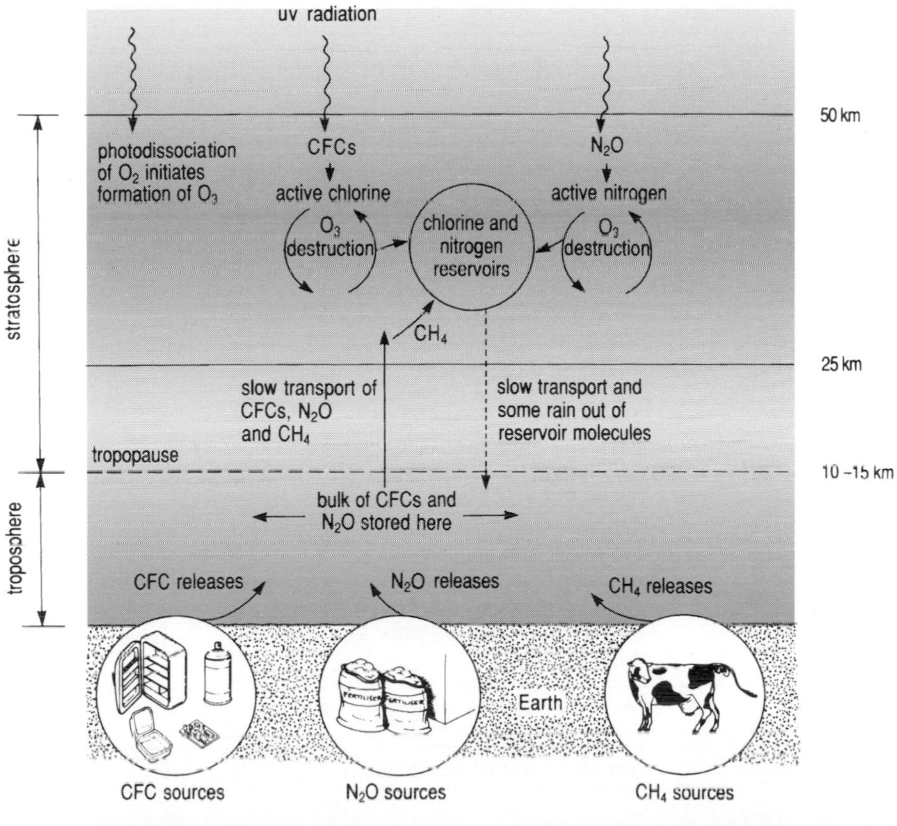

◄ Figure 2.10
Some of the key processes, both chemistry and transport, that determine the concentration of ozone in the stratosphere (as represented in a one-dimensional (1-D) model, Section 4.2)

To see what this entails, take the characteristic *vertical distribution* of ozone – or ozone profile (Figure 2.2b) – as a first example. Here, the chemical balance that determines the concentration of ozone at any particular altitude depends not only on the local conditions (temperature, intensity of solar radiation, and so on) – but also on the supply of key catalytic species and other trace gases. This supply is itself driven by atmospheric motions. Source gases have to be transported into the stratosphere, and then move up through it. And there is evidence that long-lived reservoir molecules (like HCl, for example) can move down – and be washed out in rain in the troposphere. In short, the stratosphere is host to a constant traffic of ozone and the many gases that interact with it – as suggested in Figure 2.10. To simulate the ozone profile a model must capture this interplay between chemistry and dynamics.

On top of this, large-scale atmospheric motions also play a critical role in moving stratospheric ozone around the globe – and so determining its *global distribution*. There is evidence to this effect in Figure 2.11. Here, the 'contours' record measured values of the **total ozone column** (also known simply as '**column ozone**') – that is, the total amount of ozone above a unit area (usually 1 cm²) of the Earth's surface. Notice that values of column ozone contain no information about the vertical distribution of ozone: just the total number of molecules in a column stretching up from the surface to the top of the atmosphere. The data in Figure 2.11 are typical of years before the ozone hole started to appear – recorded as a function of both *latitude* (from the north pole at the top to the south pole at the bottom) and *season* (from January to December across the figure).

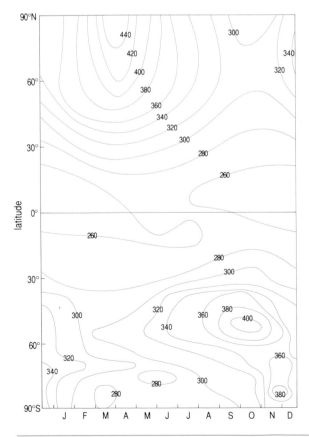

◀ *Figure 2.11*
*The global distribution of ozone as a function of latitude and time of year. The numbers are average values of the total ozone column in Dobson Units\* (DU = 2.7 x 10¹⁶ molecules per square centimetre), based on data from a network of ground-based measuring stations around the world – for years prior to 1974.*

\**An odd unit, named after Gordon Dobson, the Oxford University Reader in meteorology who developed one of the techniques used to measure ozone.*

*Q*   Where and when do the highest amounts of ozone occur? Is this what you would have expected?

*A*   According to Figure 2.11, the column ozone is greatest at high latitudes in late winter and early spring – around the end of March in north polar regions (top left), and in mid-October at around 60° south (bottom right). This is surprising. Since the production of ozone is driven by sunlight, one might expect maximum amounts at low latitudes, where there is the greatest amount of solar radiation: in practice, minimum amounts are found in equatorial latitudes.

In fact, the main ozone factory *is* at tropical latitudes (at altitudes of some 30–40 km), but it is transported from here downward and toward the poles in a circulation that is biased in favour of the 'winter' hemisphere – as suggested by the highly simplified picture in Figure 2.12.

To capture what is known about atmospheric circulation strictly requires the sort of three-dimensional global circulation model (GCM) described in the next chapter. If detailed ozone chemistry could be incorporated into a GCM, then the resulting model would be ideal for simulating all the subtleties of ozone distribution – with altitude and latitude (and with longitude too, although this is less marked), and with the changing seasons. This remains a major challenge and to keep such a task to manageable proportions many such forecasts have in practice used models that incorporate a much simplified treatment of atmospheric dynamics – one-dimensional or two-dimensional models.

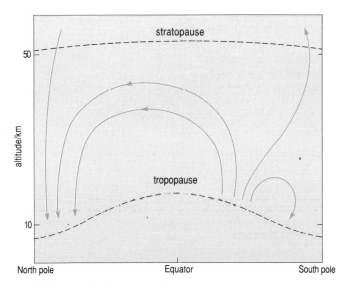

▲   *Figure 2.12*
*Stratospheric circulation, shown highly schematically, for winter in the northern hemisphere. The flow is reversed during winter in the southern hemisphere – but does not travel further south than about 60°S because the circulation encounters 'resistance' at that latitude (see Section 5.1). As a result, global ozone levels do not peak in the tropics – where most ozone is produced – but near the north pole and at 60°S. (Notice that the height of the tropopause varies with latitude.)*

## 4.2 Models and scenarios: changing perceptions

*One-dimensional (1-D) models* effectively treat the atmosphere as a single vertical column. Averaging techniques are used to reduce *all* transport processes to simple vertical movement (up and down, along the lines portrayed in Figure 2.10). Such models can simulate the vertical distribution of ozone, but not any 'horizontal' (i.e. north–south or east–west) or seasonal variations. So the ozone profiles (and hence values of column ozone) they produce are necessarily *average* pictures – averaged around the globe, and usually over one or more simulated years as well.

The great strength of 1-D models is that they are computationally fast, even when they include very detailed ozone chemistry. As a result, they have been widely used to predict changes to the total (average) ozone column. A typical study involves introducing the perturbation of interest (effectively a change in 'input' conditions such as a specified increase in CFC emissions), and then running the model for many simulated years until the modelled atmosphere settles into a new steady state. The ozone column (and/or altitude profile) is then compared with a similar model calculation that did not contain the perturbing influence.

Since 1974, 1-D model forecasts of ozone depletion by CFCs have fluctuated widely: Figure 2.13 presents a fairly typical record through to the early 1980s. These fluctuations reflect the many uncertainties associated with the modelling process, but the marked shifts recorded here can all be largely attributed to changes in the scientific description, specifically the chemistry, incorporated into the model – to revised reaction rates and to the inclusion of reactions which had been overlooked earlier.

1-D models have allowed studies of the interactions between the different ozone 'catalyst' families (nitric oxide, hydrogen and chlorine). They have also allowed scientists to ask 'What if?' questions: for example, what might be the effect of cooling the stratosphere? So, although the modelling appears crude, it can be useful. However, these models gave no warning of the ozone hole over Antarctica (see Section 5); instead the trend was to more reassuring forecasts (see Figure 2.13). The chemistry dealt with the gas phase, and did not anticipate the role of reactions between gases and liquid or solid particles (heterogeneous reactions) that take place in

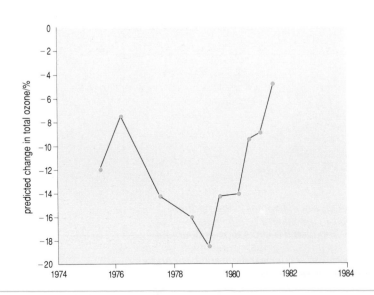

◀ *Figure 2.13*
*Record of calculations of the expected change in total ozone (at steady state) from CFC release, as predicted by a particular 1-D model. The horizontal axis represents the year in which each calculation was made.*

*polar stratospheric clouds (PSCs)* (see Section 5.1 and Box 2.3 below) when temperatures are low. *2-D and 3-D Models* are now employed which can reproduce many of the features of stratospheric ozone that are observed, both globally and during the formation of the Antarctic ozone hole. Unfortunately they still tend to underestimate ozone depletion by a factor of two due to an incomplete description of PSCs and other sources of particles, and the complexity of the dynamics and its interaction with the chemistry.

On one point, however, the models are all in agreement: a chlorine atom has the same effect whatever its source. The array of chemical compounds that serve as sources for chlorine (and bromine) in the stratosphere is indeed bewildering, but they can be grouped into distinctive types which often show very similar properties. Known collectively as **halocarbons**, the major types are listed in Table 2.2 (above) and their main properties and uses are summarised in Box 2.1.

*Activity 3*

Q1  The ODPs of hydrofluorocarbons, given in Table 2.2, are expressed as zero. What does this mean?

Q2  HFCs have been suggested as a possible alternative to CFCs. Assuming they have the right properties to act as substitutes, what is the objection to their use?

Q3  The use of HCFCs, developed as a 'transitional' substitute for CFCs, is due to be phased out by the year 2020. Why is there concern about their manufacture and use, given their low ODPs and short lifetimes?

---

*Box 2.1   Halocarbons*

1    The *chlorofluorocarbons* (CFCs) are the main group of the ozone-depleting substances (described in the introduction to this chapter). They are stable, non-toxic and non-inflammable and have a range of applications as refrigerants, propellants, blowing agents for insulation, and industrial solvents.

2    The miscellaneous group consists of the *carbon tetrachloride* ($CCl_4$) used in dry cleaning and as a chemical feedstock, and *methyl chloroform* ($CH_3CCl_3$) used as a solvent and also a cleansing agent. These are both artificial chemicals, while the short-lived chloromethane, or *chloromethane* ($CH_3Cl$) produced in the oceans, is the main natural source of organic chlorine.

3    *Hydrochlorofluorocarbons* (HCFCs) were developed as substitutes for CFCs with ODPs of roughly one-tenth the value of CFCs. The most common representative of this group is HCFC-22 and its production and atmospheric concentration are increasing rapidly.

4    *Hydrofluorocarbons* (HFCs) contain neither chlorine nor bromine and therefore do not deplete ozone (the ODP is effectively zero). They were also developed as possible substitutes for CFCs but may be controlled because they are significant greenhouse gases.

5    Sources of bromine are: (a) the *halons* – similar in chemistry to CFCs but containing bromine. They are used as non-corrosive fire extinguishers and have other specialised applications, but are very strong ozone depleters. (b) *Methyl bromide* is produced naturally by algae in the ocean, but over a third is anthropogenic: its main use is as a soil fumigant but it is also produced inadvertently from the burning of biomass.

If we wish to model future chlorine burdens in the stratosphere to assess the effect of introducing CFC substitutes or to compare different control scenarios, a common denominator is needed. One such measure which allows a comparison of different chemicals, the ozone depletion potential (ODP), was introduced in Section 3.1. A look at Table 2.2 suggests that each chemical type shares common properties: for example, CFCs have long lifetimes and ODPs mostly close to unity; HCFCs have shorter lifetimes, and ODPs between 0.04 and 0.1; and the halons have the highest ODPs.

Q   Ignoring for the moment bromine compounds, can you see which properties of the compounds in Table 2.2 contribute to a high ODP?

A   The two key factors seem to be: the compounds must contain chlorine (a higher number of chlorine atoms seems to help) and the atmospheric lifetime needs to be reasonably long.

Because there is only a slow exchange of air between the lower and upper atmospheres (see Section 2.1), on average it takes gases in the troposphere five years to reach the stratosphere. If the atmospheric lifetime is much shorter than this, most of the compound will be destroyed before it can reach the ozone layer, whereas with lifetimes of fifty years or more almost all of the compound will be effective.

Q   Most of the compounds with short lifetimes, less than twenty years, say, have a common feature to their chemical formulae. Can you identify what it is?

A   With the exception of the HFCs, all the chemicals with the shortest lifetimes contain hydrogen in their formulae.

This is because the carbon–hydrogen bond (shown in Figure 2.7) is attacked by a natural oxidising agent found in small quantities in the troposphere, the hydroxyl radical (OH). The hydroxyl radical, which also plays an important part in stratospheric ozone chemistry, is very reactive and is the main tropospheric 'sink' for all trace gases with a carbon–hydrogen bond, including methane. This property of the carbon–hydrogen bond has been exploited in the development of HCFCs – to produce shorter lifetimes and hence lower ODPs – but these compounds still deplete ozone and their release in the atmosphere is still damaging.

Compounds containing bromine, the **halons** and **methyl bromide**, have received increased attention in recent years. Although organic bromine compounds tend to be less stable than their chlorine counterparts, when the bromine they contain reaches the stratosphere, its effect on ozone is much greater – by a factor of 40 at least. Once there, the bromine is far less likely to be bound up as a reservoir molecule but remains available in an active form that can react with ozone. As you saw in the answer to Activity 2, the basic catalytic reaction of the bromine atom with ozone is the same as for the chlorine atom:

$$Br + O_3 \longrightarrow BrO + O_2 \qquad\qquad\qquad \textbf{6}$$

$$Cl + O_3 \longrightarrow ClO + O_2 \qquad\qquad\qquad \textbf{7}$$

However, a third reaction can take place *between* ClO and BrO:

$$ClO + BrO \longrightarrow Cl + Br + O_2 \qquad\qquad\qquad \textbf{8}$$

The net effect of reactions 6, 7 and 8 is to convert two molecules of ozone (O₃) to three molecules of oxygen (O₂), with the original chlorine and bromine restored. This cycle of reactions is thought to be effective at temperatures above those at which PSCs form. In the Arctic, where winter temperatures are not always low enough for the formation of PSCs, the chlorine–bromine cycle may play a major role in ozone depletion.

The relative contributions of chlorine and bromine-containing compounds can be expressed in an alternative form that measures their concentration or 'loading' of the atmosphere based on their known emissions. This can be done separately for chlorine and bromine loading, but if a scaling factor is used for the increased effectiveness of bromine, just as it is for ODPs, the two figures can be combined. The resulting *equivalent effective chlorine* then provides a simple summary of the contributions from

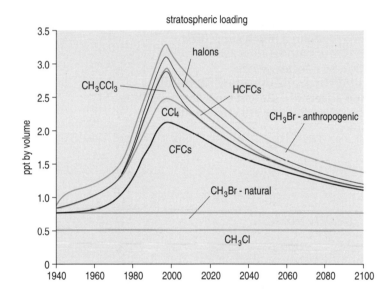

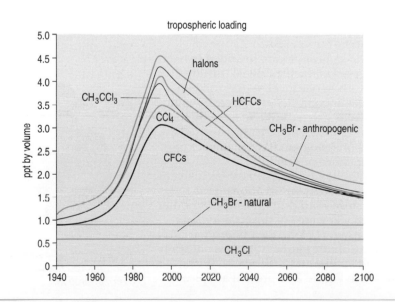

◀ *Figure 2.14*
*The contributions from the halocarbons to the chlorine/bromine loading of the troposphere (bottom) and stratosphere (top) from 1940 to 2100, under compliance with the Montreal Protocol and the London and Copenhagen Amendments. The chlorine/bromine loading is projected to peak in the troposphere in the mid-1990s and approximately five years later in the stratosphere.*

any particular family of ozone-depleting substances to the loading of the atmosphere. (As the term 'equivalent effective chlorine' is a bit of a mouthful, in this chapter we will use the phrase **chlorine/bromine loading** to indicate the combined effect of the two molecules in the atmosphere.) Figure 2.14 shows one such projection of future loading based on compliance to the Montreal Protocol (and the subsequent Copenhagen amendments). The technique can be used to investigate the effects of a given control strategy on any of the wide range of chemical families known to harm the ozone layer.

## 4.3   Summary of Section 4

To simulate all aspects of the variation of stratospheric ozone – spatially and with the seasons – strictly requires a sophisticated 3-D GCM that incorporates detailed ozone chemistry. Because of computational limitations, many of the assessments of the threat to ozone posed by human activities have used models (1-D or 2-D) that contain a much simplified treatment of atmospheric dynamics. Even current model forecasts underestimate the effects of ozone depletion by a factor of two. Nevertheless the comparative effects of different ozone-depleting chemicals can be expressed in terms of ODPs, and the effects of different emission (or control) scenarios can be explored with the concept of the effective loading of the stratosphere by chlorine and bromine.

## 5   What is happening to the ozone layer?

Large losses of total ozone in Antarctica reveal seasonal $ClO_x/NO_x$ interaction.* (J.C. Farman, B.G. Gardiner and J.D. Shanklin, *Nature*, 16 May 1985)

* These symbols are often used to represent the chlorine and nitrogen families, respectively.

The measurements from the British Antarctic Survey (BAS) station at Halley Bay reported by Joe Farman and his colleagues are included in Figure 2.15. There is no doubt that this report marked a crucial turning-point in the long-running saga about CFCs. On one level, it provided an image – of a 'hole in the sky' (through which 'pour' deadly cancer-causing 'rays', in the more lurid press accounts), later given dramatic force by the publication of NASA's satellite 'pictures' (see Plate 6) – that has exerted a powerful hold on the public and political imagination. But the impact on the scientific community was, if anything, more profound.

Up to this point, the debate about CFCs had turned, almost exclusively, on the results from model calculations – *not* actual observations. True, monitoring programmes had revealed a steady build-up in CFCs and other important trace gases. But there remained a key area of uncertainty: was anything actually happening to the ozone layer? Here, the underlying problem is that column ozone – like atmospheric temperature – is naturally highly variable, and over a wide range of spatial and time scales. Within the tropics, the average figures we looked at in Figure 2.11 present a typical

*Joe Farman, whose team first detected the ozone hole over Antarctica.*

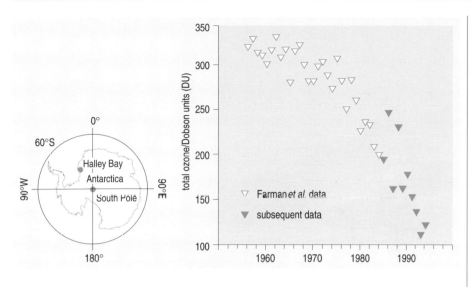

◀ *Figure 2.15*
*The October average total ozone observed over the BAS station at Halley Bay from 1956 to 1994. Before the mid-1970s, average October values were around 300 DU, but a rapid decline since then has led to values close to 100 DU.*

◀    *A balloon launch at Halley. The instruments carried aloft signal back information about the meteorological state (temperature, pressure, humidity etc.) of the lower stratosphere. Other balloons carry instruments to measure the ozone concentration at different altitudes.*

▲   *Halley 4, part of the British Antarctic Survey station in Antarctica, where ongoing measurements of column ozone first detected the ozone hole. Like its predecessors, this station has now nearly disappeared under the snow!*

picture – with small, but well-defined, variations with either latitude or season. Beyond the tropics, however, the detailed picture is much more complex. Here, there can be such large fluctuations in column ozone from day to day and year to year, and from place to place, that it becomes very difficult to detect any underlying 'signal' of a small consistent long-term downward trend that could be attributed to human activities. As with global warming, the basic requirement is for an extended, and reliable, data set.

The longest data record of ozone measurements comes from a network of ground-based stations referred to in Figure 2.11 – the so-called **Dobson network**, of which Halley Bay is a part. Although this network was extended in 1957, the distribution of stations remained quite uneven, with 42 out of 67 in northern temperate latitudes. Since 1979 the problem of spatial coverage had been much alleviated by instruments borne aloft on satellites (like the **Total Ozone Mapping Spectrometer – TOMS** – on board NASA's Nimbus-7 satellite until 1992 and, since then, on Meteor-3), but here the data record was too short to permit the identification of any statistically significant trends. Overall, then, the early 1980s saw little convincing evidence that CFCs had *already* caused damage to global ozone levels.

Set against this background, the results from Halley Bay carried a double-edged message. On the one hand, they provided the first unequivocal signal of a real change to column ozone – albeit apparently restricted to Antarctica in springtime (September/October in the southern hemisphere, remember). On the other hand, the signal was so large and unexpected that its immediate effect was to challenge the validity of existing models: why did they fail to predict this dramatic decline?

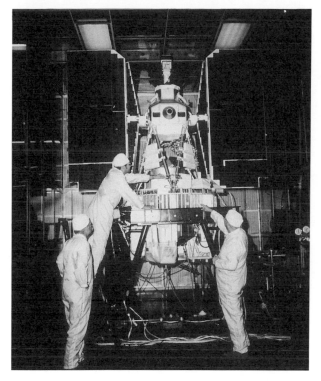

▲   *NASA's Nimbus-7 meteorological satellite carries a sensor from which global 'maps' of atmospheric ozone can be constructed.*

▲   *The Dobson spectrometer, which is used to measure ozone concentrations.*

## 5.1   *The ozone hole: why Antarctica?*

### *Introduction*

The results from Halley Bay have since been amply confirmed by other workers. Particularly telling are the computer-processed images from the TOMS data (Plate 6), which show that the depleted region (less than 300 DU) extends over the entire Antarctic continent – and beyond. Other measurements, from satellites and balloons, have revealed that the depletion is mainly concentrated between 12 and 24 km in altitude – spanning much of the lower stratosphere at these latitudes. So the 'hole in the sky' is actually more like a *slice* out of the heart of the ozone layer: a striking example from 1987 is shown in Figure 2.16(a). Whatever happens to Antarctic ozone in the future, 1987 will undoubtedly remain a landmark year because it saw the first hard evidence for a link between ozone loss and chlorine chemistry.

The evidence came from a huge US-led experimental campaign during August and September – the period during which the ozone hole develops each year. The project involved some 150 scientists and technicians from several nations, based at Punta Arenas on the southern tip of Chile. Crucially, the team was equipped with two aircraft – a DC-8 and a modified U2 spy plane (an ER-2) – both capable of flying *into* the depleted region, but the latter at altitudes of up to 18 km. As indicated in Box 2.2, it was instruments aboard the ER-2 that caught chlorine monoxide red-handed: the 'smoking gun' of the CFC debate had finally been found.

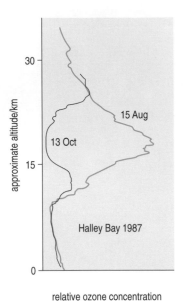

relative ozone concentration

▲   Figure 2.16(a)
*Between mid August and mid October 1987, some 95%
of the ozone between 14 and 23 km altitude was
destroyed over Halley Bay.*

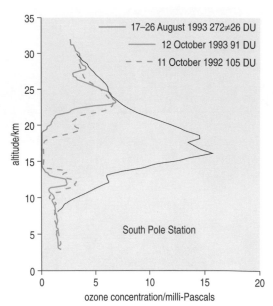

ozone concentration/milli-Pascals

▲   Figure 2.16(b)
*In the Octobers of 1992 and 1993 ozone in the lower
stratosphere (below about 19km altitude) was almost
totally destroyed at the South Pole Station.*

## Box 2.2   Finding the 'smoking gun'

The schematic record of observations from the
ER-2 showed that at ~18 km altitude, poleward of
~65°S, the composition of the vortex was highly
perturbed. As this chemically perturbed region
(CPR) was entered, the concentration of chlorine
monoxide (ClO) increased sharply over several
hundred kilometres, reaching values some 10 times
greater than those observed immediately outside
and some 100 times greater than those at lower
latitudes. In late August ozone concentrations were
roughly constant across this boundary.

However, by the middle of September there was
a sharp decline in ozone as the chemically
perturbed region was entered … As well as the
gross changes in ozone and ClO seen as aircraft
flew into the CPR, there were also smaller-scale
changes in ozone and ClO which mirror one
another. This decline of ozone only in the region
where the ClO concentrations were high provides a
strong indication that chlorine chemistry is
responsible for the ozone depletion.

Source: SORG, 1988, p. 9 and Figure 1.7

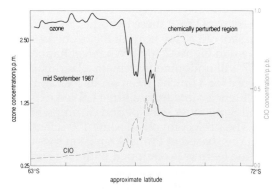

▲   *NASA's ER-2 – described as a 'rocket engine
with glider wings' – carries instrumentation
packed into two wing pods – and a lone pilot.*

▲   *Schematic record of observations from ER-2
of ozone and chlorine monoxide concentrations.*

## What causes the hole?

Several factors come together during the long dark Antarctic winter to prime the stratosphere with the 'reactive' chlorine that destroys much of the local ozone layer so dramatically each spring.

First, there is the unique meteorology of the polar stratosphere. At the autumn equinox (the end of March), the Sun sets for six months at the south pole, and an area of darkness spreads over the polar cap. It gets very cold. Up in the stratosphere (*not* at ground-level, note), this rapid cooling sets up a pattern of very strong westerly winds that swirl around the pole – the so-called **polar vortex** (Figure 2.17), extending out to around 60°S. This 'structure' is extremely stable: it endures throughout the polar winter, weakens slightly with the return of sunlight in September, but does not finally break down until summer – usually some time in November. An important effect of the vortex is to isolate the air within it. In particular, it effectively blocks the circulation of ozone-rich air from the tropics, preventing it from moving further south than about 60°S for the 'dark side' of the year – the 'resistance' referred to in the caption to Figure 2.12 (Section 4.1). Now refer again to Figure 2.11. These circulation patterns explain – in part, at least – why the amount of ozone *within the vortex* used to hold steady (at around 300 DU) throughout most of the winter and spring. Thereafter, the final dispersal of the vortex in early summer allowed a rapid influx of air from lower latitudes – carrying in 'fresh' ozone and thus increasing the amount, to some 400 DU. The vital difference is that now the ozone column is almost constant through much of the winter, but falls rapidly in the spring – usually to less than 150 DU – before recovering again in the summer. The dramatic growth in the size and duration of the ozone hole during the southern spring is illustrated for recent years by Figure 2.18.

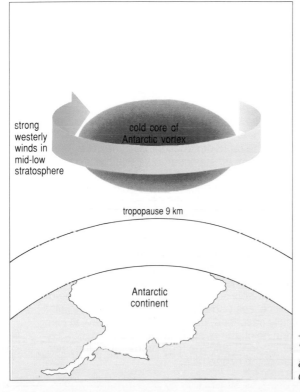

strong
westerly
winds in
mid-low
stratosphere

cold core of
Antarctic vortex

tropopause 9 km

Antarctic
continent

◀   *Figure 2.17*
*The Antarctic vortex forms*
*as air cools and descends*
*during the winter months.*

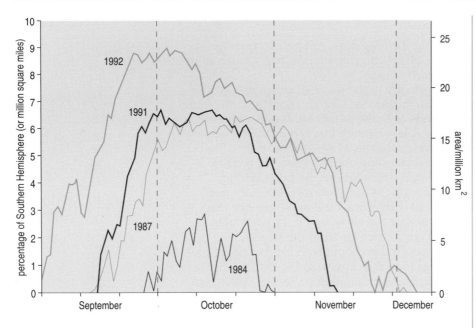

◀  *Figure 2.18*
*The seasonal growth and decay of the Antarctic ozone hole during the southern spring for the years 1984, 1987, 1991 and 1992. In 1992 the maximum areal extent of the hole (defined as the area enclosed by the 212 DU contour) covered nearly 10% of the Southern Hemisphere; the hole has generally started earlier and persisted longer into the spring.*

The second key feature of the polar stratosphere is its unusual chemistry. Once again this stems from the isolation of the Antarctic vortex. Eventually, its frigid core becomes cold enough for icy particles to form – known as **polar stratospheric clouds**, or **PSCs**. They have an ethereal beauty (see Plate 7), but are deeply implicated in polar ozone destruction. The detailed action of PSCs is described in Box 2.3, but the essential point is simple and dramatic: at temperatures of –80°C or lower, chemical reactions

---

### Box 2.3    Polar stratospheric clouds

Two types of PSC are distinguished as the temperature drops below –70°C down to –90°C each winter. Type I starts to form at the 'higher' temperature as supercooled droplets of water and nitric acid condense, then slowly freeze into solid hydrates of nitric acid. Below about –80°C condensation of water molecules on these hydrates leads to the formation of particles consisting predominantly of ice (Type II). Although the formation, classification and reactions of the cloud particles are complex, and still being unravelled, only two aspects of their behaviour need be considered here.

Firstly, the central role of PSCs is to release chlorine from their 'holding tanks' – remember those reservoir molecules, hydrogen chloride (HCl) and chlorine nitrate ($ClONO_2$), described in Section 2.2 – that normally keep its destructive effects in check. A variety of chemical reactions is initiated at the surface of the PSCs which has the effect of releasing active chlorine molecules ($Cl_2$, HOCl) and of removing nitrogen compounds which react to

lock them up again as chlorine nitrate.

As long as the PSCs are present, the polar vortex is *primed* to release the chlorine radicals (Cl and ClO) *as soon as sunlight returns* in late winter, which then, along with similar bromine radicals, catalyse the rapid destruction of ozone. The second, dramatic aspect of most PSC reactions is that they speed up as the temperature falls, and, if present in sufficient quantity, the colder Type II PSCs are thought to be able to 'prime' the whole polar vortex in the lower stratosphere within a matter of days.

Neither the formation of PSCs, nor their role in perturbing the chemistry of the stratosphere was included in the early 1-D and 2-D models (mentioned in Section 4), so their failure to predict the ozone hole is not surprising. Although some of the processes underlying PSC behaviour are still not fully described, current 3-D models are all clear on one point: the overwhelming role of anthropogenic chlorine (and bromine) as the main cause of the ozone hole.

at the surface of PSCs become very effective so that in a matter of days the stratosphere is primed for the rapid destruction of ozone when sunlight appears.

### Activity 4

Based on the record of ozone depletion over Halley Bay (in Figure 2.15), what other features of this record do you think the explanation summarised above needs to address?

You may have noticed several features – other than the clear evidence of increasing ozone depletion – but I would like to draw your attention to two:

*   Why was there a sudden onset of ozone depletion in the early 1980s?
*   What caused the fluctuations of ozone levels in the late '80s?

To these, I would add a third, related, question:

*   What of the future?

Although the answers to these questions cannot be complete, it is clear that the late 1970s saw the Antarctic stratosphere cross some critical threshold. Part of this may have been induced by natural changes – in large-scale global circulation patterns, say – coming to favour a stronger, colder polar vortex, well-laced with PSCs. But without the inexorable build-up in anthropogenic emissions, there would have been little chlorine around to take advantage of this change.

This interplay between chemical and dynamic factors also drives the year-by-year variability of the ozone layer. The Antarctic is primed each winter for ozone destruction throughout the lower stratosphere, but what actually happens in a given year depends on the meteorological conditions that prevail: the stronger the vortex, and the longer its frigid core is maintained after the return of sunlight, the deeper and more sustained will be the ozone loss. A well-known feature of the stratospheric circulation is the **quasi-biennial oscillation** (or **QBO**), which causes the winds in the equatorial stratosphere to flip from easterlies to westerlies and back every two to three years. The QBO is known to influence both the temperature and intensity of the polar vortices: westerly phases were associated with deep vortices in 1987, 1990 and 1993. In the late '80s this effect appeared to play a critical role in deciding which of two states of the Antarctic stratosphere prevailed, a strong circulation with heavy ozone depletion, or a weak, distorted vortex – as in 1988 – that dissipated early with less severe ozone loss.

From 1989 to 1991 much of the ozone in the region of the stratospheric clouds was destroyed each spring, just as it was in 1987 (Figure 2.16a). The reduction of the total ozone column to 150 DU at Halley each October, about half the historical value, seemed to have reached the limits of destruction. Then, in 1991, came the eruption of Mount Pinatubo in the Philippines which injected massive amounts of sulphur dioxide into the stratosphere, probably more than any other volcano this century.* In the course of a year the stratospheric winds had spread the 20 million tonnes of **sulphate aerosols** in a veil from pole to pole. Record low total ozone values were observed in 1992 and 1993 throughout the Antarctic (and elsewhere): the ozone hole was both deeper and wider than before (Figures 2.15 and 2.18). At Halley the total ozone column fell to 110 DU, and over much of the

* Once again, the difference between troposphere and stratosphere is crucial. Sulphur dioxide emitted into the troposphere is usually washed out by rain within a few weeks – and is the main source of acid rain. In the stratosphere it lies beyond the reach of rain clouds and persists for several years, affecting the chemistry of the ozone layer. In both parts of the atmosphere sulphate aerosols are formed. Their effects on climate are described in Chapter 3, Section 3.5.

Antarctic continent in the lower stratosphere (between 14 and 19km), 99% of ozone was destroyed (see Figure 2.16b).

▲  *'There's a hole this big, I tell you ...'*

It is likely that these increases can be attributed in part at least to tiny sulphate particles or aerosols from Pinatubo in the lower stratosphere. Sulphate particles can act as catalytic sites for the 'priming' of reactive chlorine and bromine and hydrogen $(HO_x)$* compounds, in a manner similar to PSCs, but at temperatures above those at which PSCs are most effective. The sulphate particles were able to reach parts of the ozone layer, between 10 and 15kms for example, that PSCs, centred about 17kms were 'unable to reach' (compare Figures 2.16a and 2.16b). That does not explain the enhanced losses observed above 18km as the sulphate layer was not found above this height. For an explanation of this we must look to another cause. Many models of the Antarctic stratosphere have predicted that the springtime ozone losses should cool the lower stratosphere. Since the late 1980s, observations at Halley have frequently found temperatures more than 5°C below the historical range of values associated with October, November and December.

*Q*  Why might a cooling of the Antarctic stratosphere in springtime be a cause for concern?

*A*  At temperatures below –80°C, PSCs remain highly active and, as long as this condition persists, ozone destruction can continue. If the cooling is sufficient to keep temperatures below –80°C, ozone destruction will continue later into the southern Spring. There is evidence that this has been happening in recent years (see, for example, Figure 2.18).

The main action of the volcanic sulphate aerosols was probably to extend the limits of the ozone hole, increasing its depth, area and persistence into spring and summer. It now seems likely, though, that the ozone hole has passed another threshold: as the volcanic materials faded away in 1994 and 1995 the hole appeared to be nearly as deep as the two previous years and continued to cover an area of between 20 and 25 million square kilometres, maintaining the trend seen in Figure 2.19. The effects of Pinatubo may merely have brought forward by a few years new patterns of ozone destruction already set in train by increased chlorine and bromine loading of the stratosphere. Not only does the springtime ozone hole continue to deepen and widen but losses extend well into the southern summer – December and January – giving an increased risk of UV radiation at the surface of the Antarctic and its borders.

* Here the subscript 'x' refers to a variable number of oxygen atoms.

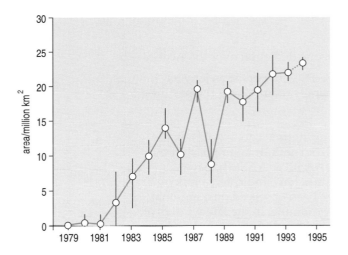

◀   Figure 2.19
*The growth in area of the
Antarctic ozone hole in the
southern spring. Since 1992,
the hole, if circular, would
reach uniformly to 65°S.*

*(Source: adapted from
information provided by
Goddard Space Center, NASA.)*

What of the future? Current computer models of the Antarctic vortex
are now able to reproduce the events described for the 1980s and 1990s and
the good news is that, although the ozone hole is expected to 'go down
faster' each year, the hole should not get any deeper or wider. This
prediction is based, firstly, on (reasonably reliable) projections that chlorine
and bromine loading will peak by about the year 2000 then slowly subside
and, secondly, on an understanding that, even with the reduced
stratospheric temperatures, PSCs could only last until the end of October
(when temperatures normally reach –60°C). While this prediction is
comforting, it must be treated with some caution. At the moment, even
models which include PSCs still underestimate overall ozone loss and,
given the considerable uncertainty about the exact nature of PSC
composition and temperature thresholds for chlorine activation, it is
difficult to assess how late in the year these reactions may remain
important. For the next ten years at least, when the chlorine and bromine
loading in the stratosphere is expected to be at or above current levels,
Antarctic springs will be observed with some trepidation.

## 5.2   Global implications

Q   Where else might you expect to find evidence of accelerated ozone
depletion?

A   Above the north pole – the region most like Antarctica as far as climate
goes.

### An Arctic hole?

In fact the different distribution of land masses and oceans in the Arctic and
Antarctica results in the north polar winter stratosphere being generally
*warmer* than its southern counterpart. This has two important
consequences. First, although PSCs do form over the Arctic, they are
usually less abundant. Second, the circumpolar vortex is weaker, and may
be disturbed, or even disrupted, during the winter by warm air pushing in

from lower latitudes. As a result, the vortex and the PSCs within it rarely persist until sunlight returns.

Q   Why would this affect the severity of ozone depletion over the Arctic?

A   Substantial ozone depletion requires the PSCs to persist as long as possible *after* the return of sunlight – thereby maintaining the perturbed chemical composition.

At the time of writing (early 1996) a major springtime ozone hole has not occurred over the Arctic, but observations from aircraft taking part in the Airborne Arctic Stratospheric Expeditions (AASEI & AASEII), have confirmed that ozone depletion takes place and have indicated a loss of 20% in the winter of 1991/92. Similar losses have been observed on a regional scale, over Siberia and northern Europe, for example, in the late winters of 1992/93 and 1994/95. These losses have been linked to chlorine chemistry by measuring the distribution of active and reservoir species. Chlorine monoxide (ClO), used as an indicator of active chlorine (see Box 2.1), is observed to increase in December when the PSCs first form. It reaches a maximum in January and at the same time the reservoir molecule HCl is reduced. The process takes place intermittently throughout the winter months, with some recovery of reservoir molecules when temperatures are not low enough for PSCs to form. The Arctic is thus primed, if not completely, for ozone destruction just as it is in the Antarctic. In addition, it is thought that the effect of PSCs can be reinforced by sulphate aerosols, and, in the relatively warmer stratosphere of the Arctic, the reaction of bromine radicals with chlorine radicals (see Section 4.2) may point to a significant contribution from bromine.

Two factors appear to prevent an Arctic ozone hole developing at the moment. Firstly, the polar vortex tends to disappear in February, before the arrival of sufficient sunlight for the catalytic removal of ozone, leading to an influx of warmer ozone-rich air and a decline of PSCs and active chlorine. Compare this to the Antarctic which remains cold for a further two months. Secondly, less nitric acid is trapped by the PSCs in the Arctic than in the Antarctic (a process known as *denitrification*), so that, when sunlight returns, active chlorine is readily captured to form the second reservoir molecule, chlorine nitrate.

The chemistry, though complex, is essentially the same as occurs in the Antarctic: anthropogenic chlorine and bromine compounds are the main culprits. At present the weaker and more variable vortex with its warmer temperatures has prevented the Arctic from going the same way as the Antarctic, but this same variability limits our ability to forecast ozone changes in the future. However, a persistent vortex in late winter, coupled with reduced temperatures, has the potential to produce considerable ozone losses at high latitudes.

## Worldwide ozone trends

Since the late 1970s ozone measurements have been made both by satellite and from ground-based instruments. Although there have been problems of calibration and comparability between different types of instrument, there is now broad agreement on the results. Ozone losses have been observed in mid-latitudes of both hemispheres as well as in the polar regions. From 1979 to 1994 the trend for the northern hemisphere, between 30°N and 60°N, has been minus 6% per decade in winter and spring, and

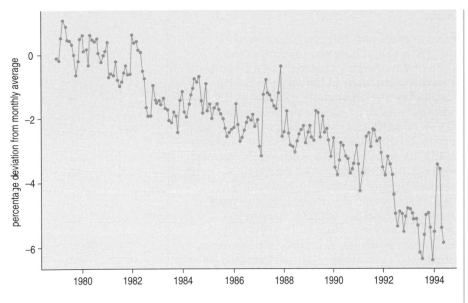

*Figure 2.20
Global ozone trends from
1979 to 1994 measured
between 60°S and 60°N.
The data for this figure do
not include the regions of
greatest ozone depletion
over the Arctic and Antarctic.
The anomalously low figures
observed during 1992 and
1993 are probably a
consequence of the
Pinatubo eruption.*

minus 3% per decade for summer and autumn. In the southern hemisphere,
between 30°N and 60°N, the annual trend of minus 4 to 5% per decade has
been similar but without strong seasonal differences. In the tropics,
between 20°N and 20°S, the trend is slightly negative, but not statistically
significant. The global trends outside the polar regions – between 60°N and
60°S – are illustrated in Figure 2.20.

It has sometimes been argued that a trend like this could be caused by
factors other than the ozone-depleting chemicals, for example the changes
in UV radiation that are known to vary over the eleven-year sunspot cycle.

Q    Why should this affect the ozone layer?

A    While a full explanation may be more complex, changes in UV
radiation from the Sun would be expected to affect the initial *production*
of ozone in the stratosphere, as described in Section 2.1.

Observations of ozone levels since the 1960s (over several cycles) show that
global ozone levels do indeed vary with the sunspot cycle, but with an
amplitude of only 1–2%. The trends in Figure 2.20 show changes much
larger than this and the solar cycle has gone from a maximum in 1979 to a
minimum in 1985 then back again to a maximum in 1991, so the
explanation must be sought elsewhere. In fact it seems likely that the losses
in ozone globally are due to a combination of transport from depleted polar
zones combined with local losses due to gas phase reactions (catalysed
mostly by $HO_x$, chlorine and bromine) and by heterogeneous reactions on
particles such as sulphate aerosols.

Two separate observations provide evidence to support this
explanation. Firstly, seasonal variations for the northern hemisphere show
greatest losses during spring (with a peak of 6–10% in February to April
1993) after depleted polar air has a chance to mix. Secondly, after the
eruption of Pinatubo in 1991 and the injection of sulphate aerosols into the
stratosphere, a 3–4% ozone loss was observed in the tropics in the following
six months while on a global scale losses of about 2% were observed in 1992
and 1993.

## Ozone levels and UV radiation

Thus far, Section 5 has concentrated quite naturally on ozone *losses* to indicate the effects of increased chlorine and bromine loading of the atmosphere and to point to some of the complex chemistry and dynamics affecting the ozone layer. The next section looks at the *effects* of ozone depletion, most of which are caused by the consequent increase in UV radiation reaching the lower atmosphere, in particular the biologically active component known as **UV-B radiation**. It may be helpful at this point to review the factors other than ozone depletion which affect the intensity of the UV radiation (including the UV-B component) reaching the ground.

The amount of UV radiation reaching the surface of the Earth is strongly influenced by three environmental factors: the angle the Sun makes with the horizon (the higher it is in the sky, the shorter the path of radiation through the atmosphere); the total ozone column in the atmosphere; and the incidence of clouds and pollution. The first factor varies in a wholly predictable manner, while the averaged column ozone has been mapped for each latitude and season (see Figure 2.11), though fluctuations on shorter time-scales do occur. In the lower atmosphere, changing weather can quickly alter the distribution of clouds and pollution, causing very large fluctuations in UV radiation over time-scales of hours or less, and over small regions, making any slow, long-term trends difficult to detect.

Q   *Before ozone depletion occurred,* where on the globe, and when, would you expect UV radiation to be the greatest?

A   Figure 2.11 shows that the total ozone column is lowest – less than 260 DU – over a broad band of the tropics between 10°N and 20°S. Maximum UV is likely to occur here at those times of the year and, just as important, at the times of day – the hours around noon – when the Sun is overhead.

This answer ignores the less predictable effect of clouds and pollution, but would be correct for clear sky conditions. What we are discussing here is the intensity of the radiation on a horizontal surface – as might be approximated by the exposed surface of an average sunbather. For a more typical person who is upright rather than prone, the reflection of UV radiation from bright surfaces such as snow or sand may be significant.

Two other factors can also affect the intensity of UV radiation. The first is height above sea-level, because much pollution is low-level, so mountaineers and skiers are likely to be exposed to high levels of UV. Figure 2.11 also indicates that the highest natural levels of ozone occur at high latitudes in the months of winter and early spring which is fortunate for skiers and others who live and take recreation in high latitudes because these happen to be the conditions under which current ozone depletion is most severe.

The second factor is distance from the Sun. The Earth's orbit about the Sun is not quite circular but an ellipse, with the result that the distance between the Sun and the Earth changes a few per cent with the seasons. At present the Earth is closest to the Sun in the northern hemisphere winter and most distant in summer. In practice this means that in the summer months, when maximum UV-B radiation occurs, the southern hemisphere receives a greater intensity of radiation than the north.

*Activity 5*

Make a summary of Section 5 for yourself by working through these questions:

Q1  Why did it prove so difficult to obtain convincing evidence that emissions of CFCs were damaging the ozone layer? How did the results from Halley Bay alter this situation?

Q2  Make a list of the key steps believed responsible for the ozone hole over Antarctica each springtime. Why does a similar hole not appear over the Arctic? (Or strictly, not as of 1995: you may be aware of developments since then.)

Q3  What is the observational evidence to support the thesis that sulphate aerosols, formed in the stratosphere after the eruption of Pinatubo in June 1991, could play a role similar to PSCs in accelerating ozone loss over Antarctica?

Q4  What factors influence the strength of ultra-violet radiation experienced at the Earth's surface?

# 6   Consequences of ozone depletion

The most obvious cause for concern about ozone loss stems from its role as a filter of the Sun's ultra-violet radiation. Figure 2.21 is a repeat of Figure 2.4 (the black curve), but compared this time with the pattern of radiation received at the Earth's surface (the green curve). Concentrate on the UV region. Notice that the band labelled **UV-C** ($2.0–2.9 \times 10^{-7}$ m) is virtually eliminated by the atmosphere. This is just as well, because UV-C is lethal to micro-organisms (whence its use in germicidal lamps), and can destroy both nucleic acids and proteins: in the range from $2.4–2.9 \times 10^{-7}$ m, protection from UV-C is due entirely to absorption by ozone.

More important as far as ozone loss is concerned is the narrow band between $2.9 \times 10{-7}$ and $3.2 \times 10^{-7}$, known as 'biologically active' UV or **UV-B**. Here the *attenuation* of the solar input evident in Figure 2.21 is again due to ozone, but the effect is less complete: a fraction of UV-B penetrates all the way to the ground. UV-B radiation is known to have a multitude of effects on humans, animals, plants and materials and indeed on the chemistry of the atmosphere itself. Most of these effects are damaging – but few are sufficiently well understood at present for the impact of enhanced UV-B to be quantified. However, given its importance, it may still come as a surprise to learn that there were few reliable measurements of this radiation band until the late 1980s. Part of the reason is its low intensity: at the edge of the Earth's atmosphere it represents about 2% of the solar spectrum, but only a small fraction of this reaches the surface of the Earth.

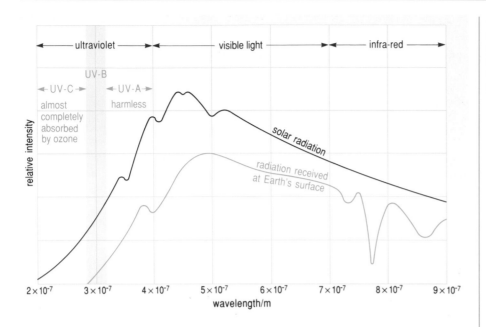

◀  *Figure 2.21*
*Incoming solar radiation
(from Figure 2.4) compared
with that received at the
Earth's surface after passing
through the atmosphere.*

Although always difficult to measure, UV radiation has been monitored at a few sites for many decades, but before the late 1980s instruments rarely had sufficient accuracy or stability to provide an historical base-line for UV-B for the period before ozone depletion started. It was only after the discovery of ozone depletion that reliable measurements started to be made (with an absolute accuracy of $\pm5\%$), but this has not given enough time to establish long-term trends. Small changes due to ozone depletion are still likely to be swamped by the effects of clouds and pollution, so the clearest pictures come from instruments taking clear-sky measurements or sited away from urban pollution.

## 6.1   Measurements of enhanced UV-B Radiation

Nevertheless, there is now overwhelming observational evidence that ozone depletion causes increases in UV-B radiation in accordance with theoretical models. As might be expected, large increases of UV-B radiation have been observed at high southern latitudes in association with the springtime ozone 'hole': see, for example, Figure 2.22. At Palmer station on the edge of the Antarctic (64°S), for instance, damaging radiation levels during spring now exceed the *summer maxima* recorded in San Diego in Southern California when the Sun is nearly overhead (see Figure 2.23). At mid-latitudes, away from the ozone hole, UV-B measurements for clear skies show relatively high levels in the southern hemisphere in summer, in Chile, New Zealand and Australia, for example, compared to corresponding latitudes in the northern summer in Europe.

Q  Why do you think UV-B measurements are higher in the southern hemisphere?

A  As solar angle and clouds are excluded, the main factor has to be the differences in total ozone column, but both tropospheric pollution in the northern hemisphere, and the difference in Sun–Earth separation during the northern and southern summers play a role.

Until recently, measurements have not shown any clear trends in UV-B levels in mid and high latitudes outside the polar regions, although there is a strong likelihood that before the 1980s pollution in urban areas reduced UV-B radiation. However, large increases in UV-B in 1992 and 1993

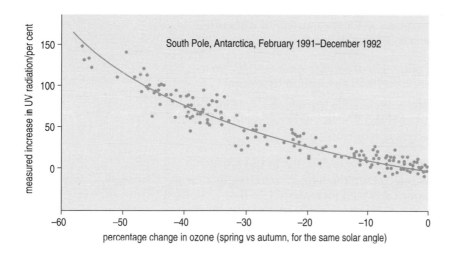

South Pole, Antarctica, February 1991–December 1992

◀ *Figure 2.22*
*The increase in the intensity of sunburning (erythemal) UV radiation at the south pole due to reductions in stratospheric ozone. The points in the diagram compare levels of springtime UV radiation, under ozone hole conditions, with UV values taken in the southern autumn, when ozone levels are 'normal'. The comparisons are made for the same solar angle, compensating for its effect.*

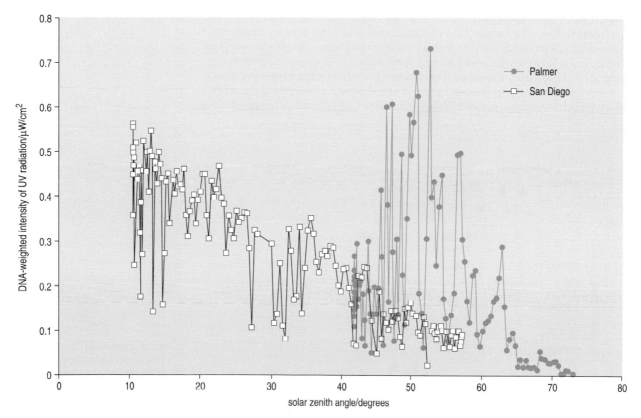

▲ *Figure 2.23*
*The intensity of UV radiation at noon for clear skies plotted as a function of the angle of the Sun above the horizon (the solar zenith angle) – which varies with the season. The figure compares Palmer Station at 65°S in Antarctica with San Diego in southern California at 32°N and illustrates the excess radiation levels found in the Antarctic.*

(resulting from the effects of the Pinatubo eruption) have been measured in middle and high latitudes in the northern hemisphere compared with previous years, despite the obscuring effects of clouds. These are the first reported examples for densely populated areas of persistent increases associated with ozone depletion.

One important feature of the recent measurements and of model predictions, and one that also contributes to the measurement difficulties, is the uneven absorption by ozone across the UV-B band.

*Q*   Using Figure 2.21, which end of the UV-B band from $2.9 \times 10^{-7}$ to $3.2 \times 10^{-7}$ m shows the greater *proportional* absorption by ozone?

*A*   The short-wave end at $2.9 \times 10^{-7}$ m shows the greater absorption. The figure shows that virtually all UV-B at $2.9 \times 10^{-7}$ m is removed, while at the other end about three-quarters is removed.

As the ozone column is depleted, the shorter wavelengths in the UV-B band show higher proportional increases than the longer wavelengths. For example, a 1% reduction in ozone causes an increase of UV-B of approximately 1% at $3.1 \times 10^{-7}$ m, but 3–4% at $3.0 \times 10^{-7}$ m and greater than 10% at $2.9 \times 10^{-7}$ m. This is significant because the shorter wavelengths are by far the most damaging to living organisms and tissues. A typical example of this sensitivity, based on experimental measurements, is shown in Figure 2.24. However, for most biological processes what counts is the total dose or dose increase, not the change of proportion; if this is considered, then the longer wavelengths still show the greater absolute increase. In order to allow for these two opposed wavelength-dependent effects – UV-B increases and biological sensitivity – a weighting factor is used which indicates the available *effective* radiation for a given process. Figures 2.22 and 2.23 in this section use the erythema (sunburning) and DNA-weighted measures of UV-B, with values of 1.2 and 2.2 respectively.

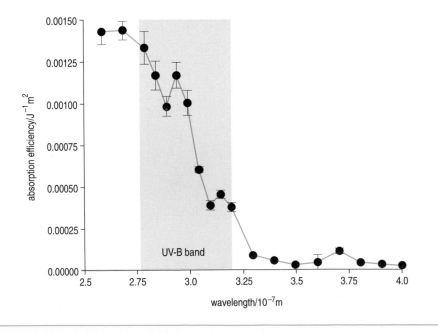

◀   *Figure 2.24*
*The action spectrum for the inhibition of photosynthesis in the cyanobacterium Nodularia spumigena shows, typically, that the shorter wavelengths within the UV-B band ($2.9 \times 10^{-7}$m $- 3.2 \times 10^{-7}$m) have the greatest biological effect.*

## 6.2   Human health

Through its contact with exposed skin and eyes, UV-B radiation has a direct effect on humans and animals. The effects can be beneficial, as in the production of vitamin D, or damaging, where there are a number of possible adverse effects – ranging from short to long term. In some cases, as with the impact on the immune system, both aspects may be in operation: suppression of immune reactions in skin may bring relief to psoriasis-sufferers but can lead to a lowering of defences against tumours and infectious agents.

Excessive exposure to UV-B radiation is known to cause short-term damage to eyes, sometimes called 'snow-blindness', particularly where there are bright reflective surfaces such as sand or snow. An extremely painful inflammation of the eyes occurs but the effects, fortunately, are temporary. Of greater significance is the increased likelihood of developing cataracts caused by *chronic* (i.e. long-term) exposure to sunlight and in particular UV-B radiation, the key factor being the accumulated, lifetime dose. The World Health Organisation (WHO) has estimated that about half of all cases of blindness worldwide are due to cataracts, a clouding of the crystalline lens of the eye. There are many other risk factors for cataract development, such as malnutrition, diabetes, heavy smoking and ageing, but while these may produce higher relative risks the number of people exposed to the risk is in each case fewer than for sunlight (with the exception, of course, of ageing). Estimates based on epidemiological evidence from the US Environmental Protection Agency (EPA) suggest that each 1% decrease in ozone will produce approximately a 0.5% increase in the incidence of cataracts.

The link between UV-B and the incidence of skin cancer is particularly emotive: here, there are two main strands of evidence. First, skin cancer is predominantly a disease of light-skinned people and the dark pigment – melanin – is known to be an effective filter of UV-B. The second strand comes from *epidemiology* – a study of the factors that influence the

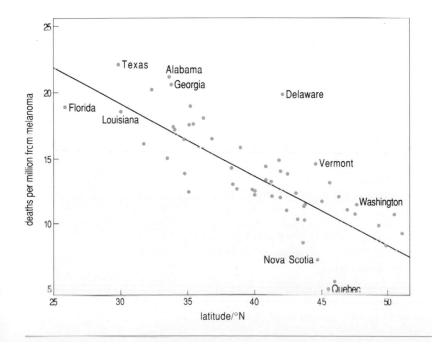

◀   *Figure 2.25*
*The variation with latitude of human death due to skin melanoma among white males in the United States and Canada. Annual averages for 1950–67.*

occurrence of the disease in human populations. These studies reveal a striking inverse correlation with latitude. Some typical data from North America are shown in Figure 2.25: similar patterns have been found within white populations in the United Kingdom, Norway, Australia and New Zealand.

Q   Why should this variation with latitude support a link between the incidence of skin cancer and the intensity of UV-B exposure?

A   The intensity of UV-B is greater near the equator, both because the Sun is more nearly overhead, so the radiation has a shorter path through the atmosphere, and because on average less ozone is found there (remember Figure 2.11).

*Melanoma*, the particular form of skin cancer reported in Figure 2.25, is actually much rarer than other types – known collectively as *non-melanoma skin cancers (NMSC)* – but it is the most serious, with a substantial mortality rate unless diagnosed early: some 40% of patients die from the disease within five years. However, its epidemiology does not indicate a simple relation to UV-B exposure: it affects relatively young people and, unlike other skin cancers, it has been increasing over the last few decades in all light-skinned populations studied. This, and the observed correlation with indoor working and social class, has led to the suggestion that melanoma is associated with intermittent, but intense exposure to UV-B – the kind that causes sunburn – particularly during childhood.

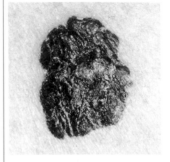

▲   *A malignant melanoma affecting the lower leg.*

NMSC is more clearly correlated with long-term UV-B exposure and light-skinned people are at most risk. NMSC is 20 to 30 times more common than melanoma, but, fortunately, its associated mortality rate is far lower: where good medical care is available, it is thought to be less than 1%. Total deaths, however, are probably comparable with those due to melanoma and for some groups – those who are immuno-suppressed, for example – NMSC is particularly hazardous. With approximately 1.2 million new cases of NMSC each year, estimates from the EPA suggest that a 1% decrease in column ozone would translate into 25 000 additional cases. Using current projections of ozone depletion, even under the Copenhagen Agreements the incidence of NMSC at a latitude of approximately 50°N will be 25% higher in the years 2030 to 2050 than in 1980.

Of course, it is possible to take precautions against the adverse effects of UV exposure – those who experience snow-blindness are unlikely to want to repeat the experience – and it is difficult to predict how such changes in personal behaviour may influence long-term trends. It is certainly to be hoped that educational programmes coupled with daily forecasts of safe exposure times – particularly in those countries of the southern hemisphere where light-skinned populations are most at risk from high UV-B levels – will lead to modified behaviour. However, it should not be forgotten that many common domestic animals – cattle, sheep, cats and dogs – are also susceptible to skin cancer and eye damage; for them, a change in behaviour is not an option.

Q   Why, if increased UV-B exposure is the major risk factor for NMSC, has there been no increase in the incidence of NMSC, whereas an increase in the number of melanomas has been observed?

A   NMSC is related to *total* dose and typically has an incubation period of twenty to thirty years, so the effects of changes in UV radiation may not appear before the turn of the century. It mostly affects light-skinned

people who have spent much of their life in the open and is more likely
to be found in older people. Changes in life-style could clearly affect the
size of the population exposed to risk. In contrast, melanomas usually
affect those who are not habitually exposed to the Sun. The risk factors
are not fully understood, but high doses of UV in early life are thought
to be damaging. Changes of behaviour have been the major cause of the
increase: the trend for young people and families to take holidays in
sunnier climates, for instance.

Finally, UV-B damage to the immune system in both animals and humans
can weaken defences not only to cancer but also to a variety of infectious
diseases, and in many cases skin pigmentation provides no protection.
The impact of these immune effects on the spread of infectious diseases
amongst the human population is not known. However, in geographical
regions where infectious diseases already present a significant hazard, and
for those sections of the population with impaired immune function, the
effect of increased UV-B may pose a significant threat to human health.
Indeed, these as yet indeterminate threats may prove to be more of a
hazard than those where a quantitative estimate, however uncertain, has
been made.

## 6.3    Terrestrial plants

Plants are adapted to present levels of radiation, but rather little is known
about their response to enhanced levels of UV-B. To date, most studies have
focused on agricultural crops typical of mid-latitudes. Of the 300 or so
species and cultivars screened for tolerance to UV-B, some two-thirds have
been found to be sensitive – although the degree of sensitivity varies
widely, even among cultivars within a given crop species. Typically,
sensitive plants show reduced growth and smaller leaves: unable to
photosynthesise as efficiently as other plants, they yield smaller amounts of
seeds or fruit. In some cases, these plants also show changes in their
chemical composition, which can affect food quality.

The limited data available so far suggest that increased UV-B levels
may also affect forest and grassland productivity. Potentially more

◀    Loblolly pine trees
irradiated with UV-B
simulating 0%, 20% and
40% ozone depletion.

important, it is possible that subtle changes in plant growth induced by UV-B could upset the delicate balance in natural ecosystems – thereby changing the distribution and abundance of plants. Quantifying this effect remains a key area of uncertainty – as indeed do the more direct impacts on food production and forestry.

## 5.4   Aquatic ecosystems

Life in the oceans is also vulnerable to UV radiation. Although not as important as visible light or temperature or nutrient levels, there is evidence that ambient solar UV-B radiation is nevertheless an important limiting factor in marine ecosystems. The potential impact of any increase in UV-B will depend critically on the depth to which it penetrates – more than 20 m in clear waters, but only some 5 m or so in unclear (silty, say) water. These estimates are crude, because underwater penetration of UV is notoriously difficult to measure, but, if confirmed, they suggest that roughly half of all marine fishes, all nearshore flora and fauna (including coral reefs), and many of the living things in estuaries and lagoons could be at risk. Certainly, enhanced UV-B has been shown to damage a range of small aquatic organisms – zooplankton, larval crabs and shrimp, and juvenile fish – as well as slowing photosynthesis in phytoplankton. Many of these latter organisms are capable of active movement allowing them to optimise light levels over the cycle of day and night. Exposure to UV-B radiation has been shown to interfere with both movement and orientation mechanisms leading to reduced productivity and survival rates.

Q   Why should damage to these small organisms be of particular concern?

A   They are at the base of the marine food chain, so any change here – in species composition, say – could have consequences higher up.

As was noted in Chapter 1, in many countries more than 50% of the dietary protein of the human population is obtained from the sea – so once again, the potential exists for a substantial *indirect* effect on human health. As with terrestrial ecosystems, too little is known for predictions of the overall biological consequences of ozone depletion to be made with any degree of confidence. However, there is a more immediate cause for concern – the ozone hole over the Antarctic (and the increasing depletion of ozone over Arctic areas). With the exception of upwelling near continental shelves and the equator, the highest concentrations of marine phytoplankton are found at high latitudes with concentrations typically 10 to 100 times greater than in the tropics and sub-tropics (see Plate 2). Although the desolate, ice-covered continent of Antarctica is home to few land plants and animals, the southern oceans are some of the richest ecosystems on Earth – teeming with life that has evolved in the presence of little UV radiation. In Spring, as the ozone hole fluctuates above these waters, the variation of UV-B radiation in the upper ocean layers has now been measured: preliminary studies have indicated reductions of 6–23% in productivity. Many marine organisms, including phytoplankton and the krill that feed on them, may be particularly susceptible because peak UV-B values occur at critical phases in their development.

## 6.5   Atmosphere and climate feedbacks

In this last section we explore some of the more unexpected feedbacks that can occur from the reduced levels of ozone in the stratosphere. Firstly, we will be looking at the main consequences of stratospheric ozone depletion: the increase in UV-B radiation in the troposphere. Then we will briefly review some possibilites of interactions between ozone loss and the topic of the next chapter, global warming.

### The effect of increased radiaiton

The lower atmosphere – the troposphere – contains many trace gases and pollutants which have a important influence on both air quality and the climate, though the manner of their interaction with each is complex. An effect of an increase in UV-B radiation is to make the troposphere more reactive by influencing the balance of tropospheric ozone and related oxidants such as hydrogen peroxide, $H_2O_2$, and the hydroxyl radical, OH.

The effect on tropospheric ozone is significant in its own right, as low-level ozone is both a serious pollutant and a powerful greenhouse gas. Changes in its distribution or concentration will thus have consequences for both atmosphere and climate.

---

**Box 2.4    Tropospheric ozone and UV radiation**

Tropospheric ozone is thought to have increased by at least 50% since pre-industrial times (from 'natural' levels of approx 20–30 ppbv) and is now recognised as a serious pollutant in cities and surrounding areas. It occurs naturally through downward flux from the stratosphere but is also formed through the interaction of sunlight with a cocktail of pollutants – hydrocarbons, carbon monoxide in the presence of nitrogen oxides ($NO_x$) – usually associated with road traffic. Just as in the stratosphere, ozone in the lower atmosphere reacts with UV but its balance is very sensitive to the concentration of $NO_x$. Where $NO_x$ levels are high, an increase in UV-B radiation is predicted to raise levels of tropospheric ozone. In practice this means most urban and rural areas, where harmful concentrations of ozone are expected to occur earlier in the day and closer to emission sources and population centres, with some rural areas suffering more urban patterns of pollution. In the relatively unpolluted or pristine areas where $NO_x$ remains low, tropospheric ozone is expected to reduce, and indeed lower levels have been measured in both the Antarctic and Canadian Arctic.

---

Q   Will enhanced UV radiation in the troposphere affect CFCs?

A   It is most unlikely. CFCs are not dissociated by the UV radiation that reaches the troposphere; this only happens in the stratosphere where the radiation is much more energetic. On the other hand the change is likely to affect some of the 'CFC substitutes'.

Enhanced UV-B levels are expected to increase the concentrations of the hydroxyl radical, OH, giving rise to some interesting indirect effects or feedbacks – not all harmful. In Section 4 we saw that molecules with carbon hydrogen bonds were attacked by OH in the troposphere. Each increase of the hydroxyl radical, the 'atmospheric cleansing agent', gives nearly proportionate reductions in concentrations of methane and the CFC substitutes, HCFCs and HFCs, by speeding the rate of their removal. In the last decade, the rate of growth of the greenhouse gas, methane, in the atmosphere has slowed; one third of this slowdown is possibly due to this indirect consequence of enhanced UV radiation. Similar changes in hydrogen peroxide concentration, the main oxidiser of sulphur dioxide to sulphuric acid, may have an impact on patterns of acid rain deposition, while in general the increase in tropospheric activity could lead to an increased production of cloud condensation nuclei, from the oxidation and subsequent nucleation of natural and anthropogenic sulphur. These examples illustrate the complex interactions now taking place in the atmosphere.

## Ozone loss and global warming

Ozone loss can have direct and indirect impacts on global warming: directly, because ozone, whether in the troposphere or stratosphere, is a greenhouse gas, as are the CFCs and most other ozone-depleting chemicals. The global and regional reductions observed in stratospheric ozone are thought to provide a partial 'offset' to the warming caused by the CFCs in the atmosphere. This topic is discussed more fully in the next chapter. Some indirect effects – on the abundance of greenhouse gases and clouds – were touched on in the previous paragraph. Another possible interaction is the effect of ozone loss at high latitudes on phytoplankton: ozone depletion may lower their productivity, in turn reducing the uptake by the oceans of the greenhouse gas carbon dioxide. All could influence the rate of global warming but the dynamics are poorly understood at the moment and hard to quantify.

Perhaps of greater immediate concern is the possible impact that global warming may have on ozone loss. The possibly serious implications of reduced temperatures in the lower stratosphere – due to feedbacks caused by the ozone losses – have already been discussed in Section 5. To these must be added a far from obvious effect of global warming, which is also predicted to reduce stratospheric temperatures. A global cooling trend has been observed in the lower stratosphere for several decades, while over the Antarctic and Arctic record low temperatures have occurred in the stratosphere in recent years. Models sugggest that the major contributor to these trends is ozone depletion itself, but global warming may also be playing a part.

## 6.6   Summary of Section 6

Rather little is yet known about the physical effects or the ecological consequences of changes to the ozone layer. Nevertheless, along with increased evidence of raised UV-B levels, awareness is growing of the types of damage that might ensue – not just directly to human health, crops and terrestrial and aquatic ecosystems, but also indirectly via possible disturbance to the climate. This, coupled with our lack of knowledge about the scale of the problem, has finally triggered international action to begin the process of healing our ozone shield.

*Activity 6*

In using epidemiological data to predict changes in the incidence of skin cancer, the argument goes as follows. Skin cancer rates depend strongly on latitude; this latitude dependence is assumed to be due to differences in the intensity of UV-B; thus the change of incidence can be calculated for a given change in UV-B, brought about by a given loss of ozone. During the early debate about CFCs, it was pointed out that there is another way of interpreting the data. This links the incidence of skin cancer directly with the fact that the ozone layer varies with latitude (Figure 2.11) – thinning by around 3 DU (i.e. some 1% of the average ozone column) for each 200 miles or so displacement toward the equator. Thus, on this basis, the chances of contracting skin cancer are increased identically by a perturbation producing a 1% depletion in the ozone column, or by moving 200 miles south. How would you respond to this argument?

# 7   Strategies for protecting the ozone layer: towards the Montreal Protocol – and beyond

The scientific basis for the original Montreal Protocol was the *theoretical* prediction that, should CFC and halon emissions (see Activity 2) continue to grow for the next few decades, there would eventually be substantial damage to the ozone layer – essentially the situation outlined at the beginning of Section 5. Here we look at the terms of the Protocol and their amendments in the face of evidence that ozone loss has already occurred. But first, a look at the historical background to this seminal international agreement is in order.

## 7.1   The 'spray-can war' in the United States

The first step toward limiting the release of CFCs into the environment came with unilateral action in the United States. In May 1977 three federal agencies – the EPA, the Food and Drug Administration and the Consumer Product Safety Commission – issued a joint statement that set in train legislation to ban the use of CFCs as propellants in spray-cans by the end of 1978. This action was the culmination of a period of heated scientific and political controversy ignited by Rowland and Molina's initial publication. The debate raged throughout 1975 and into 1976. It focused on the aerosol industry because spray-cans necessarily involve complete release of the propellant gas: at the time, some 75% of CFC emissions worldwide were attributed to this source.

Throughout this period, industry – both the giant chemical companies (notably Du Pont in the US) that produce CFCs, and the aerosol manufacturers themselves – contended that the 'burden of proof' rested with the scientists who claimed that CFCs posed a threat. In particular, they argued that a major industry should not be jeopardised – with the attendant risks of unemployment and to the economy in general (this was a period of worldwide economic recession) – on the strength of a theoretical prediction, hitherto unsupported by observations on the real atmosphere. Rather they advocated a 'wait and see' approach.

There is force to this argument – essentially a presumption of '*innocent until proven guilty*'. But there is a counter view, namely that a legitimate cause for concern about damage to the environment should evoke a verdict of *'guilty until proven innocent'* – mirroring the principle routinely applied to new drugs, for example, which are extensively tested for harmful effects before being released for clinical trials. Of course, this begs the question of what constitutes 'legitimate cause for concern', and who decides. In the case of CFCs the issue was further compounded by the unusual stability of the compounds. Those who argued for an immediate ban stressed that this, together with the natural variability of the ozone column, meant that monitoring the atmosphere could not provide an *'early* warning' of ozone loss. Rather, by the time a clear signal was detected there would already be enough CFCs effectively 'stored' in the troposphere to keep the damage going for many decades – even if emissions were then halted at once: just the circumstances we now find ourselves in.

The propaganda war was played out in the media, and in testimony before committees of the state and federal legislatures. But in the end, the US National Academy of Sciences (NAS) was effectively cast as the 'Supreme Court' in this environmental trial. Its report appeared in September 1976: in brief, it supported the scientific case against CFCs (confirming a 'legitimate cause for concern'), but stressed the uncertainties in the calculations and the importance of further research. In policy terms it recommended that 'non-essential' uses of CFCs should be drastically curtailed (widely interpreted as a ban on their use in aerosols), unless new findings emerged within two years to mitigate the threat: in short, a verdict of 'guilty until proven innocent'.

## 7.2    *The international dimension: towards Montreal*

When the scare broke in 1974, the US accounted for over 50% of world usage of CFCs. While acknowledging this position in its support for unilateral action, the NAS report also stressed the global nature of the problem, and the importance of encouraging other countries to adopt similar restrictions. This call went largely unanswered. To take the United Kingdom as an example, this was not because of any major differences about the scientific assessment of the problem. On the contrary, a 1976 report from the DoE Central Unit on Environmental Pollution (Pollution Paper No. 5) produced much the same forecast of ozone depletion as its American counterpart. But the tone of the report was far more cautious, coming down *against* the 'need for precipitate action'. This reflected an attitude prevalent among UK scientists that the United States was overreacting in the urgency of its response. This, in turn, encouraged a public debate that had a far more measured tone, and generally lower profile, than that in the US.

The debate and the control strategy adopted by the US had its effect: *worldwide* production of CFCs fell from the peak levels recorded in 1974, and continued to decline throughout the late 1970s. This period also saw further massive research programmes on both sides of the Atlantic. Once again the vehicle for providing policy-makers with a scientific assessment of the problem can be characterised as the 'expert committee' – under the auspices of the NAS (and later NASA) in the US, for example, and the DoE and Meteorological Office in the UK. Further parallel reports appeared in late 1979 – again reaching broadly similar scientific conclusions. But, equally, the disagreement about what action should be taken was again apparent.

The policy of 'cautious inaction' advocated in the UK report was broadly in line with the EC approach. Here a precautionary policy, amounting to a *voluntary* agreement to limit CFC production (but at a level allowing significant expansion), was about to be adopted. By contrast, the US report called for an urgent global ban on aerosol usage. It also pointed to the alarming growth in other uses of CFCs – a pattern that was to reverse the downward trend in worldwide production after 1982. The US report prompted a further attempt at unilateral action by the EPA. But these were the last months of the Carter Administration, and the political climate in the US was about to take a sharp turn against the environmental lobby. This not only blocked the EPA's proposals, but also sought to prevent the agency from taking any further precautionary action in the absence of *international* agreement that it was necessary.

The first steps toward achieving such an agreement were taken in 1981. Since 1977 UNEP had continued to issue regular bulletins which reviewed current research and forecasts of ozone layer depletion. In May 1981 it set up a working group of legal and technical experts charged with drawing up a draft 'Convention for the Protection of the Ozone Layer'. After long and difficult negotiations, this led to the *Vienna Convention*, signed by twenty nations in March 1985; it came into force in September 1988.

It is ironic that the negotiations for this convention took place during a period when the 1970s' pattern of ever more gloomy forecasts was reversed (see Figure 2.13). But these forecasts – and more importantly, their downward trend – featured prominently in press reports during the early 1980s. Given this perception of a reducing threat, a major row between the US and the EC that erupted early in 1985 (each effectively wanting its own preferred control strategy incorporated in the UN code) and the timing (*before* news of the ozone hole first appeared), it is not surprising that the Convention is a somewhat anodyne document. It declared the signatories' determination 'to protect human health and the environment against adverse effects resulting from modifications to the ozone layer', and provided, *inter alia*, for the exchange of information and scientific data relevant to this determination. But, it had no regulatory powers at all – although it did provide for protocols to be adopted later.

## 7.3   The Montreal Protocol

The Protocol was agreed on 16 September 1987 and came into force on 1 January 1989, having been ratified by the required 11 countries representing two-thirds of global consumption. As originally agreed, it provided for a staged control of five CFCs (those listed in Table 2.2 in Section 3.1) and three halons (organic compounds containing bromine atoms). The overall effect would be to freeze consumption of the halons at 1986 levels in 1992, but to cut consumption of the CFCs to 50% of 1986 levels by the year 2000. *Production* of the CFCs was subject to less severe cutbacks, the intention here being to provide a surplus to meet the growing domestic needs of developing countries that became parties to the Protocol. Most such countries – notably China and India – were not represented at the negotiations. This provision – together with an Article exempting developing countries from controls on consumption for ten years – was designed to encourage their participation in the Protocol, while discouraging the proliferation of CFC production capacity around the world.

The terms of the Protocol were an inevitable compromise between the conflicting national interests of the countries involved. Negotiations for a protocol began in earnest at the end of 1986 – a time, note, when the Antarctic hole had again heightened fears about the ozone layer, but no hard evidence had yet emerged linking this phenomenon with chlorine chemistry. Straight away, the old split reappeared between the US (now backed by the Scandinavian countries) and the EC (and Japan). In fact, the US had hardened its stance, going into the negotiations with a call for immediate cuts in production, as part of a longer-term strategy leading to a virtual phase-out of CFCs. This position was later softened, not least as a result of infighting within the Reagan Administration. But at the time the strategy was resisted by several European countries; even here, the UK looked increasingly isolated from its Community partners – initially refusing to countenance any cuts at all.

By May 1987 a measure of compromise had been achieved – sufficient to allow a draft protocol to be drawn up. The terms were broadly similar to (if somewhat stronger than) those eventually agreed – but at the time, further progress seemed unlikely. Here, the UK position was crucial, with the prospect of its veto within the EC blocking approval of the second round of cuts.

Britain's retreat from this position came in August 1987. This U-turn has been characterised as a somewhat cynical move by the UK government, aimed at softening Britain's deservedly poor image on environmental issues but at virtually no cost to itself. Indeed, over 50% of the UK market for CFCs then went into aerosols, so major cuts in consumption could be achieved by the simple – and comparatively cheap – expedient of phasing out this usage. In addition, ICI – by far the major UK producer of CFCs – shifted its ground at this time, coming out publicly in favour of substantial cutbacks, provided the timetable was not too tight. An important factor here was undoubtedly the recognition that a schedule to phase out CFCs *worldwide* would provide an equal opportunity for companies to market substitute chemicals – newly developed compounds, production of which would be protected by patents (unlike the long-established CFCs) and thus potentially highly profitable. Alongside these factors, a sustained campaign by environmental pressure groups, together with a hardening of attitudes among UK scientists, undoubtedly played their part.

With further compromises, the terms outlined earlier were finally agreed – but the *timing* of this agreement was extraordinary! Just two weeks later, the first results started coming in from Punta Arenas. The enormity of what is happening over Antarctica, together with the publication of the 'Ozone Trends' report in March 1988, prompted media coverage and public debate throughout 1988 and early 1989 – at a level to rival that of the mid-1970s in the US. Only this time there was little controversy, public concern and action (in boycotting aerosols in the UK, for example) being matched at last by political will – culminating in a decisive shift in the international atmosphere behind much tighter controls. In the UK the most prominent feature of this period was the so-called 'Thatcher Conference' in London in March 1989 – actually hosted jointly by UNEP and the UK government. Public declarations apart, the real business had already started, was moved on two months later in Helsinki, and came to fruition in London in June 1990.

## 7.4 Beyond Montreal

A crucial feature of the Montreal Protocol is Article 6, which provides for a periodic review of its control measures 'on the basis of available scientific, environmental, technical and economic information'. The reviews are initiated by the Conference of Parties (of governments ratifying the Protocol), the decision-making body that has the power to amend the Protocol. It meets annually, but every two or three years establishes *expert assessment panels* (covering science, impacts, and technology and economics) to help its Parties consider revisions to its core commitments. There have been three review cycles since Montreal, culminating in meetings in London, Copenhagen and Vienna. At the first two meetings, to a background of increasing evidence of ozone loss, the Parties negotiated tighter controls and brought under their remit a wider range of ozone-depleting subtances.

The years leading to the 1990 London meeting saw the dramatic deepening of the Antarctic ozone hole with evidence 'beyond reasonable doubt' that anthropogenic chlorine was largely responsible, as well as fears for the Arctic ozone layer. The reports from the assessment panels were undoubtedly influential in making the 1990 *London Amendments* to the Montreal Protocol far-reaching. Firstly, the range of ozone-depleting chemicals covered by the original protocol was extended: in addition to the original halons and CFCs (extended from 5 to 13 compounds), carbon tetrachloride and methyl chloroform were included, signalling concern about short-lived as well as long-lived species. Secondly, complete phase-outs were recommended for all these chemicals preceded on a tighter timetable by more stringent cuts: *

- CFCs, halons and carbon tetrachloride to be phased out completely by 2000, with cuts of 50% or more by the mid-1990s.
- Methyl chloroform to be phased out by 2005 with progressive cuts from 1995.

The London meeting could not reach agreement on how to treat HCFCs – not least because of warning noises from the chemical industry that early controls could make investment in their development and production difficult. The parties characterised HCFCs as 'transitional substances' and called for their 'prudent and responsible use' prior to their phasing out between 2000 and 2040. But in 1991 the science panel reported stronger evidence of mid-latitude ozone losses. It also found that bromine probably accounted for 25% of the ozone losses in the Antarctic hole, and pointed to the significance of its major source, methyl bromide, which occurs naturally and from human activities. Controls were again strengthened by the 1992 *Copenhagen Amendments* and extended to include methyl bromide and the HCFCs:

- Halons to be eliminated by 1994 (in effect, immediately) and CFCs, carbon tetrachloride and methyl chloroform phased out by 1996.
- Methyl bromide production to be frozen in 1995.
- HCFCs to be frozen in 1996 with progressive cut-backs from 2004 to 2020 (99.5% cuts) and complete phase-out by 2030.

Without the Montreal Protocol and its subsequent amendments, the continuing use of CFCs and other ozone-depleting substances would have led to at least a tripling of the stratospheric abundances of chlorine and bromine by 2050. As the projections in Figure 2.26 show, the London and

* Controls apply to production and consumption. The baseline year to which cuts refer is usually the year controls were first proposed for the chemicals: e.g. 1986 for CFCs and halons, 1989 for carbon tetrachloride and methyl chloroform, and 1991 for methyl bromide. Some temporary exemptions to the controls are allowed for essential medical uses (e.g. inhalants) and sensitive firefighting applications (e.g. aircraft).

Copenhagen Amendments are expected now to lead to a peak around the turn of the century with a slow decline over the next fifty years. This is not before time. In the years following the Montreal Protocol the Antarctic ozone hole has continued to deepen, suffering losses greater than had been thought possible, while new mechanisms for ozone destruction have been discovered in the sulphate aerosols from the eruption of Pinatubo. And in the first half of the 1990s ozone levels in mid-latitudes and over the Arctic have reached all-time lows. In contrast, the first signs that the controls were taking effect – the rise of CFC concentrations in the troposphere (shown in Figure 2.8) appears to be levelling off – provide more encouraging news. These trends are consistent with manufacturers' figures for CFC production, shown in Figure 2.27, and adds confidence to model projections that chlorine/bromine loading of the lower atmosphere should peak shortly, followed five years later in the stratosphere.

Since the early 1980s, with the steady increase in the chlorine/bromine loading of the stratosphere, a series of thresholds seem to have been passed, none of which has been anticipated with any accuracy and none of which (as yet) has been reversed. At fairly low loadings (of about 2 ppb by volume), in the early 1980s the Antarctic ozone hole made its first appearance. In 1987 the extent of the hole increased. This was repeated in 1989 and all subsequent years. Since the early 1990s the hole suffered

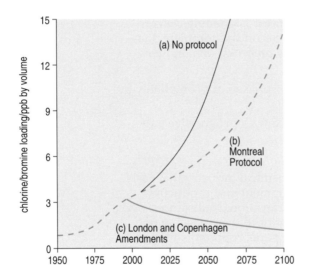

◀ *Figure 2.26*
*Projections of future chlorine/bromine loading of the stratosphere for different regimes of the Montreal Protocol: (a) without the Protocol; (b) with the original provisions of the 1987 Montreal Protocol; (c) under the Montreal Protocol with the London and Copenhagen Amendments in force.*

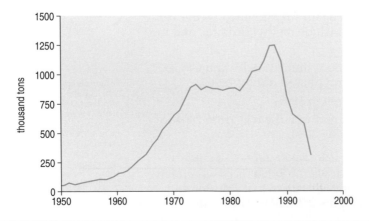

◀ *Figure 2.27*
*Estimated world production of all chlorofluorocarbons from 1950 to 1994.*

almost total destruction of ozone in the lower stratosphere and losses have increased in the northern hemisphere. It is clear that as long as the atmospheric loadings remain at or near their current levels of 2.5 to 3 ppb – until at least 2005 according to the projections from Figure 2.14 – the ozone layer may sustain further shocks and must be considered to be in extreme jeopardy.

The projections shown in Figure 2.14 also indicate that chlorine/bromine loading of the stratosphere will not return to 1980 levels until approximately 2050, a concern which was taken up by the 1994 science panel. It stated that further controls on ozone-depleting substances were not likely to have any significant effect on the timing or magnitude of the expected peak loading in the stratosphere, and concentrated instead on proposals to speed the *decline* of chlorine loading of the stratosphere in the early decades of the next century. These included, for the first time, proposals to control emissions (as opposed to controls on their production and use). The recommendations from the 1994 science panel to the 1995 Vienna Conference are listed below. Each change could each lead to a cumulative reduction of ozone loss over the next fifty years of between 5 and 13%:

- Eliminate methyl bromide emissions from agricultural, structural and industrial activities by the year 2001 (leading to a 13% reduction).
- Eliminate HCFC emissions by the year 2004 (5% reduction).
- Prevent halons, currently contained in existing equipment, from release into the atmosphere (10% reduction).
- Prevent CFCs, currently in existing equipment, from release into the atmosphere (3% reduction).

The delegates to the 1995 Vienna meeting gathered in celebratory mood on the tenth anniversary of the original Vienna Convention. However, compared with the achievements of the last ten years, the latest revisions to the Montreal Protocol suggested a loss of momentum, falling well short of the recommendations from the science panel. The developed countries agreed to ban both HCFCs and methyl bromide by 2020, twenty years later than recommended and with certain provisos. For HCFCs a ten-year exemption of 0.5% for servicing existing equipment remains – this is no more than a restatement of the 1992 position. With methyl bromide which is widely used as a pesticide and soil fumigant for intensive cash crops, particularly those grown under glass, there are considerable commercial interests at stake. US farming interests argued that methyl bromide be exempted for 'essential agricultural use', which in practice could be very wide-ranging, although the Netherlands have reported there are satisfactory substitutes for 90% of its uses. This will not be decided until 1997. While developed countries were making slow progress with the remaining ozone depleting chemicals, developing countries expressed concerns about the operation of the multilateral fund to pay for the switch to ozone-safe technology.

## Funding and control mechanisms

Other issues remain to be dealt with at the international level and are likely to assume greater importance at future meetings. For example, as bans on production bite and the range of controlled substances is extended, problems of compliance and monitoring become more centre-stage.

Another major challenge that came to the fore at the Vienna meeting is the management of phase-outs in developing countries in a fair and co-operative manner. Developing countries have the same obligations as others for substances controlled by the Montreal Protocol and the London amendment, but with a ten-year delay. An agreement on their obligations for HCFCs and methyl bromide, left unresolved at Copenhagen, was reached at Vienna, but only with difficulty. They agreed to freeze HCFCs by 2016 and phase them out by 2040, and to freeze methyl bromide production by 2002 (at 1995–98 levels). European delegates, however, were concerned that a delay before the introduction of a freeze could encourage growth of these chemicals in the short term. From the developing countries, on the other hand, there was a growing unease that their developed partners were not fulfilling some of their promises.

Historically, the developed countries have accounted for the lion's share of the total consumption of controlled chemicals and the responsibility for existing ozone depletion clearly rests on their shoulders primarily. On the other hand, improvements in living standards in many developing countries include plans that could easily change this pattern: the expansion of domestic refrigerators, for example. Negotiations at the 1990 London meeting produced a major advance – agreement to a financial mechanism under which developed countries will meet the incremental costs that developing countries incur in complying with the Protocol. This is intended to cover all aspects of the process of technology transfer, including access to the necessary technical expertise, patents, training and so on, as well as the direct costs of converting existing production facilities or of establishing new ones, or of simply importing CFC substitutes. The heart of the mechanism is a special 'multilateral fund' – set at an interim level of $240 million for the first three years – to be financed by contributions from developed countries.

With this agreement came the crucial participation of both China and India, but the start-up of phase-out projects in developing countries has been slower than expected and tensions and difficulties remain. The fund is only getting 85% of the money promised by the richer nations and several countries, including China and India, have indicated they will only keep to the ozone agreements if sufficient funds are forthcoming to help finance their commitments. With the likelihood of a rapid growth in refrigeration use in industrialising Asia, it is clearly important that agreement on funding is reached soon. As with many other areas of the ozone agreements, since 1990 the achievements have been considerable, and a framework has been put in place, but the next few years will show how difficult and costly it is to implement the obligations.

*Activity 7*

Make a summary of Section 7 for yourself by noting down the key stages in achieving international agreement to phase out CFCs. To what extent were decisions taken at each stage influenced by evidence of damage to the ozone layer? What other influences were at work?

# 8 Conclusion

By the year 2000, a quarter of a century after the alarm was first raised, the burden of ozone-depleting chemicals in the stratosphere will have passed its peak and started a slow decline, according to current projections – as long as no more unexpected shocks occur. This has only been made possible by firm international agreement on the production and release of halocarbons. That same period will have seen a major shift in international attitudes to global environmental problems – a shift in which some would claim that the ozone hole over Antarctica has played no small part. Certainly, it is shocking to recognise that damage to this remote and pristine environment could be caused by such everyday activities as getting rid of an old fridge, or using an underarm deodorant. And once the link was established, the international community *appeared* to react quickly and decisively – as did the public at large. But had it actually waited too long?

There is no simple answer. However, the long debate about CFCs – marked by calls for a better scientific understanding of the problem, more accurate forecasts of long-term ozone depletion, and so on – surely carries an important message when set against the sudden and wholly unexpected loss that did occur. The atmosphere is a complex and finely balanced system. Tamper with it, and it may respond in a gradual and more-or-less predictable way. Or it may not. Under these circumstances, the notion of 'scientific certainty' takes on a different cast – especially when a built-in time-delay, as here, means that the problem, once manifest, is not easily reversed.

Clearly, these points apply with equal – or greater – force to the problem of global warming and climate change. Here, much remains to be understood. Nevertheless, a consensus is growing that the time for 'cautious inaction' is past. And there is an important precedent in the Vienna Convention: there, for the first time, nations agreed in principle to tackle a global environmental problem *before* its effects were felt, or even scientifically proven. Hopefully, the international co-operation manifest in then agreeing to phase out CFCs will spill over into the debate about global warming – but the measures required are likely to prove a lot more costly, and more difficult to orchestrate, as you will see in the next chapter.

# References

REDDISH, A. and RAND, M. (1996) 'The environmental effects of present energy policies', Ch. 4 in Blunden, J. and Reddish, A. (eds) *Energy, Resources and Environment*, London, Hodder and Stoughton/The Open University (second edition) (Book Three in this series).

ROWLAND, F. S. and MOLINA, M. (1974) 'Stratospheric sink for chlorofluoromethanes: chlorine atom-catalysed destruction of ozone', *Nature*, Vol. 249, pp. 810–12.

SILVERTOWN, J. (1996) 'Earth as an environment for life', Ch. 6 in Sarre, P. and Reddish, A. (eds) *Environment and Society*, London, Hodder and Stoughton/The Open University (second edition) (Book One in this series).

SORG (1993) *Stratospheric Ozone 1993*, Fifth Report, London, HMSO.

# Further reading

Two very readable books on this subject are:

DOTTO, LYDIA and SHIFT, HAROLD (1978) *The Ozone War*, London, Doubleday.

GRIBBIN, JOHN (1988) *The Hole in the Sky*, London, Corgi Books.

Rather more technical, though still fairly accessible to non-specialists, is a series of reports produced by the UK Stratospheric Ozone Review Group, all published by HMSO.

- *Stratospheric Ozone*, First Report, 1987
- *Stratospheric Ozone 1988*, Second Report
- *Stratospheric Ozone 1990*, Third Report
- *Stratospheric Ozone 1991*, Fourth Report
- *Stratospheric Ozone 1993*, Fifth Report.

UNEP has also produced a short account of its role in achieving the phase-out of CFCs:

UNEP (1989) *Action on Ozone*.

·The following book gives an excellent survey of the background and significance of the Montreal Protocol by the chief US negotiator:

BENEDICK, R. E. (1991) *Ozone Diplomacy: new directions in safeguarding the planet*, Cambridge, MA, Harvard University Press.

A readable review of the progress of the ozone agreements since 1990 and the institutions behind them is to be found in:

PARSON, E. A. and GREENE, O. (1995) 'The complex chemistry of the international ozone agreements', *Environment*, Vol. 37, No. 2, pp. 16–43.

Authoritative updates on the status of polar ozone and worldwide trends, and on further revisions to the Protocol, appear in *New Scientist*.

# Answers to Activities

*Activity 1*

The desired effect is achieved if hydrogen (H) first attacks ozone ($O_3$), forming the hydroxyl radical (OH), and then this collides with another oxygen atom (O) – freeing the hydrogen for further destruction. In symbols:

$$O_3 + H \rightarrow O_2 + OH$$
$$O + OH \rightarrow O_2 + H$$

*Activity 2*

*Q1* (a)  Since Concorde flies within the stratosphere (Section 1), nitric oxide from their engines would be expected to add to the normal stratospheric burden of NO – and hence enhance ozone destruction via the catalytic cycle in equations 2 and 3.

(b)  To be a problem, some of the NO produced at ground-level would have to be carried up into the stratosphere. But NO is a very reactive molecule (as suggested by its role in the catalytic cycle) – with a very short lifetime in the troposphere. In fact, this reactivity makes NO a pollutant of the *lower* (rather than upper) atmosphere, its presence in vehicle exhausts etc. leading to nitric acid (a component of acid rain – *Reddish and Rand*, 1996) and photochemical smog.

*Q2* From the cycles in the text and in Section 3.2, the following would be a good guess:

$$O_3 + Br \rightarrow O_2 + BrO$$
$$O + BrO \rightarrow O_2 + Br$$

Note: BrO or OBr are equally acceptable.

*Q3* Of the compounds listed in Table 2.2, halon-1301 has an atmospheric lifetime comparable with those of CFC-11 and CFC-113. Each of these carries more than one Cl atom per molecule, yet rates an ODP little different from 1.0. By contrast, the halon has an ODP of 12–13, but carries only a single Br atom. This suggests that Br atoms must be more efficient at destroying ozone than are Cl atoms. Other evidence confirms this.

*Activity 3*

*Q1* The ODP of zero must mean that the molecules (as suggested by the name) do not carry ozone-destroying catalytic species – notable chlorine and bromine – into the stratosphere.

*Q2* HFCs are powerful greenhouse gases. The same properties that make them suitable as substitutes mean they are inert and thus also have very long lifetimes in the atmospheres. As we shall see in Chapter 3, this is the main objection to their use.

*Q3* Although HCFCs have low ODPs, typically one-tenth or one-twentieth the value of CFCs, they still deplete ozone. At the moment HCFC-22 is the main CFC substitute and emissions are rising sharply. As Table 2.2 shows, its concentration in the atmosphere after only a few years of use is already significant. Although it has a low ODP, the low value largely derives from its short lifetime: it is still putting chlorine atoms into the stratosphere. Any advantage over CFCs is soon lost if enormous amounts are released continuously into the atmosphere. Precisely the same criterion applies to methyl chloroform, a common cleaning fluid, which has a similarly short lifetime and low ODP value (see Table 2.2 again). Figure 2.14 shows that, nevertheless, it has made a significant contribution to the chlorine/bromine loading of the atmosphere in recent years. For that reason, it is due to be phased out in 1996 along with CFCs, halons and carbon tetrachloride.

*Activity 5*

*Q1* The central problem was the natural variability of the ozone column which meant that long-term monitoring was required in order to detect the small, but statistically significant, downward trend in ozone levels that CFC emissions were expected (on the basis of model forecasts) to cause worldwide. By contrast, the results from Halley Bay were quite unequivocal – with ozone concentrations well outside the range previously recorded (Figures 2.15 and 2.18). Their impact was twofold. First, they generated widespread concern – sufficient to support the crucial international campaign (Box 2.2) that furnished proof of a link with CFCs. Second, they shaped the reanalysis of existing long-term data records by the Ozone Trends Panel.

*Q2* There are two key ingredients:

• First, the unique meteorology of the Antarctic stratosphere. The loss of sunlight leads to strong cooling, which in turn produces the polar vortex – a sort of containment vessel that both isolates and cools the air within it. At temperatures below about –80°C, icy particles – polar stratospheric clouds (PSCs) – can form.

• Second, the special chemistry that takes place on the surface of the PSCs. This releases chlorine from its 'inactive' reservoir molecules into a form ready to produce a burst of active radicals once sunlight returns – whence the rapid and deep ozone depletion in early spring.

• The hole is 'healed' each summer by the influx of warm, ozone-rich air from lower latitudes once the polar vortex dissipates.

The lack (so far) of a major northern hole is mainly attributed to the winter stratosphere usually being warmer over the Arctic – so producing fewer PSCs, and a weaker vortex, that is generally disrupted before sunlight returns. However, the region is pre-conditioned for ozone depletion: dispersal of this air may contribute to the unexpectedly large ozone losses observed over high northern latitudes in winter.

*Q3* Apart from laboratory results, which we have not discussed, there is, firstly, the observed global distribution of sulphate aerosol throughout the lower stratosphere within six months of the eruption. Following this, record low ozone levels were observed over Antarctica in 1992 and 1993, particularly between 10 and 15 kms in the lower stratosphere (compare Figure 2.16b with Figure 2.16a). Anomalously low global values were also observed in 1992 and 1993 (see Figure 2.20) and over the Arctic, although these losses might be attributed to other causes.

*Q4*  There are a large number of factors. I have grouped them under five headings, which may explain why small, long-term increases have been hard to detect:

(a)  The angle of the Sun above the horizon – which depends on latitude, time of day and time of year.

(b)  The total column ozone (broadly speaking the stratospheric ozone layer, but influenced to a small extent by fluctuations in tropospheric ozone).

(c)  The condition of the lower atmosphere: clouds and aerosols (mostly pollution) can both reduce ultra-violet radiation at the surface.

(d)  Reflection of ultra-violet radiation from the surface (usually the same surfaces that reflect visible light well).

(e)  The distance between the Sun and the Earth.

*Activity 6*

The argument seeks (quite deliberately at the time it was propounded) to obscure the crucial distinction between changes in the ozone column *averaged over the entire Earth* (as predicted by a given 1-D model calculation, say) – and *natural variations* – here with latitude, but equally well from month-to-month, or year-to-year, at a given location. (There is an analogy here with temperature changes discussed in Chapter 3.) Thus for example, a drop of some 5°C in the global average temperature would plunge the Earth back into a glacial period. But changes of this magnitude from day-to-night, or day-to-day, are commonplace – as are similar differences in the average annual temperature between one place and another, each deemed to be habitable.

   To return to the ozone problem – a 1% depletion over the entire globe means that *everywhere* would be subjected to higher levels of UV-radiation. The link with skin cancer would then predict an increase in such cancers unless *everyone* (or at least all the light-skinned people) moved to higher latitudes (thereby effectively cancelling the change) – or took other precautionary action.

   Note that to concentrate on skin cancer also ignores all the other possible consequences of enhanced UV – to human health, but also to crops, and terrestrial and aquatic ecosystems, which cannot so easily be shifted around. Here, overall ozone depletion superimposed on the natural variations could expose organisms to levels of UV-B higher than ever before experienced – a situation that already appears to be occurring in springtime in the oceans around Antarctica.

*Activity 7*

In brief, the chronology is as follows:

1974   Rowland and Molina publication.

1978   US bans use of CFCs as propellants.

1979   EC adopts precautionary policy – voluntary agreement to limit CFC production.

1981   UNEP sets up working group to draft a framework convention for the  protection of the ozone layer.

1985   Vienna Convention signed.

1987   Montreal Protocol agreed.

1990   The London Amendments with phase-out by 2000.

1992   The Copenhagen Amendments with phase-out by 1996.

1995   The Vienna Conference, phase-out of bromide by 2020.

Only the latter stages of this process were influenced by direct evidence of ozone loss – following news of the ozone hole (May 1985) and its subsequent deepening (especially in 1987), and publication of the *Ozone Trends* Report (1988). Since 1990, agreements have taken place against a backdrop of a deepening and widening ozone hole in the Antarctic, record low ozone levels in the northern hemisphere and the discovery of new mechanisms of ozone destruction.

To me, the major influences at work throughout the process appear to have been:

- The scientific assessment of the scale of the problem and the urgency of doing something about it – and hence the attitude of the scientific community.

- The industry – both directly (for example, during the 'spray-can war' in the US), and through its influence on policy-making both nationally and internationally (as in, for example, input to the technical and economic assessments that underpin reviews of the Protocol). More recently, there has been the involvement of a new sector – agriculture.

- Environmental agencies (e.g. the EPA in the US) and pressure groups (e.g. Greenpeace, FoE etc.).

- Governments' policy-making processes, including the reassessment of national interest (as in, for example, the UK U-turn in the negotiations for Montreal, the recent participation of India and China, and so on).

- International organisations, notably UNEP – in placing the issue on the international agenda, and keeping it there (especially during the early 1980s), achieving co-operation between developed and developing countries, and financing phase-outs in developing countries.

- The media and public education, especially during the mid-1970s in the US (but not in the UK!) and then again, worldwide, in 1988/89.

- The growing power of consumer pressure, for example in boycotting aerosols, support for local recycling schemes and so on.

# Chapter 3   Atmospheres and climatic change

## 1   Introduction

Measures aimed at protecting the natural environment on a global, rather than just a local, scale figured increasingly prominently on international agendas in the 1980s. Chapter 1 described one of the first major steps in international co-operation on environmental issues – the 'Law of the Sea' agreement that provides for the protection and management of the oceans. More recently, international concern focused on the atmosphere, as evidence accumulated that human activities are upsetting a natural balance and might result in climatic changes of a sort never before experienced within historic time. As we have seen in Chapter 2, one consequence of the breakdown of natural balances – the 'ozone hole'  prompted concerted international effort, resulting in the control of CFCs and other ozone-depleting chemicals through the Montreal Protocol and its subsequent amendments.

Another manifestation of a change in atmospheric balance – the probability of global warming arising from trace gases, such as carbon dioxide released in the atmosphere as a by-product of human activity – has given rise since the late 1980s to an explosion of (often highly sensationalised) media reports, and to much scientific, lay and political debate. Some topical cuttings from (carefully reported) articles are reproduced in Figure 3.1. While media concern and interest has fluctuated (seeming to blow hot and cold with the prevailing weather), the scientific and political communities have moved cautiously towards agreement.

In 1988 the United Nations Environment Programme (UNEP) and the World Meteorological Organisation (WMO) established the Intergovernmental Panel on Climate Change (IPCC) to provide an authoritative, international opinion on climate change. The IPCC produced its first scientific assessment in 1990 which had a major influence on the UN Framework Convention on Climate Change signed by over 160 nations at the 1992 Rio Earth Summit. In March 1994 the UN Climate Change Convention came into force, followed in March 1995 in Berlin by the first meeting of the ratifying parties. While few firm commitments have yet been made to control greenhouse gases in any of the meetings, the broad principles agreed by the parties are potentially far-reaching: it seems likely that the step-by-step approach established by the Montreal Protocol is a model to be emulated. Nevertheless, no-one underestimates the problems of achieving effective action over global warming: modern society found CFCs useful but has quickly developed substitutes. It will have greater difficulty moving away from its widespread dependence on fossil fuels for energy – the source of the main greenhouse gas, carbon dioxide.

As you read this chapter, look out for answers to the following key questions:

- How, and why, have atmosphere and climates changed in the past?
- How can climate change be forecast and what are the current expectations for the next century?
- What are the likely impacts of global warming?
- What political negotiations have taken place and what changes are needed?
- What strategies are available to reduce global warming?

## Britain could be the new Sunshine State

◀ Figure 3.1
Reports of global
warming from 1989 and
1995.

Within the lifetime of today's children, South-East England could enjoy a semi-Mediterranean climate as average temperatures increase by more than 6 degrees Fahrenheit. The garden of England could become its sunporch, and farming could boom in the North.

These are the benefits which Britain could enjoy thanks to the greenhouse effect, according to evidence being collected by a United Nations panel.

Britain will warm more than the global average, because the greenhouse effect will heat up cooler regions faster than the trop-ics. South-East England could become more like today's south-western France, with warmer, drier summers, but wetter winters. Huge golden fields of sunflowers could flourish across southern England.

And if the warming continues, Britain could grow its own baked beans: at present they all have to be imported.

But the South could also suffer much more frequent droughts and present forms of farming would suffer. There would probably be more thunderstorms.

Source: *Observer*, 26 November 1989

## Global warming gets cold shoulder

World leaders have gone cool on global warming. The climate summit that begins in Berlin tomorrow, therefore, will be an exercise in playing for time.

The most that is likely to emerge from 11 days of talks is a decision to make a decision in 1997 on what to do next about the threat of man-made climate change. And even that is far from certain.

Oil-exporting countries such as Saudi Arabia and Kuwait would prefer the environment ministers to steer well clear of any commitments that could limit the world's rapidly growing use of fossil fuels. [...] The majority of climate scientists continue to warn that climate change is coming and that it will be more rapid than at any time in the past 10,000 years.

Source: Nicholas Schoon, *The Independent*, 27 March 1995, p. 8

## Hottest year heralds global warming

Early British snow notwithstanding, 1995 is likely to be the world's warmest year since records began, say researchers at the Meteorological Office's Hadley Centre. "We have all the November data in from around the world," says David Parker, "and so far we are 0.05°C warmer than the previous record-holder, 1990."

Last week, the UN's Intergovernmental Panel on Climate Change met in Rome, and concluded that "the balance of evidence" suggests that human activities are warming the planet. The four warmest year worldwide have all been in the 1990s, with nine of the ten warmest occurring since 1980.

The record spectacularly confirms the predictions of climate modellers that global temperatures – which fell in 1991 and 1992, following the eruption of Mount Pinatubo in mid-1991 – would bounce back with a vengeance in the mid-1990s. Pinatubo sent a veil of particles into the stratosphere, which shielded the Earth from solar radiation for about two years.

Source: *New Scientist*, 23/30 December 1995, p. 5

## Global warming 'jury' delivers guilty verdict

Like Sherlock Holmes's dog that didn't bark, it is the warming that hasn't happened that has finally convinced climatologists that human activity is probably to blame for global warming.

Worldwide average temperatures have risen by half a degree in the past century – the biggest warming since the end of the last ice age. But the intensity of warming has been patchy. Nonetheless, last week the scientific working group of the UN's Intergovernmental Panel on Climate Change, in its first full report for five years, threw aside past statements that the warming was within the range of natural variability.

Instead, for the first time, it argues that current warming "is unlikely to be entirely natural in origin" and that "the balance of evidence suggest a discernible human influence on global climate". The reason for their new confidence is that they now believe they can explain the hitherto perplexing patchiness of the warming.

The chief human influence on climate is the general buildup of carbon dioxide and other greenhouse gases in the atmosphere – largely from fossil-burning fuels. But the IPCC's full 400-page report, to be published in April, argues that the patchiness of global warming is caused by a second human influence. In many parts of the world, including much of Europe and North America, warming has been partly masked by another form of pollution from burning fossil fuel – aerosols of sulphates and soot that form a thin haze which reduces solar heating.

Source: Fred Pearce, *New Scientist*, 9 December 1995, p. 6

The ultimate objective of the Climate Change Convention is to ensure stabilisation of greenhouse gases in the atmosphere at safe concentrations and at a rate that will allow ecosystems time to adapt. Scientists can advise on what needs to be done to achieve a given safe level and may offer criteria to assess what is 'safe', but while a consensus is emerging over the global effects, few scientists would claim to forecast with any degree of certainty or detail the likely effects of global warming on individual nations or societies (as claimed for Britain by one of the articles in Figure 3.1). Herein lies the core of the problem. Proposals for any kind of action require immediate costly investment as insurance against possible future changes whose exact nature is still uncertain. Deciding when to act and how far to take the actions is a matter of value judgement. In order to make such a judgement, we need to understand not only the political and economic factors involved in any particular course of action, but also the scientific evidence. A major theme of this chapter, therefore, is to present the underlying science, to distinguish areas of scientific consensus from areas of controversy, and to illustrate some of the scientific uncertainties. Our understanding of the processes of climate change remains incomplete in several important areas, but has improved significantly in the last ten years, allowing more convincing modelling of current climate patterns and of possible futures. The overall aim of this chapter is to provide a background that will enable you to follow future developments in an informed way and to assess the relevance and reliability of new research data as these are reported in the media and in popular scientific articles.

Much of the content of this chapter is scientific, embracing a range of disciplines. However, an effort has been made to keep the number of technical terms used to a minimum. If you do not have a science background, the activities within the text should help you to focus on the most important points and it is essential that you attempt these as you come to them, before reading on. Activities at the ends of sections can be used for further self-assessment or revision.

# 2   Climate: past, present and future

In the late 1980s the phrase 'the greenhouse effect' came into common usage, often in the context of global warming that had occurred during the present century and was being linked to human activities. This is not an accurate use of the terminology. The 'natural' greenhouse effect has operated for over four billion years, maintaining temperatures suitable for life on Earth. It is perhaps the main key to understanding past and present climates (as outlined in *Sarre and Reddish*, eds, 1996) and is thus an appropriate topic to begin the story of climatic change.

## 2.1   The greenhouse effect

All hot objects emit electromagnetic radiation at a range of wavelengths, the exact range depending on the temperature of the object.

*Activity 1*

Figure 3.2 shows part of the electromagnetic spectrum. (This was introduced in *Reddish*, 1996.) General experience tells us that a poker heated in a fire glows 'red hot'; if heated to a higher temperature, in an oxy-acetylene flame say, it would glow 'white hot'. Generalising from this example, and remembering that 'white light' is a combination at all the visible wavelengths, work out whether the average wavelength of emitted radiation increases or decreases as the temperature of the emitting object falls.

The Sun, with a surface temperature of over 6000°C, emits radiation with wavelengths between $2 \times 10^{-7}$ m and $4 \times 10^{-6}$ m (i.e. 0.2 μm to 4 μm), with its peak emission in the visible band. Objects, such as planets, illuminated by solar radiation *reflect* some of it (without change in wavelength) and absorb the rest, so warming up. They then *emit* radiation with a wavelength that will depend on their temperature but will certainly be *longer* than that of the incoming solar radiation, since the objects will be cooler than the Sun. Well-established laws of physics can be used to calculate what is called the *effective radiating temperature* of a spherical body of known reflectivity heated by a known amount of solar radiation. In this context, both the Earth and the Moon can be treated as spherical bodies and the appropriate calculations show that their effective radiating temperatures should be approximately −18°C. This is indeed the average temperature of the Moon, which orbits the Sun at the same distance as the Earth, and if, like the Moon, the Earth had no atmosphere, it too would be at −18°C. A body at this temperature emits radiation with wavelengths in the range $4 \times 10^{-6}$ to $10^{-4}$ m (4 to 100 μm), i.e. in the infra-red band. These spectra of emitted radiation from the Sun and the Earth are shown in Figure 3.3(a). (Ignore parts (b) and (c) of this figure for the moment.)

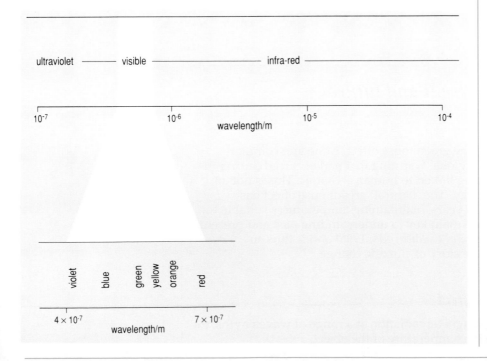

Figure 3.2
A portion of the electromagnetic spectrum.

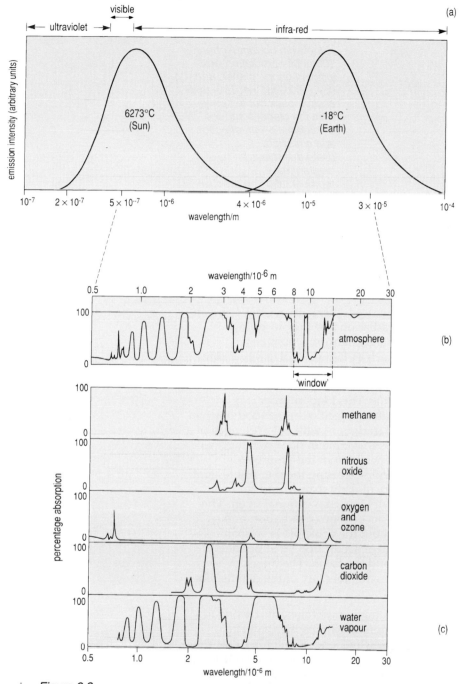

▲ Figure 3.3
(a) Specification of emitted radiation from bodies at 6273°C and −18°C, corresponding to the effective radiating temperatures of respectively the Sun and the Earth.
(b) Percentage of atmospheric absorption of radiation passing through the atmosphere, as a function of wavelength.
(c) Percentage of absorption attributable to various atmospheric gases occurring in their natural proportions.
(Note that nitrogen – the main constituent of the atmosphere – does not absorb in the infra-red waveband: it is not a greenhouse gas.)

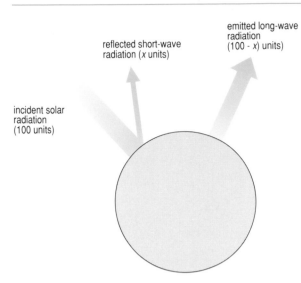

reflected short-wave
radiation (*x* units)

emitted long-wave
radiation
(100 - *x*) units)

incident solar
radiation
(100 units)

◀  *Figure 3.4*
*Schematic illustration of radiative balance for a planet with no atmosphere. 100 units of radiation are incident on the planet, and [x + (100 − x)], i.e. 100, units escape back out to space from the planet's surface, by a combination of reflection and emission.*
*(Note that this is a diagrammatic representation; in practice the solar radiation would be incident on the complete hemisphere and the planet would radiate in all directions.)*

The total radiation budget must be balanced: there always has to be an equilibrium between the incoming solar radiation and the out-going radiation. If there were no atmosphere, this equilibrium could be represented schematically by a diagram such as Figure 3.4. However, the Earth *does* have an atmosphere, and this changes the picture substantially. Gases in the atmosphere absorb radiation, but they do so selectively: different gases absorb different wavelengths. This kind of process is familiar in domestic microwave ovens: water molecules in the food happen to absorb strongly in the microwave band (radiation with a wavelength of a few centimetres), whereas solid substances that contain no water, such as china and plastic containers, do not. The energy from the absorbed radiation is transferred to the water molecules, causing them to rotate, and this heats the food from the inside far more quickly than would the conduction from the food surface that occurs in a conventional oven. The energy that atmospheric gases acquire by absorption of radiation makes their molecules vibrate, rather than rotate, but the principle is similar. Figure 3.3 (b) and (c) shows the percentage of absorption by atmospheric gases for radiation emitted from the Sun and the Earth. The diagram is complicated, and you are not expected to remember any of the details. For now the important points to note are:

1    Each peak represents absorption of a range of wavelengths by one or more gases. If a peak shows 100% absorption, the atmosphere will be completely opaque to radiation within that waveband. Remember that the radiation is of different wavelengths and comes from opposite directions in the two cases. Solar radiation of a wavelength that is 100% absorbed will never reach the Earth's surface; infra-red radiation that is 100% absorbed will never reach the top of the atmosphere.

2    Most of the gases absorb in several different wavebands, but are more or less transparent to radiation of intermediate wavelengths. Those that absorb in the infra-red are known as **greenhouse gases**.

3    There is comparatively weak absorption in the visible spectrum: most of the solar radiation passes through the atmosphere and reaches the Earth's surface.

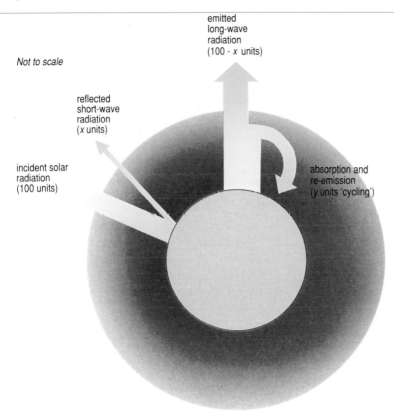

*Not to scale*

reflected
short-wave
radiation
(*x* units)

emitted
long-wave
radiation
(100 - *x* units)

incident solar
radiation
(100 units)

absorption and
re-emission
(*y* units 'cycling')

◀ *Figure 3.5
Schematic illustration of
radiative balance for a
planet with an absorbing
atmosphere. The 100 units
of incident solar radiation
are again balanced by a
total of 100 units of radiation
escaping back into space.
However, y units of long-
wave radiation also cycle
within the atmosphere.
(The same diagrammatic
representation has been
used as in Figure 3.4.)*

4    On the other hand, most of the long-wave spectrum emitted from the
Earth is absorbed. There is just one region of weak absorption, between
about $8 \times 10^{-6}$ and $13 \times 10^{-6}$ m (i.e. between 8 and 13 µm); this is the
'window' that allows some infra-red radiation to escape into space and
stops the Earth heating to temperatures inimical to life.

The simple picture of Figure 3.4 therefore has to be modified. A better
representation is shown in Figure 3.5. Some of the infra-red radiation from
the Earth's surface is absorbed in the atmosphere, so heating it. The warm
atmosphere radiates in its turn, again at infra-red wavelengths since its
temperature is roughly similar to that of the Earth. Some of this re-emitted
radiation goes out into space (the radiative balance with incoming solar
radiation still has to be maintained at the top of the atmosphere), but some
goes back towards the Earth's surface and heats it up. This phenomenon,
which keeps the Earth warmer than its effective radiating temperature, is
known as the **greenhouse effect**. The average air temperature at ground-
level is in fact about 15°C. As we have seen, without the atmosphere, the
Earth, like the Moon, would be at about −18°C. The natural greenhouse
effect thus keeps the Earth roughly 33°C warmer than it would be if there
were no atmosphere.

As an aside, it is worth noting that the greenhouse analogy is a very
misleading one. The main reason that a greenhouse heats up on a sunny
day is that the air is trapped inside it and the natural process of convection
is prevented. This is quite a different situation to the climatic 'greenhouse
warming' we are discussing here, which is a purely radiative effect.

Figure 3.6 provides a more detailed breakdown of the energy flows in
the atmosphere than Figure 3.5 and gives estimates of the quantities

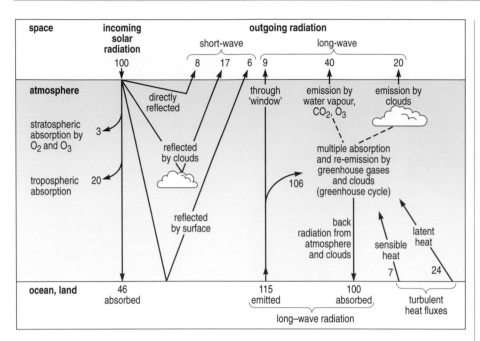

◀ *Figure 3.6   Schematic representation of the overall energy budget for the Earth and its atmosphere. Figures are in global averages, expressed as a percentage of the incoming solar radiation of 100 units (100%) which has been measured as 342 $Wm^{-2}$. The data in the figure can be derived from model calculations or measurements, so you may find slightly different values quoted elsewhere.*

involved. Let us concentrate for the moment on the radiation entering and leaving the top of the atmosphere.

$Q$   How many units of the incoming solar radiation are reflected?

$A$   31 units. For every 100 units entering the atmosphere, 8 are reflected by the atmosphere, 17 by clouds and 6 from the Earth's surface giving a total of 31 (the 'x' factor in Figures 3.4 and 3.5). The ratio of 31 to 100, or 31%, is called the 'albedo' (described in *Silvertown*, 1996).

$Q$   What happens to the rest of the incoming solar radiation?

$A$   The other 69 units, or 69%, are absorbed in the atmosphere or at the surface of the Earth. They represent the net incoming solar radiation which has to be balanced by the outgoing long-wave radiation, as illustrated by Figure 3.6.

The *total* incoming flux of solar radiation (100 units) has been measured by satellite at the top of the atmosphere: averaged for the globe over a year it amounts to 342 Watts for every square metre (342 $Wm^{-2}$). These are the units we shall use to describe and quantify the greenhouse effect for the remainder of the chapter.

$Q$   What is the value, in $Wm^{-2}$, of the outgoing long-wave radiation?

$A$   It has to be the same as the net incoming solar radiation, i.e. 69% of 342 $Wm^{-2}$. This is simply $(69/100) \times 342\ Wm^{-2} = 236\ Wm^{-2}$.

## *The enhanced greenhouse effect*

The radiation balance under natural conditions is shown in Figure 3.7(a), but what happens as the concentration of greenhouse gases builds up in the atmosphere from human activities – the so-called *enhanced* or *anthropogenic*

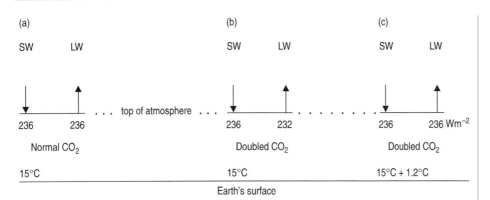

Figure 3.7   *The enhanced greenhouse effect for an average surface temperature of 15°C. Under natural conditions (a) the net short-wave radiation entering the atmosphere, 236 Wm$^{-2}$, is exactly balanced by the long-wave radiation leaving it. If the carbon dioxide concentration in the atmosphere doubles (b) the long-wave radiation escaping from the atmosphere is reduced by 4 Wm$^{-2}$. This leads to a net heating of the Earth and atmosphere until balance is restored (c) with a 1.2°C warmer Earth. If feedbacks\* are included, the surface temperature rises on average by 2.5°C.*

\*See Box 3.2

greenhouse effect? For simplicity, let us consider a sudden doubling of the carbon dioxide concentration (the effect would be similar if other greenhouse gases were chosen). An increase of greenhouse gases would mean that the atmosphere becomes less transparent to long-wave radiation; more would be re-emitted back towards the surface and a smaller fraction would escape from the top of the atmosphere into space\*\*. The outgoing long-wave radiation is reduced by an estimated 4 Wm$^{-2}$, creating an imbalance: as Figure 3.7(b) illustrates, more radiation now enters the atmosphere than leaves it. This causes both the atmosphere and the surface of the Earth to warm up until the outgoing long-wave radiation again matches the incoming radiation. The temperature change needed to restore the balance is 1.2°C (see Figure 3.7c), but when atmospheric feedbacks from water vapour and clouds are added, the greenhouse effect is further increased, and the expected rise doubles to approximately 2.5°C.

In this description of the enhanced greenhouse effect we have used a simple model of interaction between the radiation balance of the atmosphere and its temperature. The net change of radiation at the top of the atmosphere\*\*\* is a convenient and simple way to characterise the effect of a given concentration of greenhouse gas and is called the **radiative forcing** of the gas. The figure of +4 Wm$^{-2}$ for a doubling of carbon dioxide concentration (a positive value indicates an *increase* of net radiation) serves as a bench-mark and is known with reasonable accuracy, but the best estimate of the resultant temperature rise – currently 2.5°C – is less certain. The enhanced or anthropogenic greenhouse effect forms the main subject of this chapter, but, to avoid unnecessary repetition, the terms 'enhanced' or 'anthropogenic' are sometimes omitted in the discussions that follow.

## 2.2   Rhythms in the climate

Changes in climate, and other environmental changes with which they are associated, take place over a considerable range of time-scales. Over the longest time-scale – the age of the Earth – the climate has been affected by

\*\*In Chapter 2 we noted that temperature falls with height in the troposphere (at a rate largely determined by convective processes), from an average surface temperature of 15°C to -50°C or colder at the tropopause. When the outgoing long-wave radiation is intercepted by natural greenhouse gases it is re-emitted to space from higher in the atmosphere where the temperature is colder. Because the amount of radiation emitted and absorbed by a given material *decreases* when its temperature falls – as turning an electric bar fire on and off will demonstrate – the total emission from the Earth is reduced. If the long-wave radiation were visible, the Earth, seen from space, would 'glow' less brightly, wrapped in its insulating blanket of greenhouse gases. The radiation balance is only restored by raising the temperature throughout the atmosphere until emissions return to their former level. The net effect is to warm the surface.

\*\*\*Strictly speaking the top of the troposphere is used, with adjustment for the radiation from the stratosphere. This is because the temperatures of the Earth's surface and the troposphere are tightly coupled together.

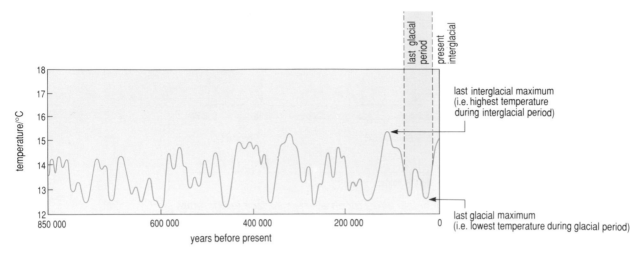

▲  *Figure 3.8*
*Globally averaged surface temperatures on the Earth during a succession of glacial and interglacial periods.*

▲  *In 1814 the Thames froze to such a depth that fairs were set up on the ice. There are considerable variations in temperature within the longer-term climatic changes.*

fundamental changes in the composition of the atmosphere. The Earth's early atmosphere was probably dominated by carbon dioxide and other gases released by volcanic activity, quite unlike its current composition. But when primitive plant life first appeared on Earth, probably in the oceans, it started absorbing carbon dioxide and releasing oxygen through photosynthesis. Added to this, the normal processes of weathering also remove carbon dioxide from the atmosphere and transport calcium and bicarbonate ions to the oceans. There they are taken up by marine organisms and eventually accumulate on the ocean floor, a mechanism often referred to as the 'biological pump'. Over geological time, as life evolved, these kinds of processes have returned carbon dioxide to the Earth's crust, locked up in rocks, where it is very slowly recycled by volcanoes. Thus our present atmosphere is in a state of dynamic equilibrium, with biological and geochemical cycles, albeit on lengthy time-scales. Stephen Schneider, an American climatologist who has been at the forefront of debates on climate change, has called this interplay 'the co-evolution of climate and life'.

Many changes, however, occur in a regular rhythm. For example, in the last million or so years, there has been a pattern in which a glacial period, that lasts about 100 000 years, is followed by a warmer, interglacial period of 10 000 to 20 000 years' duration* – the current one has lasted 10 000 years.

The rhythm of the glacial and interglacial periods evident from Figure 3.8 is linked with slight, but regular, variations in the Earth's orbit round the Sun. Over a period of about 100 000 years, the shape of the orbit changes, from elliptical to almost circular, and back again. The more elliptical the orbit, the greater the differences between the seasons. Secondly, the tilt of the Earth's axis of rotation to the plane of the orbit varies, with a periodicity of 41 000 years, causing slight shifts in the positions of the tropics and greater or lesser differences between summer and winter in any given place. Thirdly, the axis wobbles, with a periodicity of 21 000 years. As a result, there is a slow drift in the time of year at which the Earth is closest to the Sun and the pattern of the seasons gradually changes over many such cycles. A Serbian astronomer, Milankovic, first proposed in the 1920s that the pattern of glacial periods might be linked to these fluctuations in the orbital geometry of the Earth, but at that time his hypothesis could not be tested because the periods of glaciation had not been accurately dated. When this was finally done, in the mid-1970s, it became clear that the amount of ice cover over the past 800 000 years does indeed fit with three cycles of 100 000, 41 000 and 21 000 years' duration. However, one very fundamental question remained. If the Sun's output remains constant, the Earth will receive the same amount of radiation every year. The Milankovic cycles only result in slight seasonal or geographical temperature changes. How are these changes enough to trigger the onset or the end of glacial periods or indeed the rapid temperature fluctuations that are now thought to have occurred at certain times in the past?

The interglacial period in which we now find ourselves has already lasted 10 000 years and, in comparison to the preceding glacial period, appears to have been a time of unusual climatic stability. Temperature records going back 100 000 to 200 000 years have been reconstructed from ice cores several kilometres deep taken from the centre of the Greenland and Antarctic ice caps. The Greenland record, displayed in Figure 3.9, investigated by Professor Dansgaard and colleagues, shows that temperatures increased *comparatively* slowly in the transition from glacial to interglacial conditions at about 10 000 years before present (BP) – globally a rise of between 3°C and 5°C over a thousand years – and then remained stable. In contrast, the record for the glacial period before this suggests rapid temperature fluctuations of up to 5°C over periods as short as one hundred years.

It seems highly likely that feedbacks are at work in the climate system which speed up the switch from glacial to interglacial conditions and may amplify the other fluctuations. Many mechanisms have been suggested, but one in particular derives from the record of greenhouse gases contained in the ice cores. Figure 3.10, taken from the Vostok core in the Antarctic,

*You may come across popular books or articles in which the term 'ice age' is used to describe what is properly called a 'glacial period'. In scentific terminology an ice age is a very long span of time, during which there is an alternation of glacial and interglacial periods. The present (Pleistocene) ice age has already lasted for about 4 million years. The most recent previous ice age spanned the time between 270 and 310 million years ago. In the time intervening between these two ice ages (i.e. between 4 and 270 million years ago), the climate was more stable.

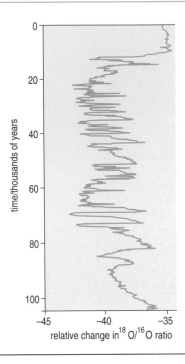

◄ *Figure 3.9*
*Variations in Arctic*
*temperature over the past*
*100 000 years as deduced*
*from measurements of the*
*isotope $^{18}O$, from the*
*'Summit' ice core from*
*Greenland. A change of*
*5 units of the isotope ratio*
*$^{18}O/^{16}O$ (see Box 3.1) in the*
*ice core, compared to a*
*standard value, corresponds*
*on this graph to a*
*temperature change of*
*about 7°C.*

## Box 3.1   Records in the ice

Small quantities of different oxygen isotopes occur naturally in water molecules, but the ratio of $^{18}O$ to $^{16}O$, normally 1 part in 500, is sensitive to the temperatures at which water evaporates from the ocean and forms ice crystals in clouds. The snow that then falls from the clouds onto the ice caps forms layers of ice that create a record of climatic conditions. With careful calibration, measurement of this ratio (or a similar ratio for hydrogen isotopes) can be used to estimate temperature changes in the surrounding regions. In addition, bubbles in the ice contain tiny samples of air which can be used to analyse concentrations of the greenhouse gases, carbon dioxide and methane. While the temperature record is typical of the polar regions, from where the precipitation is mostly drawn, the greenhouse gases are evenly mixed in the atmosphere and their record in the ice is representative of global concentrations.

*A glaciologist working on an ice core. Ice cores are the source of a great deal of information about past climates.*

illustrates the period of the last major glaciation which began about 120 000 years ago and ended 10–20 000 years ago. The trends for the carefully reconstructed carbon dioxide (and methane) concentrations are strongly correlated with the temperature record. However, we should be cautious in reaching the conclusion that changes in greenhouse gases have caused the temperature changes. In fact, the reverse may be the case, at least during the cooling phase of the glacial cycle when it appears that changes in carbon dioxide concentration lagged behind the temperature change.

One final feature of the Greenland record, shown by Figure 3.9, is worth noting. At approximately 12 000 years ago, the warming at the end of the last glacial period was abruptly interrupted and the temperature in the Arctic dropped by at least 5°C. This colder period – the Younger Dryas, named after an Arctic flower which spread at that time – ended just as abruptly 10 700 years ago. The cooling event seems to be linked to the break-up of the massive ice-sheets over eastern Canada, but the speed of the transitions was unexpected and led scientists to look for new mechanisms. Professor Broecker has suggested that the melting of large quantities of ice and its release into the ocean switched off at least part of the deep water circulation in the North Atlantic (see Chapter 1) – the 'conveyor-belt' that normally aids the northward flow of warm water – leading to the sudden cooling. The return of the conveyor-belt led to the equally sudden reappearance of warmer conditions.

While we do not yet understand many of the details of what happened during the ice ages, or even during the last transition to our present

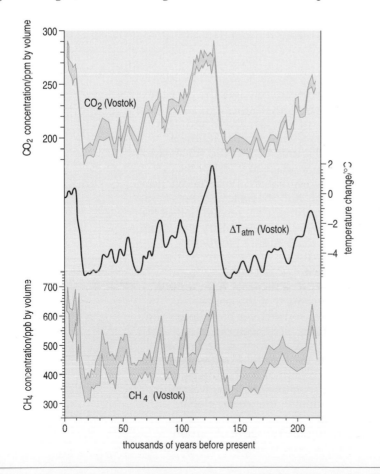

◀  Figure 3.10
Temperature anomalies and methane and $CO_2$ concentrations over the past 220 000 years, derived from the ice-core record at Vostok, Antarctica.

interglacial climate, it is clear that the greenhouse effect has played an important role in the major climatic shifts of the past. The additional question which scientists are now addressing is whether human activities are altering our climate on a much shorter time-scale, bringing to an end a period of comparative stability. Our perspective on this will be informed by what we know of rapid climate changes in the past, including the possibility that changes in the deep ocean circulation may be induced by global warming.

## 2.3   *Climatic evidence*

Climatologists have no doubt that the Earth has become warmer in the last century. In this subsection we shall examine this warming trend in some detail and look at the evidence that links the rise in temperature to human activity.

Q   Looking back to Figure 3.1, what evidence was presented that global warming accelerated in the 1980s and early 1990s? Is such evidence necessarily indicative of global warming?

A   As reported in the *New Scientist* article of 23/30 December 1995, nine of the ten warmest years recorded have occurred since 1980 and the four warmest have been in the 1990s. The recent run of warm years does not necessarily imply that temperatures will continue to rise – the period may simply be anomalous. It is *consistent* with present projections of global warming, but by no means a proof that it is happening.

A better idea of current temperature trends is given by Figure 3.11 which shows the variations that have occurred on land and sea over the past 130 years. Once the annual fluctuations are averaged out, these cumulative data show an increase in global mean temperature of between 0.3°C and 0.6°C over the past century. Records of the sea-surface temperature follow a similar pattern to the land-based records, as Figures 3.11(a) and (b) show. Other types of measurement confirm these general trends. Temperature-sensing balloons were not launched routinely until about 1950, but we now have forty years of data from such instruments and they show quite dramatic warming of the troposphere, accelerating in the second half of the period. Measurements of sea-level changes over the past century show that mean levels have risen in that time, in a way that roughly parallels temperature trends and suggests thermal expansion of sea-water as the principal cause. Additional evidence is beginning to accumulate from observations of the cryosphere (snow and ice). The majority of mountain glaciers have been in recession worldwide for much of the last century. In the northern hemisphere Arctic ice and winter snow cover have both been observed to decrease since satellite observations have been made: sea-ice by 5.5% between 1978 and 1994, a loss twice the size of Norway, and snow cover by 8% since 1973. Although the records are too short to draw definite conclusions, they are interesting because the region of the high northern latitudes is one of those expected to be most sensitive to global warming. Together, all these changes provide independent evidence for global warming, not based on temperature readings.

   Is this a case of cause and effect? Certainly the presence of carbon dioxide in the atmosphere gives rise to greenhouse warming. But has this

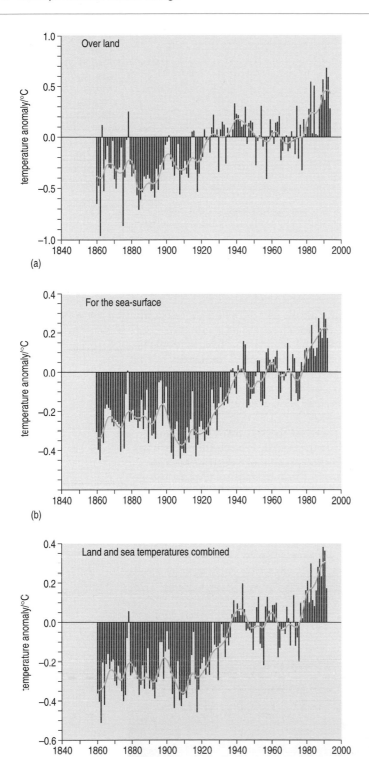

▲   *Figure 3.11    Changes in global average temperature from 1860 to 1992 relative to the period 1951–80, (a) over land, (b) for the sea-surface and (c) land and sea temperatures combined. The solid curve has been produced by filtering the data to highlight longer-term trends.*

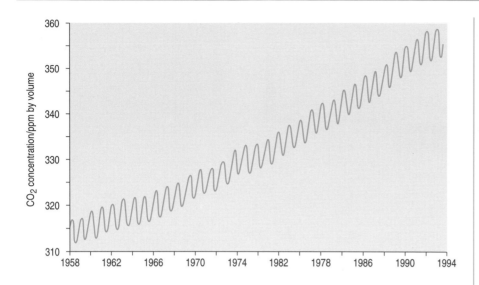

◀ *Figure 3.12 Concentration, in parts per million by volume, of atmospheric carbon dioxide measured at Mauna Loa station, Hawaii, geographically remote from any sources of pollution. Annual oscillations arise from the seasonal changes in photosynthesis and other biological processes.*

*particular* trend of rising temperatures come about *because* of an increase in the atmospheric concentration of carbon dioxide?

The longest-running continuous monitoring of carbon dioxide levels is that carried out at the Mauna Loa Observatory in Hawaii. The results of this study, illustrated in Figure 3.12, show that the concentration of carbon dioxide in the atmosphere has increased steadily over the past thirty-five years: the 1994 concentration (356 ppm) is over 10% greater than that recorded for 1958 (315 ppm). However, before we can attach any significance to this result, we need to know what the natural variation in carbon dioxide levels has been over various time-scales: have there perhaps been 10% fluctuations in the concentration in the past as a result of some natural cyclical process? For some answers to these questions we can return to ice-core data shown in Figure 3.10.

The record of the Vostok core shows that at the end of both the two most recent glacial periods the atmospheric concentrations of carbon dioxide increased sharply, from around 190 ppm to about 280 ppm, with a

◀ *The Mauna Loa Observatory in Hawaii, site of the longest-running project to monitor levels of atmospheric carbon dioxide.*

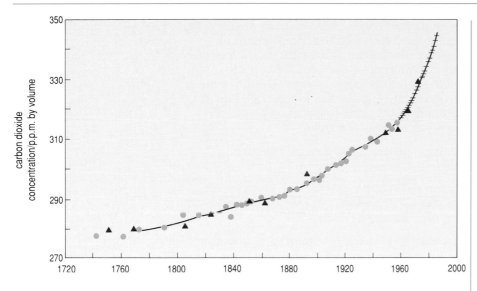

◄  *Figure 3.13*
*Increase in the atmospheric*
*concentration of carbon*
*dioxide, determined from*
*ice-core measurements of*
*Siple station, Antarctica*
*(triangles), isotope analysis*
*(circles) and direct*
*observation at Mauna Loa*
*(crosses).*

probable maximum of 300 ppm. It does indeed seem that the Mauna Loa
data are showing up something other than natural variation; even when the
observations began in 1958, the levels of carbon dioxide were higher than
they had been at any time in the current or preceding interglacial. Analysis
of another Antarctic ice-core has given more detailed information about the
build-up of carbon dioxide over the past 250 years, as shown in Figure 3.13.
The increase in carbon dioxide levels at the start of this period was linked to
deforestation and agricultural activity. Since the beginning of the industrial
age, it is undoubtedly the burning of fossil fuels that has made the greatest
contribution to the change in atmospheric concentrations of carbon dioxide.
The 1994 level of 356 ppm represents an increase of 27% over the
pre-industrial level of 275 ppm. In the 1980s global emissions of carbon
dioxide from fossil fuel burning and cement manufacture averaged 5.5 GtC
(i.e. 5.5 gigatonnes or $5.5 \times 10^9$ tonnes of carbon), each year, with an
additional 1.6 GtC mostly from changes in tropical land-use.

◄  *STEC cement factory*
*Bellevue near Tunis: cement*
*production is an increasing*
*source of carbon dioxide*
*emissions.*

There is no doubt at all that human activities are profoundly affecting the natural balance of gases in the atmosphere, leading to concentrations of greenhouse gases far above these experienced for at least 200 000 years (see Figure 3.10), and little doubt that the build-up of carbon dioxide from anthropogenic sources will lead to some degree of enhanced global warming.

So far, the discussion about greenhouse gases has concentrated on the role of carbon dioxide, but it is not the only greenhouse gas whose concentration is increasing as a result of human activities

Q   Looking back to Figure 3.3, what are the other naturally occurring greenhouse gases?

A   Methane, nitrous oxide, ozone and water vapour. From the preceding chapter, you will also recall that CFCs and many of their substitutes can also play a role in causing global warming.

These trace gases (excluding for the moment water vapour) are present in the atmosphere at concentrations far lower than carbon dioxide, but are nevertheless important as greenhouse gases for two reasons:

(a)  their atmospheric concentrations are building up more quickly than that of carbon dioxide; and

(b)  molecule-for-molecule they are more effective as greenhouse gases than carbon dioxide.

*Methane*, like carbon dioxide, is a gas that occurs naturally in the biosphere. It is produced by bacteria under anaerobic (i.e. oxygen-free) conditions, for example in the intestines of ruminant animals and in waterlogged soils, hence its common name of 'marsh gas'. Rice paddies, which in that respect are artificial marshes, are major sources of methane, and so are herds of cattle. While such sources are undoubtedly biogenic, they also clearly have an anthropogenic element, as Figure 3.14 demonstrates: the increase in

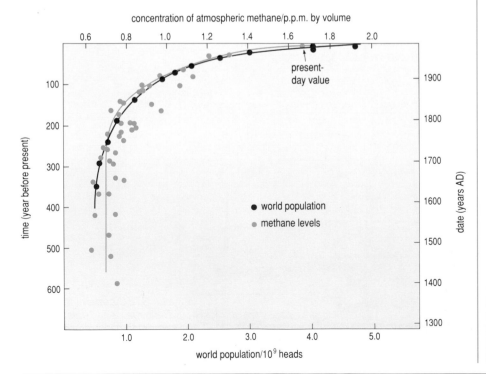

◀   *Figure 3.14 Correlation of the increase in the atmospheric concentration of methane with the growth of world population.*

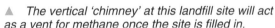
▲   *The vertical 'chimney' at this landfill site will act as a vent for methane once the site is filled in.*

▲   *An environmental scientist testing for methane and carbon dioxide gases at a vent in a landfill site where domestic and industrial waste is buried.*

atmospheric methane over the last 300 years shows good correlation with the growth of world population, suggesting that the rise in concentration of the gas is indeed related to human activities, in this case agricultural production. The rotting of any organic matter (for example, on waste disposal sites) releases methane, and leakage from fuel-extracting operations (such as natural gas pipelines and coal-mines) adds further to the atmospheric burden of this gas.

*Nitrous oxide* is another naturally occurring greenhouse gas, cycled by biological processes. However, substantial quantities of nitrous oxide are also being generated as a result of human activities, such as the industrial production of adipic acid (used for making nylon) and nitric acid, the cultivation of newly exposed soils and increasing use of nitrogenous fertilisers. In addition, combustion of fossil fuels (or indeed the combustion of anything in air, 80% of which is nitrogen) produces nitrous oxide and other compounds of oxygen and nitrogen.

*Chlorofluorocarbons (CFCs)* were discussed in Chapter 2 in relation to their role in the destruction of the ozone layer. However, they and many of their substitutes are extremely efficient greenhouse gases. They do not occur naturally, so all the atmospheric burden of CFCs is anthropogenic in origin.

◄ *Photochemical smog, in which ozone is produced by the action of sunlight on vehicle exhaust gases, in Oakland, Claifornia.*

*Ozone* occurs naturally in the upper atmosphere, where it plays a vital role in shielding the surface of the Earth from harmful ultraviolet solar radiation. Some of this ozone does descend to the lower levels of the atmosphere, but its natural concentration there is low, which is just as well, since ozone is toxic to both plant and animal life. Unfortunately, in industrial societies ozone is now being generated at ground-level by the action of sunlight on gaseous pollutants – such as the hydrocarbons, oxides of nitrogen and carbon monoxide emitted by vehicle exhausts. The so-called 'photochemical smog' that results is itself a hazard to environmental health and the ozone produced in this process also acts as a greenhouse gas in the lower atmosphere.

*Water vapour*, as Figure 3.3(c) showed, plays a very important role in the radiation balance of the atmosphere. Because it is generally more abundant than the other greenhouse gases, it has the largest contribution to the greenhouse effect. However, it is unique among greenhouse gases in that its atmosphere concentration is dependent on temperature and very little else. It is not, therefore, normally treated as a separate greenhouse gas but, as the discussion on the enhanced greenhouse effect in Section 2.1 indicated, it plays a crucial role in amplifying the effects of other gases.

The life-span of molecules of most of these greenhouse gases in the atmosphere is long: nitrous oxide persists for about 160 years, methane for about 15 years, and CFCs for around 100 years (depending on type). This allows the gases time to become well mixed by the atmospheric circulation of the lower atmosphere. They are uniformly distributed in the troposphere and their contribution to global warming through 'radiative forcing' can be calculated with some confidence (see Figure 3.15). In addition, their efficiency as greenhouse gases coupled with rising emission rates means that they have already played a significant role in climate modification. Figure 3.15 shows that in the 1980s greenhouse gases other than carbon dioxide and water vapour were estimated to account for nearly half the warming due to the 'anthropogenic' greenhouse effect. The contribution from tropospheric ozone may also be significant, but because it only

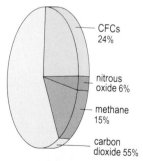

▲ *Figure 3.15 Relative contributions of various gases to the anthropogenic greenhouse warming of the 1980s. The contribution from tropospheric ozone may also be significant, but cannot be quantified at present.*

persists in the atmosphere for a matter of hours, its concentration varies considerably over time and space, depending on where it is formed, and is difficult to quantify accurately.

## 2.4  Questions about the greenhouse effect

Since the middle of the 1980s discussions about climate change, in particular the connection between increasing levels of greenhouse gases and global warming, have increasingly been aired in public and have sometimes themselves become heated. The normal process of scientific research and debate about the findings has been opened up to an unusual level of scrutiny by pressure groups, politicians and the media. While this is surely to be welcomed, the climate system is one of the most complex challenges facing scientists and, as the brief discussion above about water vapour may have indicated, for each example of a given effect there appear to be exceptions or complications. This inevitably raises questions in people's minds. Is the greenhouse effect real? Might there be other causes for the warming observed? Might not other effects lead to cooling?

### Is the enhanced greenhouse effect real?

There is little dispute about the nature of the greenhouse effect and its importance as an element of natural atmospheric systems. However, the argument has been advanced that because there is already so much carbon dioxide in the atmosphere, any extra will have no additional greenhouse effect. The scientific question is 'Are the carbon dioxide absorption bands saturated?' To which the short answer is 'No, they are not'.

It is true that at the wavelength of strongest absorption for carbon dioxide, 15μm, the band is close to saturation – the atmosphere is nearly opaque – and extra carbon dioxide will have little effect, but on either side of the peak at 14 and 16μm there is room for further absorption and enhancement of the greenhouse effect (though Figure 3.3b does not show this clearly). Indeed, as carbon dioxide concentrations increase there will always be other regions of the spectrum where currently weak peaks will start to have an effect.

The practical effect of all this is that the relationship between carbon dioxide concentration and radiative forcing is strongly 'non-linear'. In fact it is approximately logarithmic: if a doubling of carbon dioxide increases the air temperature by say 2.5°C, each further doubling leads to an additional 2.5°C – rather than the 5°C then 10°C that would be expected from a linear relation. This is perhaps fortunate for us, as it means we are unlikely to suffer the runaway greenhouse conditions found on Venus. For the greenhouse gases with smaller concentrations in the atmosphere – CFCs, for example – the relationship with radiative forcing *is* linear, which partly explains their comparatively strong effect. Methane and nitrous oxide, which are present in sufficient quantity to cause significant absorption, fall somewhere between these two examples. Models of global warming take these relations into account as well as the possible overlap of absorption between different greenhouse gases. The non-linearity of the carbon dioxide absorption, for example, is evident in the scales of Figure 3.25.

### Could other factors be responsible for the changes in climate, changes in the energy from the Sun for example?

Reliable measurements of the energy output of the Sun have only been available since 1978 when satellite-based instruments were able to monitor the solar radiation from above the Earth's atmosphere. A small change of less than 0.1% (0.2 $Wm^{-2}$) in this radiation has been recorded over the 10–12-year solar sunspot cycle. The strength of this solar cycle has been known to vary and attempts have been made to link the 'little ice-age', a cooler period around the seventeenth century, with the 'Maunder minimum' when sunspots were not observed. Recently some strong correlations have been produced between the length of the solar cycle and temperature over land in the northern hemisphere for the last 130 years. However, the correlations may be coincidental. A more compelling reason for caution is that the mechanisms of the greenhouse effect and changes in the output of solar radiation are the same: they both change the radiative forcing at the top of the atmosphere. The increase in forcing from greenhouse gases to date can be estimated with some confidence as 2.4 $Wm^{-2}$; an explanation of climate change based on solar variation alone has to explain that away. The best estimates of the variation in solar output since the seventeenth century suggest a maximum increase of between 0.1 and 0.5 $Wm^{-2}$ – significant, but no more than the effect of ten years' increase in greenhouse gases at current rates.

## *Are there factors which could lead to a cooling rather than a warming?*

Volcanic eruptions often inject columns of ash and aerosol* high into the atmosphere and have been put forward as a possible cause of temporary global coolings. There is now good evidence (from sulphate deposits in ice-cores matched to years of poor growth in tree rings) that large eruptions have been associated with many of the coolest summers of the last few centuries. The notorious 'year without a summer' of 1816 in eastern parts of North America, following the eruption of the Indonesian volcano Tambora a year earlier, was a notable example. The key factor for an effect on climate is not so much the size of the eruption, but the amount of **sulphate aerosol** – produced from the oxidation of sulphur dioxide gas – that is pushed into the stratosphere where it is no longer washed out by rainstorms. The aerosol spreads round the globe over a period of months to form a veil that then persists for a year or two. The veil reduces the solar radiation reaching the surface by scattering incoming sunlight, an effect which provides a sideshow of spectacular sunsets.

As we saw in Chapter 2, Section 5.1, the eruption in June 1991 of Mt Pinatubo is thought to be the most significant this century for climate modification, due to the massive amount of sulphate aerosol injected into the stratosphere – estimated at between 20 to 30 million tonnes. (See Plate 5.) It produced a peak radiative forcing of more than –4 $Wm^{-2}$ and averaged –2 $Wm^{-2}$ for a year, declining to half this value the following year (the negative sign indicates *reduced* forcing, i.e. a cooling effect). In other words, in the two years after its eruption Pinatubo produced a radiative forcing of comparable size to but in the opposite direction to the current greenhouse forcing of approximately 2.4 $Wm^{-2}$. A global cooling of between 0.3°C and 0.5°C was observed in 1992, and the climatic effects of the eruption have been simulated successfully by global circulation models, increasing the confidence scientists have in them. Volcanic eruptions can clearly have a significant effect on climate but the effects are short-lived and there is no evidence of systematic trends in either their frequency or intensity. Sulphur dioxide is also released into the lower atmosphere as a by-product of fossil fuel and biomass burning, where it can be a major pollutant and source of acid rain. The aerosols it forms can also influence the climate, a subject returned to in the next section.

*Aerosols are tiny droplets of liquid or small particles suspended in the atmosphere. Sulphur dioxide is readily oxidised in the atmosphere to sulphuric acid. The sulphate aerosol it forms is particularly effective at scattering solar radiation.

◄   The eruption of Mount Pinatubo in the Philippines, seen from the Clark Air Base on 13 June 1991.

## 2.5   Summary of Section 2

Greenhouse warming, an effect arising from the successive absorption and emission of infra-red radiation by molecules of certain atmospheric gases, has occurred on Earth throughout the planet's history and is crucial in maintaining the surface at a temperature suitable for life. In the past two hundred years, anthropogenic sources of greenhouse gases, particularly carbon dioxide and methane, have begun to disturb the atmospheric balance. By 1990 carbon dioxide concentration had risen by nearly 30% above its pre-industrial level, a level not seen for several hundred thousand years. There has also been an increase in the mean global surface temperature of approximately 0.5°C over the past hundred years. There is growing evidence that this is a case of cause and effect, that human activities are starting to change the global climate as well as the composition of the atmosphere.

*Activity 2*

The following questions may be used to revise some of the key concepts of this section.

Q1  Radiation from the Sun is continually incident on the Earth, so why doesn't the Earth continually get hotter and hotter?

Q2  Describe briefly the heating mechanisms in the troposphere and stratosphere and explain why overall the temperature varies differently with respect to height in the two regions.

Q3  What factors other than greenhouse gases could contribute to or mitigate global warming?

# 3   *Greenhouse warming: projections and scenarios*

## 3.1   *Introduction*

Predicting the future is an altogether trickier business than interpreting past and present data. It should come as no surprise, then, that there is still controversy surrounding assessments of the possible anthropogenic contributions to the greenhouse effect, in particular:

(a)  the likely future levels of greenhouse gases in the atmosphere arising from human activities;

(b)  the amount by which the Earth's surface temperature may rise as a consequence of the increased atmospheric burden of greenhouse gases; and

(c)  the time-scale and geographic distribution of any such temperature increase.

In this section we will examine these areas of on-going research and debate, and explore some of the difficulties and uncertainties associated with them. Concern over global warming has prompted a tremendous concentration of scientific effort in this field over recent years, but scientists still have only a partial understanding of the way in which the Earth's climate systems work. In the next decade we can expect many hypotheses and conclusions to be put forward. Quite a lot of this work will be reported in the popular media, with variable degrees of exactitude and objectivity. A major aim of this section and the next one is to enable you to follow future developments and to assess the importance and reliability of the research data on which they are based.

   The problem of trying to predict the Earth's climatic future may be broken down into a series of steps:

1   'Projecting' (i.e. predicting) the *emissions* of carbon dioxide and other greenhouse gases into the atmosphere.

2   Projecting the build-up of the *concentration* of these gases in the atmosphere.

3   Forecasting the resulting climatic change.

Because each of these steps feeds into the next one, the process becomes increasingly more difficult and uncertain from step 1 to step 3.

   We will tackle each of these stages in turn in Sections 3.3–3.6, and our conclusions will then inform the discussion in Section 4 about the impact of possible climatic change on natural processes and human societies. Before starting on this set of interwoven arguments, however, it is worth taking a brief look at the basic method underlying all the forecasts – the technique of 'modelling'.

## 3.2   *Models and scenarios*

In everyday language we use phrases such as 'architect's model' or 'model railway' to mean a scaled-down replica of the real thing. As we saw in Chapter 2, Section 4.2, a scientist uses the term 'model' somewhat differently. Scientific **modelling** may be described as an attempt firstly to break down the workings of a complex system into a series of well-defined processes, and then to find a way of simulating these processes so as to be able to predict how the system will react in a variety of circumstances. In constructing a model of the atmosphere, for example, one might start from the basic physical law that states that the temperature of an 'ideal gas' is directly proportional to the product of its pressure and its volume. The notion of an 'ideal gas' is a scientist's model if ever there was one, since no real gas ever conforms exactly to this notion of ideality (although at ordinary pressures and volumes most gases approximate to it), but the usefulness of the law lies in the fact that it is easily expressed in terms of a simple equation. There are well-established equations, too, to describe phenomena such as the change in atmospheric pressure with height, the rate at which a gas released at one point will diffuse through the atmosphere, and the radiative heating effect associated with particular gases as a function of their atmospheric concentration. It is the complete set of equations describing all the components of the system that finally constitutes 'the model'. The equations are the rules that a computer will follow so as to simulate the behaviour of the real system in response to various changes in conditions ('input') that are given to it as basic data.

The modelling process thus involves three major elements – the model itself, the capacity of the computer on which it is run and the input conditions. All three have their limitations and it is often necessary to read between the lines of various reports in order to appreciate these limitations and assess the reliability of claims made by the modellers.

Firstly, one has to consider the complexity of the model. Returning to the example of a model of the atmosphere, it is possible to construct simple models on the basis of equations expressing physical laws such as those described above. In the real world, though, a great many other factors also operate. Some are quite well understood in an empirical way (i.e. their effect is known by observation but there is no theoretical basis on which prediction of their future effect can be made): an example of such a factor in atmospheric modelling might be the influence of clouds. Still other factors are known to be important to the system but are as yet very poorly understood: the whole ocean–atmosphere–biosphere interaction falls into this category at present. In such a complex system there are also almost certainly factors that we have not even identified. Our models are therefore incomplete, over-simplified and full of approximations and assumptions. This is not to imply that simplification is necessarily a bad thing. Indeed it is the essence of modelling: there is no point in constructing a model that is just as complicated as the real word. On the other hand, a good model should include all the important features of the real-world system. The art of the modeller is to strike a balance between complexity, utility and accuracy.

Secondly, the demand for computing power still imposes limitations. Even in a very simplified model of the atmosphere, a number of processes must be included. Each process might be described by a fairly simple equation, but because all the processes interlock with one another, a great deal of computation is required to produce simultaneous solutions to all of the equations. In order to keep the task to manageable proportions, trade-offs often have to be made between the detail with which conditions are covered and the sophistication of the model. Most early computer models of the effects of doubling carbon dioxide, for example, assumed a fixed level of the gas rather than the more realistic scenario of gradually increasing the concentration, and only recently have models been used to 'couple' the atmosphere with the ocean circulation.

The third limitation of the modelling process arises from the inputs used – the numerical values or conditions the computer is told to apply to the model. Some conditions are difficult to specify in a way that a computer can use. For example, in the real atmosphere, cloud formation is an important process, but on a model in which the atmosphere is represented by large latitudinal bands, a cloud is a phenomenon on too small a scale to be simulated directly. Some quantity – perhaps a proportionality constant in an equation – therefore has to be included to give an estimate of the average cloudiness corresponding to other sets of conditions, such as temperature and humidity, that can be specified or calculated more easily. This technique, called *parameterisation*, by which small-scale phenomena and feedback loops are incorporated into a model in an average way, is widely used, but applied slightly differently in each individual case. As we shall see, models can be very sensitive to small changes in parameterisation; in evaluating models or in comparing the predictions of one against the other, it is important to bear this in mind.

Other types of input would be relatively easy to specify to a computer but are simply impossible to estimate. The future atmospheric concentration of carbon dioxide is an example of such a quantity. Under these circumstances, modellers fall back on constructing **scenarios** – pictures of what to expect given certain basic premises. A scenario is not an image of a definite future; it is an image of what *could* happen if particular conditions are fulfilled. Some of the scenarios put forward by climate modellers are frightening but may never come to pass: the fear engendered by a scenario may be enough in itself to make us take steps to ensure that the assumptions underlying that particular scenario are never fulfilled. The rest of this chapter is all about constructing scenarios, examining the models and assumptions on which they are based, and learning from them about the consequences of future courses of action.

## 3.3  Projecting carbon dioxide emissions

Table 3.1 provides an estimate of the main anthropogenic sources of carbon dioxide in the early 1990s. However, forecasting the rate at which carbon dioxide will be released to the atmosphere in the future is a complex problem in social science. In outline it requires predictions of population

Table 3.1  Annual average anthropogenic carbon budget for 1980 to 1989. $CO_2$ sources, sinks and storage in the atmosphere are expressed in GtC/yr.

| $CO_2$ sources | |
| --- | --- |
| 1 Emissions from fossil fuel and cement production | $5.5 \pm 0.5$ |
| 2 Net emissions from changes in tropical land-use | $1.6 \pm 1.0$ |
| 3 Total anthropogenic emissions = (1) + (2) | $7.1 \pm 1.1$ |
| Partitioning amongst reservoirs | |
| 4 Storage in the atmosphere | $3.2 \pm 0.2$ |
| 5 Ocean uptake | $2.0 \pm 0.8$ |
| 6 Uptake by northern hemisphere forest growth | $0.5 \pm 0.5$ |
| 7 Additional terrestrial sinks ($CO_2$ fertilisation, nitrogen fertilisation, climatic effects) = [(1) + (2)] − [(4) + (5) + (6)] | $1.4 \pm 1.5$ |

Source: Houghton, J. T. *et al.* (1995) *Climate Change 1994*, Cambridge, Cambridge University Press, Table 1, p. 18

trends, economic growth, policies of world trade, the relationship between economic development and energy consumption, and the proportion of those energy demands that will be met by burning fossil fuels. Fossil fuel usage will in turn depend on many factors, such as the price of oil, the availability and cost-effectiveness of alternative energy sources, energy conservation measures and even possible social pressures for counter-measures to anthropogenic greenhouse warming.

Even in the short term (of a decade, say), there are very large uncertainties associated with any projection of energy requirements and fossil fuel consumption. These uncertainties are different in nature from those that scientists attach to experimentally measured quantities or to the predictions of mathematical models. Demographic and economic factors are not amenable to modelling in quite the same way as quantities governed by natural laws, and economic factors are further complicated by political decisions. Predictions of fuel consumption for the 1970s based on the pattern set in the 1960s turned out to be badly wrong, because they obviously had not been able to take into account the major oil crises that in fact occurred in 1973 and 1979. In the same way, policy decisions regarding fossil fuel use could in the next decade substantially modify current patterns of carbon dioxide emissions. Faced with the impossibility of quantifying all these elements, we can only fall back on the construction of possible scenarios.

In 1992 the IPCC sponsored the development of six greenhouse gas emission scenarios, commonly referred to as the IS92 scenarios (IS92a–f), to provide inputs for climate analysis. They provide projections of carbon dioxide emissions (and those of other greenhouse gases) from 1990 to 2100 and the resulting atmospheric concentrations. Figure 3.16 shows the emissions scenarios for carbon dioxide. The scenarios are based on a range of assumptions from the World Bank and the UN about population growth and economic development, use of energy supplies, compliance with the Montreal Protocol and controls on pollutants such as sulphur dioxide and nitrogen dioxides. The highest emission scenario, IS92e, assumes moderate population growth, high economic growth and fossil fuel availability and a phase-out of nuclear fuel. A central case is represented by IS92a, which assumes a moderate growth for the population and economy, increasing use of coal as a fuel and little replacement of fossil fuels by renewables. Finally, IS92c, which produces the lowest $CO_2$ emissions, assumes low population growth (6.4 billion by 2100 compared to 5.8 billion today), low economic growth and severe constraints on fossil fuel supplies. Given the nature of the inputs, forecasts of economic growth for example, uncertainty

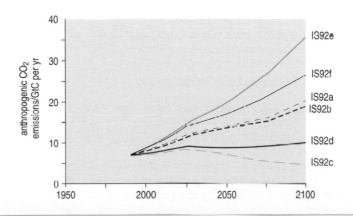

◀ *Figure 3.16*
*Emissions of carbon dioxide, in gigatonnes of carbon per year, due to human activity, for the IS92 scenarios.*

for a given scenario must be high and increases as the time horizon lengthens. For this reason the scenarios are presented together to reflect the diversity of views in the assumptions. It should be emphasised that these scenarios have been established to help evaluate the consequences of non-intervention, of taking no action to reduce greenhouse gases. Other scenarios have been designed to investigate the options of reducing emissions; these are discussed in Section 5. Assuming, then, no climate change policies, by 2100 carbon dioxide emissions are projected to be in the range 6 GtC to 36 GtC per year. Cumulative emissions for the period 1990–2100 would be in the range 700 to 2080 GtC – compared with an estimated total of 240 GtC emitted between 1860 and 1994.

## 3.4   *Projecting atmospheric concentrations of carbon dioxide*

Given a particular scenario for the amount of carbon dioxide emitted into the atmosphere, the next step is to estimate how much of it will actually remain there. Unlike the issue of emission rates, this question is amenable to scientific modelling of the type described in Section 3.2, even though the additional flux of $CO_2$ due to human activity is small compared to the natural flows. The basic features of the global carbon cycle, in which carbon circulates through the biosphere, the atmosphere, the hydrosphere (oceans and freshwater) and through soils and rocks, are well understood (see *Silvertown*, 1996).

Current estimates suggest that about half of the anthropogenic emissions of carbon dioxide are absorbed in 'sinks' by various processes; the rest remains in the atmosphere (refer to Table 3.1). Although our understanding of the proportions which can be attributed to the sinks on land and at sea is steadily improving, there are still uncertainties about many of the processes. Because of this, scientists have not yet been able to balance the carbon budget completely – a missing carbon sink of about 1.5GtC/yr exists in the northern hemisphere land mass which has not been fully explained. A complicating factor is the likely existence of **feedback loops**. For example, increased levels of carbon dioxide are likely to stimulate an increased take-up of carbon dioxide through photosynthesis. This effect, known as $CO_2$ *fertilisation*, along with possibly similar effects from increased nitrogen from the atmosphere (though acidification may also damage ecosystems) and responses to changed climate, may account for much of the absorption. On the other hand, greenhouse warming of the climate could also speed up the decomposition of dead organic matter held in soils and result in further release of carbon dioxide. In spite of these complications, the uncertainties inherent in estimating carbon dioxide concentrations for a particular emission scenario are not nearly so great as the uncertainties associated with the emission scenarios themselves. Thus the upper and lower limits on estimates of carbon dioxide concentration are effectively set by the upper- and lower-bound emission scenarios discussed in the previous section.

The IPCC IS92 emissions scenarios of Figure 3.16 are translated into projections of carbon dioxide concentrations (Figure 3.17) by incorporating a mid-range model of the carbon cycle (known as the Bern model). Although this new stage introduces additional uncertainties, shown by the area shaded green for the example of IS92a in Figure 3.17, they are small compared to the differences between scenarios. All the scenarios indicate large increases in the atmospheric burden of carbon dioxide by the year

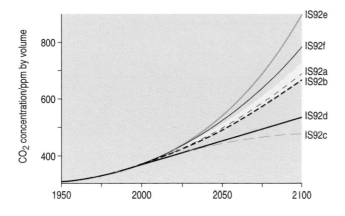

*Concentrations of atmospheric carbon dioxide, in parts per million by volume, calculated from the six IS92 scenarios, using the Bern model of the carbon cycle. The typical range of results from different carbon cycle models (for IS92a) is indicated by the green shaded area.*

2100, ranging from 75 to 200%. None of the scenarios shows levels stabilising over this time-period, the ultimate objective of the UN Climate Change Convention, although for the low emission scenario, IS92c, the rate of growth slows. The bench-mark doubling of pre-industrial carbon dioxide to 560 ppm is reached by 2070 in four of the scenarios including the mid-range IS92a. Such concentrations are unprecedented for carbon dioxide, which remains the dominant greenhouse gas in these scenarios. Even so, we need to turn briefly to a consideration of the other trace gases to complete the picture.

## 3.5    *The effects of other greenhouse gases and aerosols*

If it is difficult to predict future levels of carbon dioxide emissions, it is even more difficult to forecast emissions of other greenhouse gases. In the case of the naturally occurring gases, especially methane, the sources and sinks are not well understood. In the case of the industrial gases, particularly CFCs, future emissions will be totally dependent on technological and political developments.

Methane is in some ways the most important greenhouse gas after carbon dioxide. At present, human activities have increased the concentration of carbon dioxide in the atmosphere by slightly less than 30% above its pre-industrial level, but analyses of air bubbles trapped in ice cores show that the amount of methane present in the atmosphere has more than doubled in industrial times. There are also several positive feedback loops linking natural methane emissions to climatic change: methane, like carbon dioxide, is released from decaying organic matter. Higher atmospheric concentrations of methane cause greenhouse warming; higher temperatures increase rates of decay. Large quantities of methane are also locked away in peat bogs below frozen Arctic tundra, so the melting of areas of permafrost could lead to further release of methane.

### *Making comparisons*

Table 3.2 summarises the current concentrations and trends for key greenhouse gases; it also provides a comparative measure of the 'effectiveness' of each greenhouse gas, the *global warming potential (GWP)*. In essence, this concept is straightforward: it provides a comparison of the radiative forcing from a unit mass of given gas, well mixed in the

*Table 3.2   A summary of key greenhouse gases affected by human activities*

|  | $CO_2$ | $CH_4$ | $N_2O$ | CFC-12 | HCFC-22 (a CFC substitute) | $CF_4$ (a perfluoro-carbon) |
|---|---|---|---|---|---|---|
| Pre-industrial concentration | 280 ppmv | 700 ppbv | 275 ppbv | 0 | 0 | 0 |
| Concentration in 1992 | 355 ppmv | 1714 ppbv | 311 ppbv | 503 pptv[1] | 105 pptv | 70 pptv |
| Recent rate of concentration change per year (over the 1980s) | 1.5 ppmv  0.4% | 13 ppbv  0.8% | 0.75 ppbv  0.25% | 18–20 pptv  4% | 7–8 pptv  7% | 1.1–1.3 pptv  2% |
| Atmospheric lifetime (years) | (50–200)[2] | (12–17)[3] | 120 | 102 | 13.3 | 50 000 |
| GWPs[4], 100-year time-scale | 1 | 24.5 ± 0.75 | 320 | 8500 | 1700 | 6300 |

Notes:
[1] 1 pptv = 1 part per trillion (million million) by volume.
[2] No single lifetime for $CO_2$ can be defined because of the different rates of uptake by different sink processes.
[3] This has been defined as an adjustment time which takes into account the indirect effect of methane on its own lifetime.
[4] Although GWPs are given as single figures, the typical uncertainty relative to carbon dioxide is ±35%.

Source: Houghton, J. T. *et al.* (1995) *Climate Change 1994*, Cambridge, Cambridge University Press, Table 3, p. 25 (with additional information).

atmosphere, with respect to carbon dioxide. It is similar, both as a concept and in its uses to the ozone depletion potential discussed in the previous chapter. GWPs provide a comparison over a given time-scale, but for emissions scenarios such as IS92, it is often convenient to express the radiative forcing in terms of an equivalent quantity of carbon dioxide.

You may be wondering what happened to CFCs which have very high global warming potentials. If we consider the IS92 emissions scenarios, there is a good reason why CFCs may not play a very large role.

Q   What do you think it is?

A   The IS92 scenarios look at emissions of greenhouse gases *after* 1990. CFC emissions are stringently controlled by the Montreal Protocol: concentrations in the troposphere are now at their peak and expected to decline slowly in the next century.

The same consideration applies to another greenhouse gas, ozone in the troposphere, though for slightly different reasons: this pollutant may have doubled since pre-industrial times, but with measures to control the precursor gases, its atmospheric concentration has not increased significantly in the last decade, nor is it expected to in the future. However, the emissions and concentrations of CFC substitutes are rising rapidly (though also subject to current or future control) and will need careful monitoring. One possible family of substitutes, the perfluorates, of which $CF_4$ is an example, is noteworthy. It is present in the atmosphere in small quantities, and mostly emitted during the manufacture of aluminium. It is a potent greenhouse gas, but more importantly it is not destroyed in the atmosphere: once there, it becomes virtually a permanent resident. Because of this, it is not acceptable as a CFC substitute (see Chapter 2, Section 4).

## A few complications

Before concluding this section there are two concepts that need exploring in a little more detail: the difference between *direct* and *indirect* radiative forcing, and the effects of aerosols in the atmosphere. We will then be in a position to compare the effects of the main contributors to climate change which are summarised in Figure 3.18.

### Direct and indirect radiative forcing

As we have seen, the 'radiative forcing' due to a greenhouse gas is a measure of the net change the gas has on the radiation balance at the top of the atmosphere, and hence of its potential contribution to global warming. The forcing due to the increase in concentration of greenhouse gases (carbon dioxide, methane, nitrous oxide and the halocarbons) since pre-industrial times is currently put at 2.4 Wm$^{-2}$. These trace gases are greenhouse gases in their own right and produce 'direct' radiative forcing. Some of these, however, take part in additional chemical or physical reactions in the atmosphere which can also influence the radiative forcing: the halocarbons are an example. Halocarbons, and in particular CFCs, as we know from Chapter 2, have been responsible for global ozone loss in the lower stratosphere in addition to being greenhouse gases. The direct radiative forcing since pre-industrial times ascribed to halocarbons, mostly arising in the last few decades, is calculated to be 0.3 Wm$^{-2}$, but there is also an *indirect* forcing due to the recent ozone losses estimated at –0.1 Wm$^{-2}$, i.e. partially offsetting the warming effect. However, it is probably not appropriate simply to combine the two effects, because the pattern of stratospheric ozone loss has not been spatially even – the greatest losses have been over the poles and at mid-latitudes.

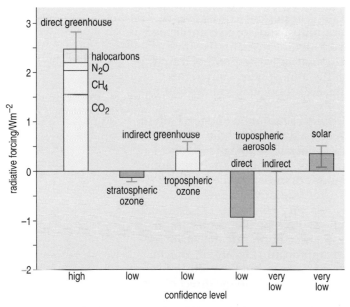

▲   *Figure 3.18   A summary of the main greenhouse 'players'. Estimates of the globally averaged radiative forcing due to changes in greenhouse gases and aerosols from pre-industrial times to the present day, and changes in solar variability from 1850 to the present day. The height of the bar indicates a mid-range estimate of the forcing whilst the green lines show the possible range of values. An indication of relative confidence in the estimates is given below each bar. The negative values for aerosols should not necessarily be regarded as an offset against the greenhouse gas forcing.*

Other trace gases such as the nitrogen oxides – *collectively termed 'NO$_x$' and not to be confused with nitrous oxide (N$_2$O)* – are not greenhouse gases in their own right but react with other components of the atmosphere to produce greenhouse gases. The nitrogen oxides are mostly anthropogenic in origin – the main sources are transport including aircraft, biomass burning and the soil. Many of the pollutants from road transport – the nitrogen oxide family (NO$_x$) together with carbon monoxide and hydrocarbons – react with sunlight under certain conditions to produce tropospheric ozone, another pollutant and also a greenhouse gas. The concentration of ozone varies considerably over space and time, and although estimates of its global warming contribution are sizeable at 0.2 Wm$^{-2}$ to 0.6 Wm$^{-2}$, there is considerable uncertainty about the figures, as is the case with most examples of 'indirect' forcing.

*Aerosols*
Aerosols are simply suspensions of tiny particles or drops of liquid in the atmosphere. In the troposphere they are short-lived, because they are swept out of the atmosphere within days or weeks by rain (by contrast, in the stratosphere, which lies above the clouds, they can persist for years). Because of this, their concentrations tend to be greatest close to, or down wind of, the major sources. They arise from a variety of sources: soil dust blowing from the land, spray from the sea, but also from smoke and chemical reactions in the atmosphere. The principal anthropogenic sources of aerosols are smoke particles from biomass burning and the release of sulphur dioxide from fossil fuel combustion (which is rapidly oxidised by hydroxyl radicals to form sulphate particles). These cause 'industrial haze' but also, crucially, affect the radiation balance in the atmosphere. The climatologically significant aerosols are those with diameters in the range 1 to 10 micrometers but particle composition is important as well as size. The sulphate aerosols are the most important groups, and are also implicated as the main source of acid rain. There is also a significant natural source of sulphate aerosol – dimethyl sulphide, a biological waste-product from the sea, which is also oxidised to sulphate in the atmosphere – but in this chapter we are mainly concerned with the anthropogenic sources and the changes they are causing.

Aerosols can have two effects on the Earth's radiation budget, both of which are thought to lead to a net *cooling*: (a) a direct effect through absorption and scattering of solar radiation back to space, and (b) an indirect effect by acting as condensation nuclei for clouds, affecting cloud density and longevity, and hence the amount of radiation they reflect.

Globally averaged estimates for the direct radiative forcing due to increases in anthropogenic sulphate aerosols vary between –0.25 and –0.9 Wm$^{-2}$ (the negative values indicate cooling) with a smaller contribution from aerosols caused by biomass burning. These figures demonstrate the current uncertainty but they also suggest significant, if localised, effects. The study of indirect forcing is still at an early stage, but the indications are that it is also negative, possibly of a similar magnitude to the direct effect.

Q   Where would you expect to find the highest concentrations of sulphate aerosols, and why?

A   You would expect to find them close to the main sources of sulphur dioxide emissions, because of their short lifetimes. The largest single sources are power stations, but in general they are found close to the industrialised areas of the world.

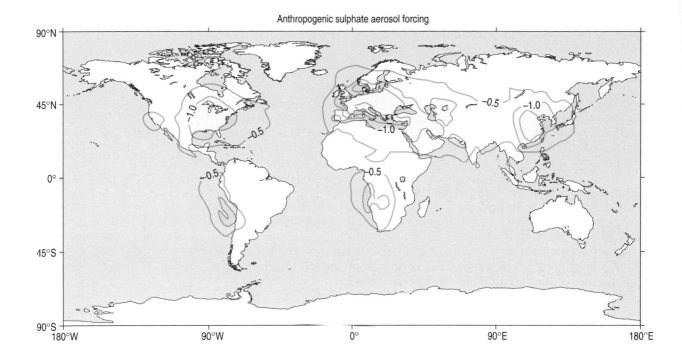

Figure 3.19 shows the estimates for current values of direct radiative forcing due to anthropogenic sulphate aerosols. The highest values are seen over the eastern US, much of Europe and eastern China and in these regions it is thought that cooling due to the increase of aerosols may have offset much of the greenhouse gas forcing. In the southern hemisphere radiative forcing from these aerosols is much weaker and there the greenhouse gas effect is more dominant. However, care must be taken in interpreting a figure for radiative forcing where the distribution of aerosols is not uniform: the global figure cannot be simply subtracted from the 2.4 Wm⁻² estimated for greenhouse gases and probably overstates its cooling contribution. Nor does it mean that global warming will go away, although it may account for the slowness in detecting it. Indeed, the opposite may be the case. Global emissions of sulphur dioxide have increased rapidly from the 1950s onwards but, because it is a damaging pollutant, in most industrialised countries emissions of the gas have been controlled, and are now falling.

▲ *Figure 3.19*
*Sulphate aerosol forcing: geographic distribution of the average, annual, direct radiative forcing ($Wm^{-2}$) from anthropogenic sulphate aerosols in the early 1990s. The negative values indicate a cooling effect. The contour interval is 0.5 $Wm^{-2}$ and values below −1 $Wm^{-2}$ are shaded green. For comparison, the average radiative forcing due to greenhouse gases is about 2.4 $Wm^{-2}$.*

Q    What is the likely effect on climate of reducing sulphate emissions?

A    The aerosols and their radiative effect will disappear rapidly, within days or weeks. In the regions of strongest emission, notably eastern US and Europe, global warming may proceed more rapidly as the 'masking' effect of the aerosols is removed.

This does, indeed, seem to be a strong possibility, but statements of this kind can only be justified when the effects of the radiative forcing have been properly simulated in a global circulation model (described in the next subsection). Figure 3.18 summarises the radiative forcing, from pre-industrial times to the present day, due to the main influences on climate change discussed in the last few pages. It is worth noting that the only estimate given with a 'high' level of confidence is the forcing from the well-mixed greenhouse gases, 2.1 Wm⁻² to 2.8 Wm⁻², of which carbon dioxide contributes 60% of the total. The uncertainty in the estimates of the other factors is indicated by the error bars.

## 3.6    *Forecasting climate change*

A great deal of research effort has been – and is still being – invested in the development of computer models for climate simulation. The main tools used by researchers are the **global circulation models (GCMs)** based on well-tested numerical weather forecasting models but designed to simulate changes over much longer periods of time, ranging from years to decades. Weather forecasting models, however, lose their forecasting 'skill' – their ability to predict events – after about twenty days, due to their extreme sensitivity to the initial conditions entered, a reflection of the chaotic nature of the atmosphere. How, then, is it possible to think of forecasting even average conditions for decades ahead? Climate models are set up differently: they are not sensitive to initial conditions, indeed it is usual to run them for many years to remove dependence on the initial data; they respond instead to the changing conditions of the Earth's surface and radiation balances. Ideally, GCMs should incorporate coupling between the atmosphere and the oceans, land, ice (cryosphere) and biosphere.

The basic principles on which computer models are constructed were described in Section 3.2. In a GCM, the troposphere is divided into a three-dimensional grid, with horizontal spacing between grid lines of about a hundred kilometres and typically twenty layers in the vertical, spaced one kilometre apart. There are usually five basic variables whose behaviour is described by equations: north–south and east–west components of winds, atmospheric pressure, temperature and humidity. These simultaneous equations are solved for each box on the grid. The better the 'spatial resolution' of the model (i.e. the finer the grid), the greater the requirement for high computer capacity and speed. In addition, a variety of phenomena that occur on scales smaller than the grid, such as clouds, precipitation and radiative heating, and exchanges of heat, moisture and momentum (friction) between atmosphere and ocean, are brought into the model using parameterisation techniques.

Once a model has been constructed, it has to be tested, a process called *validation*. Climate models can be tested in three ways that cover a range of time-scales and events. In each case we are looking at how they respond to different patterns of radiative forcing at the top of the atmosphere, just as they would to greenhouse gases. The first method is to run the model for a number of years, then compare its predictions of suitable parameters such as temperature, rainfall and pressure with the average values and seasonal variations of the global climate. Current climate models usually perform well in this test. A second type of test is to use the model to stimulate past climates which have had different seasonal patterns of solar energy reaching the Earth due to the Milankovic cycles (discussed in Section 2.2). For example, about 10 000 years ago changes to the Earth's orbit meant that solar energy in the northern hemisphere summer was 7% higher than now (and correspondingly less in winter). Under this regime, models predict a stronger monsoon over southern Asia and more summer rainfall over the Sahel. The admittedly limited 'paleoclimate' (i.e. prehistoric climate) data available supports these predictions. A third way to test the models is to get them to simulate large-scale perturbations in the climate. Some models are now able to stimulate with some accuracy the progress of El Niño events (see Chapter 1) and the climatic anomalies that accompany them (see

Figure 3.24). Others have taken the reduced input of solar radiation caused by the dust veil from the 1991 Pinatubo eruption to simulate successfully the short-term reduction in global temperatures and regional anomalies that have been observed.

In the absence of any other effects, a doubling of carbon dioxide concentration in the atmosphere from pre-industrial levels would be expected – eventually, that is, when the climate system had returned to balance or *equilibrium* – to cause a rise of temperature at the Earth's surface of 1.2°C. However, this figure makes no allowance for the feedback effects in the climate system due to water vapour, clouds, the oceans and ice which are described in Box 3.2. When these feedbacks have been incorporated, current models indicate a range of possible temperature rises from 1.5°C to 4.5°C, *with the 'best' estimate probably about 2.5°C*. The range is due to different treatment of the feedbacks by models, but their response to doubling is used as a bench-mark to compare models and often termed the 'climate sensitivity'.

---

## Box 3.2   Feedbacks and coupling in the climate system

The many interactions that take place in the climate system are illustrated in Figure 3.20. All need to be considered by a model of global climate change, in particular those interactions which serve to amplify or reduce the warming effect of greenhouse gases.

*Water vapour* is an important greenhouse gas in its own right, as Figure 3.3 shows, but its atmospheric concentration is strongly dependent on temperature – the higher the surface temperature, the more evaporation occurs from water and wet land surfaces and the more water vapour there is in the air. In turn, the increase in water vapour is likely to further increase the temperature. Thus, in

response to warming from greenhouse gases, water vapour provides a *positive feedback cycle* which amplifies the original warming. The cycle is illustrated by Figure 3.21. This action of water vapour increases the warming due to greenhouse gases by at least 60%, taking the doubling figure from 1.2°C to nearly 2.0°C, and is the major single feedback in the climate system. But note that, because its role is to amplify the effects of other greenhouse gases, it is usually discussed in the context of the 'climate sensitivity' of GCMs.

*Ice and snow* cover also provide a positive feedback mechanism to amplify global warming,

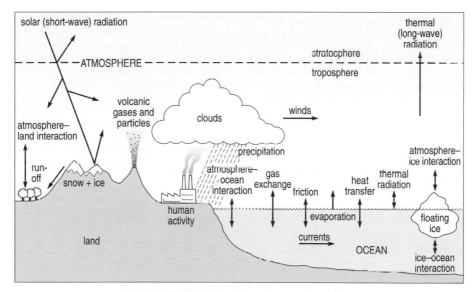

▲   *Figure 3.20   A schematic diagram of the many interactions between the atmosphere and the other 'spheres' that would be included in an ideal global circulation model.*

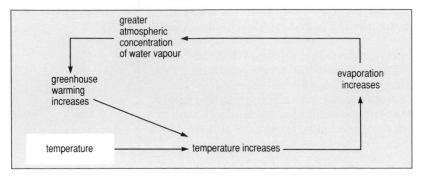

◀ *Figure 3.21*
*A positive feedback cycle illustrating the role of water vapour in amplifying the warming induced by greenhouse gases.*

particularly at the regional level. Ice and snow reflect solar radiation well (they have high albedos), but when they melt under the influence of warmer temperatures, the surface they covered will then absorb more solar radiation and increase the warming.

*Oceans* act in several ways to modify the climate. Firstly, they are the main source of water vapour and, through their 'latent heat', the largest source of heat for the atmosphere. Secondly, they have a very large heat capacity – a three-metre depth of water corresponds to the heat capacity of the atmosphere – which serves to slow down the rate of any temperature change in the atmosphere. Finally, as we saw in Chapter 1, the ocean currents transport large quantities of heat around the globe. Even small changes in this transport could have large effects on the climate.

*Clouds* are inherently difficult to model in GCMs

because they often occur on smaller scales than the typical grid size of a model. An added complication is that they combine several processes which tend to act in opposite directions. They reflect solar radiation efficiently back to space, reducing the energy available to the Earth, but they also act as a 'blanket' to thermal radiation emitted from the Earth's surface, absorbing it and reflecting part of this back to Earth, tending to reduce the energy loss to space. These effects are examples of positive and negative feedback respectively – the *negative feedback* acting to reduce any changes of temperature. Which effect predominates depends on the height and thickness of the clouds but, overall, clouds probably cool the Earth. However, the climate is very sensitive to changes in cloud type and cover: a few per cent either way can be comparable to the changes due to the doubling of carbon dioxide.

Many of the early results from GCMs that were given prominence in the media described the equilibrium response of the climate system to a *sudden* doubling of carbon dioxide, as this is the simplest simulation to carry out. Any coupling with the oceans used by the models was only of a very basic form. These simulations generally agreed on the nature and geographic distribution of the predicted warming: it would be greatest at high latitudes, particularly during winter, less in the tropics, and also enhanced for northern mid-latitude continents in summer (see, for example, Figure 3.22). While these predictions showed important agreements with recent global temperature trends, seen in Figure 3.11, there were several inconsistencies, the main being that the 'best' predictions of the increase of temperature by the 1990s, taking into account the other greenhouse gases, was double the observed value of approximately 0.5°C. In the early 1990s a new generation of GCMs was introduced which fully coupled the atmosphere to the slower circulations in the oceans – the *coupled GCMs or CGCMs*. These reflect in a more realistic manner the important impact the oceans have on climate, although the sizes of the ocean–atmosphere fluxes still have to be 'adjusted' to obtain reasonable results. When they are used with the more realistic scenario of increasing carbon dioxide at 1% a year, roughly the current rate of increase of greenhouse gases, two important changes are seen. Firstly, warming over parts of the oceans – the North

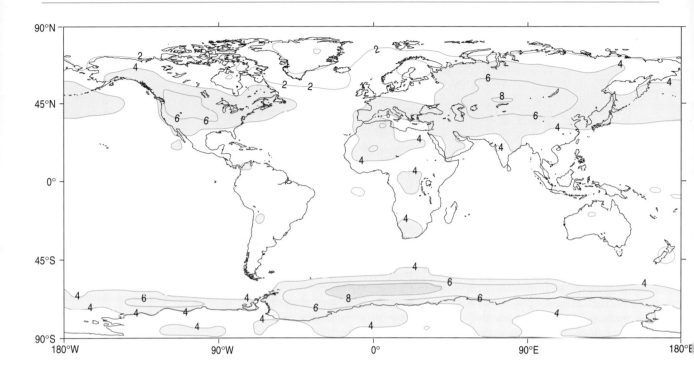

Atlantic and the southern oceans around the Antarctic – is reduced by at least 60%. These are the main areas of deep water formation, but also zones where considerable vertical mixing occurs with deeper waters: both factors tend to slow down atmospheric warming in these vicinities. Secondly, after seventy years of increase the carbon dioxide in the atmosphere has doubled, but the temperature increases are typically about 1.6°C or only two-thirds of the 'equilibrium value'. The missing heating will still occur but the rate of increase has been slowed by the thermal capacity of the oceans. These two changes bring the simulations in closer agreement with observed trends of surface temperatures in recent years. However, there are still inconsistencies:

- The models would predict land masses to warm more quickly than the sea: global trends show less difference than expected.

- Given this, the northern hemisphere should warm more rapidly than the south. This has not happened and, in particular, since the 1950s the southern hemisphere has warmed more quickly (+0.37°C) than the north (+0.14°C).

The potential impact of anthropogenic sulphate aerosols has been recognised for some time, but it was only in 1995 that its effects on the radiation balance were combined with a CGCM by workers in the Hadley Centre at the UK Meteorological Office. The results of one such simulation, using a CGCM with greenhouse gases and sulphate aerosol concentrations to reconstruct the climate changes over the last century, are shown in Plate 3. The changes of surface air temperature over the last century (Plate 3a) are compared with the predictions from the CGCM, with (Plate 3c) and without (Plate 3b) the inclusion of aerosols. When the effect of the sulphate aerosols is added, the model provides a better match of the global distribution of temperature change, and of the regional patterns over

▲   *Figure 3.22*
*The change in surface air temperature due to the doubling of carbon dioxide as forecast by the high resolution UKHI model for the summer months of June, July and August.*

the northern land masses – although the match is by no means perfect. The results help to explain why warming in the northern hemisphere has been less than predicted by earlier models. They also show a better match with the global rate of temperature change as shown in Figure 3.11(a), particularly in later years when the aerosols have had more effect. Plate 4 illustrates the latter point.

This is encouraging: here at last is a model which incorporates all of the major influences on climate change and now appears to reproduce many of the observed trends in the climate.

These results must still be treated with caution; there remain major uncertainties and deficiencies in the understanding of the behaviour of clouds and the interaction of oceans with the atmosphere. 'Biofeedback' (the effect of climate–biosphere interaction) has hardly started to be considered in models. The example presented here is also the first of its kind and will certainly be improved. _Nevertheless, it is developments such as this that has led the IPCC in 1995 to put forward a guarded statement that we are at last beginning to detect, if only in part, the effect of human activities in producing global climate change._ The main effect on the bench-mark global warming figure of including sulphates in the climate model is to widen the temperature range expected to 0.8–4.5°C, to reduce projected sea-level rises (see Section 4.2) and to slow the rate of temperature increases from 0.3°C to 0.2°C per decade. However, there is little cause for complacency: the results also suggest that global warming could accelerate once greenhouse-gas forcing begins to dominate over the effects of sulphate aerosols.

### Activity 3

In Box 3.2 the blanketing effect of clouds was described as providing a positive feedback: reinforcing the greenhouse effect. Following the example given in Figure 3.21 for water vapour feedback, we can put forward a sequence: higher temperature, more evaporation, more cloud formation, more thermal radiation reflected back to Earth, higher temperature. Write down similar sequences for:

(a) clouds acting to produce negative feedback, and

(b) ice as positive feedback.

## 3.7   Summary of Section 3

Current assessments, based on the assumption of no action to combat global warming, put an upper bound on carbon dioxide emissions of about 36 Gt of carbon per year in 2100, and a lower bound of 6 GtC. The implications of these figures are that in the worst case the pre-industrial atmospheric concentration of carbon dioxide could double before 2050, although it is more likely that this doubling will not occur until the end of the next century. The concentrations of many other greenhouse gases are, however, building up rapidly and will add to the effects of carbon dioxide. Scenarios produced by GCMs suggest that increases in concentration of all greenhouse gases equivalent to a doubling of carbon dioxide levels will lead to a mean global temperature rise of between 0.8°C and 4.5°C with a current best estimate of 2.5°C. The projections for carbon dioxide concentration and the speed of global temperature changes are without precedent. GCMs that couple the circulations of the oceans to those of the atmosphere and make allowances for tropospheric aerosols have improved modelling of climate change.

---

*Activity 4*

This activity will help you to consolidate some of the key concepts of this section.

Briefly outline the types of uncertainties associated with:

(a)  projecting carbon dioxide emissions;

(b)  projecting the atmospheric burden of greenhouse gases;

(c)  developing scenarios for climatic futures;

Explain how they are different in nature.

Do you expect these uncertainties to be reduced by more research?

---

# 4   *Living in a warmer world*

---

## 4.1   *Introduction*

There is virtually no aspect of the environment that would be unaffected by a change of a few degrees (either way) in the global average temperature, but our understanding of the mechanisms is in its infancy. While it may be reasonably easy to consider the direct effects of temperature change or a rise in sea-level, the response of natural ecosystems and human communities is inevitably more complex. Their ability to adapt – and, for managed systems, the cost of adaptation – needs to be considered, as well as their sensitivity to rates of change and extremes of climate and the total magnitude of change. Climate change may not be the most severe challenge they face but will often add to existing pressures caused by human actions. One global change which has received media prominence is the potential rise in sea-level which could put many low-lying areas in jeopardy. The risk is real, though some nations will be in a stronger position than others to take preventative action and the actual rise predicted varies considerably between scenarios.

A rise in sea-level will, of course, not be the only, nor indeed the most potentially damaging, consequence of global warming. Changes in both temperature and rainfall are likely to result in spatial shifts in patterns of agriculture and in crop yields; any such large-scale changes in food production systems are bound to have considerable social and economic impacts. Supply of, and demand for, fresh water may also be significantly affected. Natural ecosystems, on the other hand, may be less able to respond to the changing environment than the managed systems such as agriculture. There is not enough space here to explore all the possible problems in detail, and anyway that would not be appropriate since projections and concerns are constantly changing as the models are refined and our ability to forecast climate changes at the regional level is still limited. Instead, the aim of this section will be to look at just a few of the possible consequences of global warming scenarios, highlighting controversial issues and areas of current research. You will then be able to appraise future developments as these are reported.

## 4.2   Changes in sea-level

Sea-levels are normally measured relative to bench-marks on the shore and recorded by tide gauges which are operated over long time-scales. When changes in sea-level are discussed in these contexts, the changes referred to are those of *mean* (i.e. average) sea-level. To obtain a global picture, the averaging must clearly be done using data from many recording stations around the world. It must also be carried out over a reasonably lengthy period of time so as to eliminate short-term changes due to tides or to adverse weather conditions such as on-shore winds or storm surges. The total *volume* of sea-water is the most important factor determining the sea-level observed at any given site over the short term, but there are others. Over longer time-scales, deposition of sediments, local subsidence, and rearrangements of the Earth's crust either along boundaries between lithospheric plates ('tectonic' movement) or as a result of adjustments to changes in ice loading (so-called 'isostatic' movement) will also play a part, by changing the shape of the sea-water container.

   Figure 3.23 summarises data collected between 1900 and 1980. It is scaled, but presents no actual value for the sea-level at any particular date, since what is of most interest is not the absolute sea-level relative to some fixed standard but the *change* in that level with time.

*Activity 5*

(Note: you will find it helpful to attempt this activity, and read the comments on it at the end of the chapter, before continuing with the text.)

As well as changes in the global mean sea-level over the first eighty years of this century, Figure 3.23 also shows changes in the average global surface temperature over the same period.

(a)  Given that at the time these results were first presented (the mid-1980s), the global temperature rise this century was usually quoted as $(0.5 \pm 0.2)°C$, estimate the mean increase in sea-level over the same time-span.

(b)  Can you see a correlation between the two curves and, if so, is it simple or complex?

(c)  If on the basis on your answers to (a) and (b) you were asked to predict the sea-level change that might result from a future global warming of 3°C, what value would you give and on what assumption(s) would it be based?

This activity demonstrates that the construction of scenarios for future sea-level changes is fraught with difficulty. At the heart of the problem is the

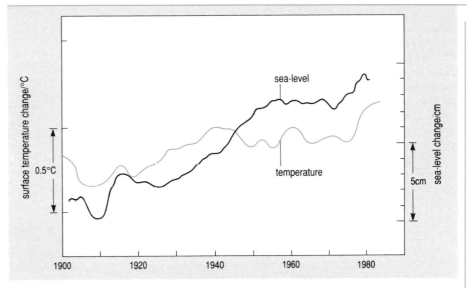

▲   Figure 3.23   Changes in global mean sea-level and average surface
temperature between 1900 and 1980.

fact that the relationship between sea-level and global surface temperature
is not a simple one. As the global temperature changes, different
components of the hydrological cycle react differently with varying
response rates and time lags. To illustrate the complexity of the problem, let
us look at just a few of the processes involved.

Broadly, changes in the total volume of sea-water resulting from
increasing temperature can occur as a result of four different processes:

- thermal expansion of sea-water
- ice melt and run-off from mountain glaciers
- changes in the extent of the ice caps (for example, in Greenland)
- disintegration of the West Antarctic ice-sheet.

Most substances expand on heating, and sea-water is no exception. The
expansion process for salt-water does respond quickly and in an
approximately linear fashion to temperature change. It is thought that
thermal expansion has probably been the major contributing factor so far in
sea-level changes associated with anthropogenic greenhouse warming, and
this helps to justify the linear correlation we used in the activity above.
There are complications, however. For example, the amount of expansion for
a given temperature change depends on salinity. The process is also difficult
to model on a global scale, because it requires knowledge that we do not yet
have on temperature changes and mixing processes in the oceans.

The melting of small glaciers as a result of temperature increase is a
process with a fairly short response time, but for larger glaciers and ice-
sheets response times may be in excess of 100 years. It does seem probable
that the Greenland ice-sheet would significantly decrease in extent if the
global temperature were to rise by several degrees. However, it is also
thought likely that there would then be an increase in the amount of snow
falling in Antarctica. The effect of the melting of ice in the northern polar
regions would thus be counterbalanced by an accumulation of snow and ice
in the southern ones. Current projections of future sea-level rises, taking

into account the first three factors listed above, and based on the IS92 emissions scenarios, suggest a value of between 20 cm and 80 cm by the end of the next century, or 2 cm to 8 cm for each decade. The contributions from the individual processes can be illustrated by using figures for the central IS92a scenario, though it must be emphasised that the uncertainty in all the values is a factor of two either way. Of an estimated total rise of 55 cm by the year 2100, thermal expansion accounts for 35–40 cm, melt from mountain glaciers 15cm and ice-caps only 5 cm with a small net loss from the Greenland ice and gain from Antarctica. There is a possibility that the West Antarctic ice-sheet could become unstable and push enormous numbers of icebergs into the oceans; this could raise the sea-level by as much as 5 metres. Scientists now regard imminent disintegration of the West Antarctic ice-sheet as extremely unlikely; even if such a collapse were eventually to occur, it is estimated that the process would take place over a period of several hundred years. Ice is melting and calving off icebergs at the periphery of the Antarctic peninsula but that is a quite different matter.

Compared to the sea-level rises of many metres that would accompany the collapse of the West Antarctic ice-sheet and the larger fluctuations that have occurred over glacial cycles, the projection of rises of tens of centimetres do not sound too threatening. The risk is very real, though, for people of many countries: the hazards include flooding and threat to life from storm surges, and loss of fresh water as well as loss of land. Currently 50 million people experience flooding from storm surges each year. If the sea-level rose by 50 cm or more this figure would double, even without allowing for population growth. Many of the world's major cities – Bombay, London and Shanghai for example – are built virtually at sea-level and many of the Earth's richest agricultural regions – such as the delta regions of the major rivers in Bangladesh and Egypt – are particularly at risk. Small flat islands might disappear altogether or become too vulnerable to storms for their population to remain. The Maldives, where the highest land is just 2.5 metres, is an often quoted example: a 50 cm sea-level rise

◀ *After storm surges in 1953, the Dutch started to strengthen their sea defences to withstand a 5-metre surge and to allow for a sea-level rise of 25 cm over the next century. The necessary construction work was a massive undertaking, planned on a 30-year programme. It is technically possible to keep the sea at bay, but the work requires long-term planning and is very expensive – probably prohibitively so for the less developed countries.*

would lead to sea-water contamination of half their ground-water supply, while a one-metre rise (at the high end of the IS92 scenarios) would inundate 80% of the land. Continental countries do at least have the option of building sea-defence systems as well as using more passive methods that use tides and winds to encourage stable barriers of sand and silt. As the Dutch have shown, it is technically feasible to keep the sea at bay, but such projects are hugely expensive. Such options are thus unlikely to be within the means of Bangladesh. In 1970 and again in 1991 it suffered devastating storm surges that together killed between 300 000 and 400 000 people. As in other densely populated areas its problems are compounded by high population densities, one result of which is land subsidence caused by the withdrawal of ground-water. The relative sea-level rise may reach 1 metre in 2050 and 2 metres by 2100 (70 cm from global warming and 1.2 m due to land movement and ground-water loss). Seven million of its population live below the one-metre contour, and 30 million below its three-metre contour; in a crowded subsistence-agriculture country the displaced population would have great difficulty relocating and adapting.

The speed at which anthropogenic greenhouse warming occurs is a crucial factor in planning for the consequent sea-level rise, as indeed it is with other effects of increasing global temperature. If the oceans rose too quickly, we would simply have to abandon the inundated land and cities. It has been estimated that a sea-level rise of one metre would threaten about 5 million square kilometres of land – a relatively small proportion (about 3%) of the world's total land area, but (under current climatic conditions) a very substantial proportion (about 30%) of its productive cropland.

## 4.3    *Effects on ecosystems, agriculture and forests*

In considering sea-level changes, we were concerned with two effects of the anthropogenic greenhouse effect, namely global temperature change and alterations to climatic patterns. In thinking about ecosystems, we also have to take into account a third factor – the direct response of plants to increased concentrations of atmospheric carbon dioxide.

The effect of enhanced carbon dioxide levels on plant growth and development (in the absence of any climatic change) has been quite extensively studied in controlled experiments, especially on crop species. In general, increased carbon dioxide concentrations stimulate photosynthesis and decrease transpiration, so that plants grow more quickly and use water more efficiently. Typical experiments in glasshouses suggest that a doubling of carbon dioxide concentration in the atmosphere can lead to increases in yield of between 10 and 50% for crops such as wheat, rice and soya beans, known as C3 plants, with the exact figure depending on the crop strain and growing conditions. However, there are no data on the effects of increased carbon dioxide concentrations for plants in field situations (for example, in large plots covered by glasshouses with artificially controlled atmospheres). Scenarios of the 'carbon dioxide fertilisation effect' are therefore based on a scaling-up from data for individual and isolated plants to whole ecosystems. The dangers of such a procedure are obvious: not every organism will respond in the same way, for example another group, the C4 plants, which include maize, millet and sugar-cane and many pasture grasses, are less responsive to these changes. Some plants may indeed grow faster, but the competition between plant species and the whole complex web of interactions within the ecosystem will change too. Under real conditions, where the availability of nutrients and water will also vary, increases in yield are likely to be much smaller than experiments suggest.

When it comes to the effect of climatic change on ecosystems or agriculture, predictions are scarcely less uncertain. A sensible analysis of the impact of temperature changes on agricultural practice would require a model that could accurately simulate those changes over distances of just a few hundred kilometres. As discussed in Section 3, the GCMs have nowhere near this degree of spatial resolution. Scenarios have been constructed on the 'broad brush' canvases of GCMs, but what they tell us is fairly obvious. If it gets warmer in the temperate latitudes, the growing seasons will be longer and the climatic zone suitable for a particular crop or ecosystem will shift polewards. On this basis it has been suggested that the cereal-growing belts of North America might shift northwards by several hundred kilometres for every centigrade degree rise in temperature. Even so, they may well suffer more heat stress: yields drop when temperatures exceed 35°C for more than a few days in a row, an event which is likely to occur more frequently in the future.

However, temperature is far from being the only climatic variable to determine the distribution of vegetation. The total quantity and seasonal variation of precipitation, air humidity and soil moisture are all vitally important too. Uncertainties in GCM predictions of precipitation patterns are even greater than in their predictions of temperature variation.

While most models produce broadly similar projections of temperature change for a given scenario, there is far less agreement on the patterns of change for precipitation. In general they agree on a global increase in rainfall over land masses: all year at high latitudes and for winter at mid-latitudes with some increase in the tropics and sub-tropics, but here there may be shifts in rainfall zones with some drier zones. But this is only half of the story: evaporation of moisture also increases worldwide, so the all important pattern of soil moisture is less easy to predict and requires another level of modelling. Large discrepancies exist here between models, which appear to be very sensitive to the way they deal with the run-off of surface water (for example, the extent to which snow melt will be absorbed by the soil, or will run off if the ground is frozen, which in turn depends crucially on the temperature predictions). There is nevertheless the likelihood that soil moisture will be reduced over northern mid-latitude land masses in summer, southern Europe for example, while areas already susceptible to drought may become more so. Given the uncertainty we cannot necessarily assume that in a warmer Britain we would be able to grow the crops that currently flourish in the south of France (compare with the *Observer* article in Figure 3.1). As a final cautionary note of warning, many (but not all) GCMs produce enhanced monsoon precipitation, but this may be reversed if the regional radiative forcing is reduced by the effect of aerosols.

As far as diversity of ecosystems and global agricultural productivity are concerned, it is not yet clear whether the overall effect of anthropogenic greenhouse warming is likely to be adverse or beneficial (although for obvious reasons most reports concentrate on the more negative aspects). What *is* clear, however, is that the distribution of growing patterns will shift, with incalculable social and political consequences. Many studies suggest that in aggregate the effects of global climate on world food production are likely to be small, given a certain amount of adaptation from farmers over crop varieties and planting dates. However, they also indicate greater disparities emerging between rich and poor areas. Developed countries may benefit overall, but tropical regions are more likely to suffer negative impacts with dry land agricultural systems in semi-arid areas likely to be particularly at risk.

Once again, it is not only the extent of climate change that matters but the pace. Where rapid changes occur, natural ecosystems may be more vulnerable than agriculture. On the basis of current models for temperate latitudes, we can expect a shift polewards of vegetation zones of nearly two hundred kilometres for each degree rise in global temperature and major changes of vegetation types over at least a third of forested areas (see Chapter 4). Trees are long-lived and take time to reproduce; they cannot easily respond to rapid changes of climate. The fate of forest ecosystems is uncertain, but many will be unable to adjust fully and some species may become extinct. And, of course, the 'migration' of a whole natural community is governed by the pace of its most slowly moving component. If a key species in a food chain gets 'left behind', then all the animals higher up the chain will be affected. In times of change, adaptability pays; this is a principle we need to apply to agricultural research. As was discussed in *Grigg* (1996), many modern strains of crop plants have been chosen to give maximum yields under ideal conditions. It might be wiser now to develop strains that are more adaptable to a wide range of climatic futures and less susceptible to occasional extremes of weather, even if their maximum yields are lower. Far more frequent extreme weather events will be part of the scene in a warmer world; it is to this part of the scenario that we will turn our attention in the next subsection.

## 4.4   Extreme events

Many of the years in the 1980s and 1990s (up to 1995) have been the warmest ever recorded. The period has also seen some remarkable severe weather events – storms, floods and droughts. The wind storm that destroyed 15 million trees in south-east England on 16 October 1987 also hit much of northern France, Belgium and Holland; these areas had not witnessed such a storm for 300 years. The storm surge that devastated Bangladesh in 1991 has already been mentioned, while unusually strong storms have also been seen in the Atlantic, for example Hurricane Gilbert caused severe damage to Jamaica in 1988, and Hurricane Andrew to Florida in 1992. In recent years there have also been record flood levels in many of the world major river systems: Bangladesh in 1988, the Yangtze in China in 1991 and the Missouri and Mississippi rivers in the US in 1993. To cap it all, there have also been unusually long and severe droughts in susceptible areas of Australia and Africa.

Are severe events like this going to increase? Are these the first signs of global warming? Strange as it may seem, it is difficult to answer these questions. We can certainly say that insurance loses due to natural disasters have increased rapidly in recent years, but this may be due to increases in population and affluence. Equally, the intensity of floods and droughts may be increased by poor management of the environment. Modelling studies do not yet agree about increased storminess in a warmer world. Even for tropical cyclones there has been little evidence of an increase in either frequency or intensity as sea-surface temperatures increase. Though there is potential for tropical cyclones to become more intense there are clearly other factors at work. There is more agreement that floods and droughts are likely to become more widespread in a warmer world. While seemingly paradoxical, this follows quite logically from the prediction of higher evaporation rates and more intense patterns of rainfall. Finally, we should expect extremes of heat to become more commonplace – in the words of Stephen Schneider, 'the dice are loaded'. August 1995 in the UK was the hottest recorded, with an average temperature over 3°C above normal, though the year may be remembered as much for the drought that developed during the summer and autumn. During August exceptionally high day-time temperatures – 10°C above normal – were experienced for nearly two weeks, but this may be the average summer weather we bequeath to our grandchildren and their grandchildren.

It may be more productive to look to the oceans for a foretaste of the future. As explained in Chapter 1, they have a large capacity to store moisture and heat, so that patterns of persistent, unusually warm or cool sea-surface temperatures are often associated with anomalies of temperature or rainfall in many parts of the globe. In fact, the pattern of extreme events in recent years has often been associated with the El Niño events in the Pacific Ocean (see also Chapter 1, Section 2). Normally the El Niño/Southern Oscillation (ENSO) switches 'on' for about a year every 3 to 7 years, bringing in its train widespread floods and droughts, as illustrated in Figure 3.24. Since the mid-1970s, however, it has been more 'on' than 'off', and from 1990 to 1994 was locked in the 'on' position. What scientists find interesting is that the pattern of global warming, prominent since the 1970s, looks very similar to the 'signature' observed with El Niño events. Is this the cause of the recent warming, or, conversely, is a more persistent El Niño itself a sign of global warming? At the moment we just do not know, but hopefully improved modelling of the ENSO will help to unravel the truth.

In the Atlantic Ocean a feature called the North Atlantic 'conveyor-belt', is responsible for the mild climate of western Europe. Salt-water at high latitudes cools and sinks to form part of a deep water circulation, helping to maintain the returning flow of warm surface water (see Chapter 1, Section 2). Under greenhouse-induced global warming, precipitation is

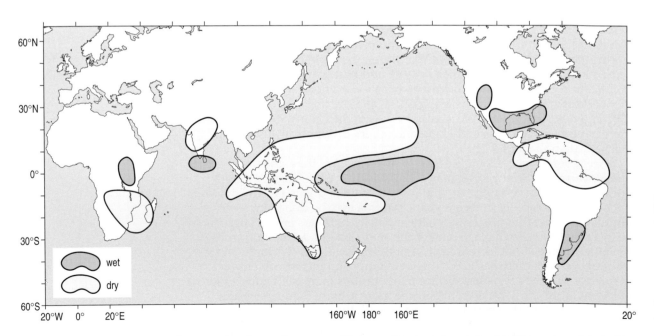

▲ *Figure 3.24   Schematic diagram showing major areas with anomalously wet and dry conditions associated with a typical ENSO event.*

expected to increase globally and in particular at high latitudes. The effect of this influx of fresh water would be to 'dilute' the ocean, making the surface waters less dense – stopping the water sinking and possibly shutting off the conveyor-belt. Several studies have been made of this possibility using models of the ocean and coupled ocean–atmosphere systems which have suggested that the conveyor-belt is very sensitive to quite small increases of fresh water, either slowing down or 'flipping' abruptly into a slower state – rather like a cyclist changing gear as the gradient gets steeper. If the conveyor-belt did slow down significantly or part of its circulation ceased, it would have severe effects on the weather of western Europe and probably the ocean environment itself. There is evidence from the ending of the last glacial cycle, reinforced now by observation of recent changes in the North Atlantic, that the conveyor-belt may have several stable states, but more studies are needed before we can say how close it is to changing gear.

## 4.5   Social, economic and political impacts

With the recognition that human activities are increasingly upsetting the natural balances of our planet, a new concern has been thrust upon the decision-makers. It has been dubbed *environmental security*. Threats to national and international peace and stability no longer come only from weapons of war; environmental changes and decline are poised to become a major source of tension both within and between nations.

Water resources, for example, are coming under increasing pressure from agriculture, which accounts for two-thirds of global consumption, and this situation may be exacerbated by the effects of climate change. Half of the world's water basins (by area) are shared by two or more countries and worsening shortages may become the focus for dispute or conflict. To take another example, the climate-induced changes in the distribution of food production, mentioned earlier, may well lead to an increase in food prices: here, as is so often the case, developing countries might suffer the most. Crop failure, whether due to flooding, drought or extreme events, in a region where hunger is already an endemic problem could cause large-scale migrations. Subsistence-level production offers little scope for adaptation and is therefore particularly vulnerable to external events. In the worst scenario, many millions of people become 'environmental refugees'.

Projections of global warming into the next century suggest that the implications of climatic change will be far-reaching indeed for national economies and international relations. Patterns of agriculture, industrial production, water resource and usage, fuel and electricity demands will all be affected. However, there will be winners as well as losers. The truly difficult questions are those of equity: not only who wins and who loses, but how those who suffer the most severe losses might be compensated. The issue is, of course, greatly complicated by the fact that it spans national borders. The activities of one country in contributing to the build-up of greenhouse gases can be seen to cause significant damage in another, quite remote, country that may have added far less to the atmospheric burden. The detrimental effect on international relations of 'exported pollution' was first highlighted by the problem of acid rain. Beside the global nature of the anthropogenic greenhouse effect, we may yet come to view the consequences of acid rain as a little local difficulty.

Finally, human health may be at risk, not only from the effects of natural disasters but also an increased incidence and spread of vector-borne diseases, such as malaria and river-blindness, mostly at the margins of currently affected areas. There is already evidence of this starting to happen, and estimates of a 20% increase in malaria would lead to an extra 50–80 million cases. While the populations of the tropics and sub-tropics would be most at risk, those in temperate regions may also be affected. The latter would also suffer most from an increase in severe heat events, particularly where the population is not acclimatised to heat. The short, but intense, heat wave in the Midwest of the US (12–15 July 1995) produced record temperatures accompanied by high levels of humidity. The Mayor's Commission of the City of Chicago has estimated that this short event resulted in the death of about 700 people.

## 4.6    Summary of Section 4

Model projections of future sea-level rises for a range of possible scenarios suggest a figure of between 20 and 80 cm, by the end of the next century, for an average global temperature increase of 2.5°C, but the slow response of the oceans means that sea-level rises will continue for centuries thereafter. A doubling of the atmospheric concentration of carbon dioxide would certainly have major impacts on ecosystems and agricultural practice. Further research into the response of plants to the enhanced levels of carbon dioxide, and GCMs capable of producing scenarios on a regional rather than global scale, are required before these impacts can be predicted in any detail, although it is clear that natural ecosystems will be less capable of adapting than managed systems. It is also difficult to provide reliable projections of regional changes in precipitation and the incidence of severe events. We mainly know that there will be changes, but the current uncertainty only adds to the difficulty of mitigating the effects. It is likely that the effects of climate change will hit hardest many of the poorest nations and populations, and while global warming may not always be their most pressing issue it will undoubtedly add to existing human pressure on their ecosystems, and social and economic resources.

*Activity 6*

The following questions may be used to revise some of the key concepts of this section.

Q1   A policy often suggested as a counter-measure to anthropogenic greenhouse warming is afforestation. Explain very briefly why afforestation helps to control the build-up of carbon dioxide, but cannot provide a long-term solution.

Q2   Discuss briefly the validity of attributing individual extreme events (such as the 1987 'hurricane' over southern England or the 1989 drought in the US Midwest) to the effects of anthropogenic global warming.

# 5   Meeting the challenge

## 5.1   Introduction

As realisation grew through the 1980s of the tremendous consequences of
atmospheric pollution, it also became apparent that completely new types
of global co-operation and policies would have to be developed to cope
with the problem. Initially, the debate focused on the uncertainties in the
projections and scientific models. It was possible simply to point to the
need for more research and to postpone any other responses while the
research was underway. Politicians were naturally cautious about
committing themselves to costly legislation or capital investments as
insurance against potential future changes the nature of which still seemed
very uncertain. As the scientists moved towards a greater degree of
consensus, however, the 'wait-and-see' response became less tenable. After
all, the whole point about insurance is that you have to have paid the
premiums before disaster strikes; no-one issues insurance retrospectively.
The political debate is now becoming more heated than the scientific one.

Many different responses have been suggested to counter the threat of
global warming, but broadly they fall into three categories:

- counter-measures – the so-called 'technical fixes'

- adaptation – adjustment to environmental changes as they occur,
without too much investment being made to anticipate or prevent such
changes

- preventative action – legislation or incentives to reduce the emission of
greenhouse gases from human activity.

*Technical fixes* have been suggested seriously by only a very small minority
of the scientists researching in this area. Proponents of such schemes argue
that since we have radically altered the balance of certain natural processes,
perhaps we should now use technological means to redress that balance by
deliberately engineering climate modifications.

One of the few suggestions that has claimed serious attention is the
attempt to enhance the uptake of $CO_2$ by phytoplankton in the oceans. At
present there is slow transfer of a portion of the carbon fixed by
photosynthesis – a steady 'rain' of organic particles known as the biological
pump – to the deep ocean, where it is effectively removed from the carbon
cycles. In most parts of the oceans primary productivity is limited by the
availability of common nutrients – phosphates and nitrates – but there are
other areas, for example the southern oceans and north Pacific, where these
nutrients are in excess and the limiting fact appears to be the micro-nutrient
iron. The proposal, then, is to use iron compounds as a 'fertiliser' to
enhance photosynthesis. While the idea of providing a quick fix to global
warming with a super-tanker full of iron filings is appealing, modelling
studies of 'iron fertilisation' of these oceans suggest that the net impact will
be slight. Small-scale experiments of seeding the ocean with iron have been
carried out – with mixed results. In fact it is likely that inadvertent
fertilisation of both land and marine biota, by nitrate deposition and carbon
dioxide, has had a much greater effect. An additional consideration is that
tampering with a complex ecosystem may well produce unexpected results

or undesirable side-effects. Fortunately there are other ways to combat global warming.

*Adaptation* is the strategy favoured by many economists. Their position is that it is unnecessarily expensive to anticipate climatic changes that may not in the end occur. Passive adaptation – adjusting to environmental change as it takes place – is cheaper, at least for the developed countries. Infrastructures such as coastal defences, or water reservoirs and distribution systems, are constantly being maintained or replaced anyway. It is not a major additional expense to modify existing refurbishment plans so as to cope with climatic change on an on-going basis. More active types of adaptive strategies are built on the recognition of the likelihood of future environmental change, without commitment to particular predictions or scenarios. One example of such a strategy would be the development and more widespread planting of new crop strains that would be less susceptible than current ones to extremes of weather. However, although managed ecosystems such as agriculture can respond well to adaptation strategies, natural ecosystems cannot. Adaptation measures have their place, but should not form the sole strategy because the most vulnerable are left least protected: rich countries can afford expensive coastal protection, whereas the very survival of poor coastal communities may be at risk.

Considerations like these have led the international community to declare for a strategy of *preventative action*. Following on from the Second World Climate Conference in 1990, negotiations led to the UN Framework Convention on Climate Change being signed in June 1992 at the UN Conference on Environment and Development at Rio.

---

*Box 3.3   Umbrella and framework conventions*

Two approaches have been used to establish effective international action on environmental problems. The first is to negotiate an *umbrella convention*, of which the Law of the Sea is an example. This approach aims to establish a comprehensive regulatory structure together with clearly defined responsibilities for the major areas covered by the treaty. An umbrella convention for climate change, for example, would have reached agreements on limiting emissions of greenhouse gases, on distributing the burden amongst nations and providing a methodology for monitoring the agreements. However, negotiations for this kind of agreement can easily get bogged down in disputes and delay, particularly where there is both technical complexity and a wide range of economic interests affected.

The alternative approach is the *framework convention*, exemplified by agreements to limit stratospheric ozone depletion. In this approach an initial convention establishes the basic objectives and principles for the international regime, together with a framework for regularly reviewing and negotiating obligations, which on agreement are written into protocols. The ozone agreements started formally with the 1985 Vienna Convention – discussed in Chapter 2 – which established the importance of protecting the ozone layer and made the case for limiting halocarbons, particularly CFCs, but had no firm targets for controls on their production or emission. These were developed later by the Montreal Protocol and subsequent amendments arising from the regular review of scientific understanding and policy responses. This approach allows time for expertise to develop and consensus to build; it also encourages national flexibility in developing detailed proposals to meet the agreed general objectives. It is, of course, the route chosen by the Climate Change Convention.

In fact the Framework Convention on Climate Change signed at Rio in 1992 disappointed many who considered the need for immediate action urgent and were concerned that the most important decisions were left for the future. But if there is not sufficient certainty in the projections of greenhouse gas emissions and the feedbacks in their life-cycles or detailed understanding of their impacts, it may not be possible to assess either the seriousness of the impacts and prevention costs or the most effective long-term course to take.

## 5.2   The United Nations Framework Convention on Climate Change

The UNFCCC came into force in March 1994, becoming international law for the ratifying 'Parties to the Convention', i.e. the nations endorsing the Convention. The main objectives and principles are set out in Articles 2 and 3 of the Framework Convention, while Article 4 deals with a structure for reviewing and negotiating future commitments.

In its preamble the Convention recognises that human activities will alter the Earth's climate, which in turn 'may adversely affect natural ecosystems and humankind'. The ultimate objective of the Convention is stated in its Article 2 and is well worth restating here:

> To achieve … stabilisation of greenhouse gas concentrations in the atmosphere at a level that would prevent dangerous anthropogenic interference with the climate system. Such a level should be achieved within a time-frame sufficient to allow ecosystems to adapt naturally to climate change and to ensure that food production is not threatened and to enable economic development to proceed in a sustainable manner.

The key issues are what constitutes 'dangerous' concentrations of greenhouse gases, and what rate of accumulation should be permitted. These are difficult policy judgements on which the parties to the Convention will have to reach agreement (difficult because of uncertainties, and conflicting economic interests). Scientists from the working groups have several tasks to perform: to reduce uncertainty by improving the modelling of the greenhouse gas life-cycles and their impact on global and regional climates; to advise on the likely impacts of any particular scenario; to explore the implications for emission patterns of meeting the stabilisation objective of Article 2; and to carry out the review process.

Article 3 sets out the guiding principles for achieving the objectives of the Convention, some of the key points of which are summarised here.

* Consideration should be given to the special needs of developing countries, particularly those vulnerable to climate change, for example countries with low-lying areas at risk from sea-level rise and semi-arid countries.

* Developed countries, responsible for most of the historical burden of greenhouse gases and having the economic resources, should take the lead in combating climate change.

* All parties have a right to sustainable development, but should also promote a supportive and open international economy. (Note: there may be some conflict of interest between these two principles.)

* Finally, the Convention embraces the precautionary principle: the need to act when threats are serious before complete scientific understanding is available. It also exhorts the parties to act comprehensively and co-operatively in their policies and to adopt measures to prevent the causes of climate change and minimise their effects.

Following tough negotiations over Article 4 in the lead-up to Rio, no specific commitments to reduce or limit greenhouse gas emissions survived, although the injunction to developed countries to return 'individually or jointly' to their 1990 levels of emissions by the year 2000, remained, if in non-binding form. However, the Convention does allow the

Conference of Parties to review the adequacy of their commitments, including the one above. This review process is a key feature of the framework approach. The Convention allows nations the freedom to decide, within broad limits, what measures to adopt in order to meet their obligations, but they are required to draw up national inventories of greenhouse gas sources and sinks, according to agreed methodologies, and to update them regularly. They are also required to communicate the results of both sets of actions to the Conference of Parties: in effect these measures are subject to international review.

## 5.3   The stabilisation of greenhouse gases

As we have noted, the main objective of the Climate Change Convention is to achieve stable levels of greenhouse gas concentrations in the atmosphere at a rate that does not threaten natural and social systems. An initial point to make is that this is not achieved by freezing emissions at current levels, at least not for several hundred years. This is illustrated in Figures 3.16 and 3.17 by the IS92d scenario which is a reasonable approximation to a constant emissions scenario. The IPCC Working Group I, dealing with scientific assessment, has looked at the implications for emission patterns of stabilising carbon dioxide at a range of illustrative levels between 350 and 750ppm by volume, shown in Figure 3.25, a methodology which can be extended to other greenhouse gases.*

*Note that by starting with scenarios for concentrations of greenhouse gases and working back to emissions that cause them, we are reversing the order in which we have so far analysed scenarios.

Q   Can you recall, by way of comparison, the pre-industrial concentration of carbon dioxide in the atmosphere and its current level?

A   The pre-industrial concentration was 280 ppm by volume and the current level, quoted for 1994, is approximately 356 ppm by volume.

Figure 3.26 shows the patterns of anthropogenic carbon dioxide emissions that give rise to the concentration profiles of Figure 3.25. Most scenarios show emissions continuing to rise for several decades, reflecting the time taken for society and the carbon cycle to respond, before falling sharply at a later stage. These patterns of emission are not unique; to a first approximation the increase in atmospheric concentrations is related to the total emissions accumulated over the time-period: thus a larger rise now

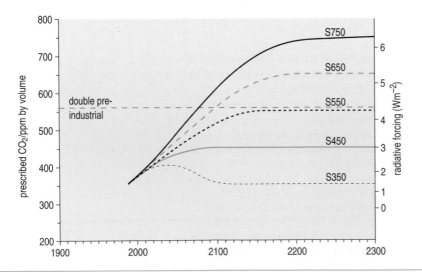

◀   Figure 3.25
Profiles of atmospheric $CO_2$ concentration leading to stabilisation at 350, 450, 550, 650 and 750 ppm by volume. Doubled pre-industrial $CO_2$ concentration is 560 ppm by volume. The radiative forcing resulting from the increase in $CO_2$ relative to pre-industrial levels is marked on the right-hand axis. Note the non-linear nature of the relationship between $CO_2$ concentration change and radiative forcing.

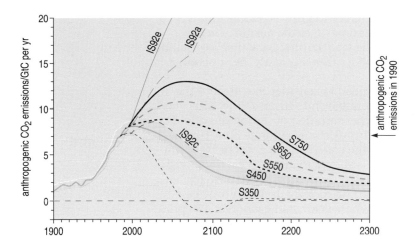

◀   Figure 3.26
*Illustrative patterns of
anthropogenic emissions of
$CO_2$ (in gigatonnes per year)
leading to stabilisation at
concentrations of 350, 450,
550, 650 and 750 ppm by
volume, following the
profiles shown in Figure
3.25. The range of results
from different models is
indicated on the 450 ppm by
volume profile. The
emissions for the IS92a, c
and e scenarios are also
shown on the figure.*

means steeper cut-backs in the future. This is a burden we will be placing
on future generations, for, as Figure 3.25 shows, stabilisation is not reached
in these scenarios for 150 to 250 years.

Q   Why do you think it takes carbon dioxide so long to reach stabilisation?

A   Because of its long atmospheric 'lifetime' (50–200 years), carbon dioxide
persists in the atmosphere. It takes many decades, centuries even, for a
balance to be reached between the carbon dioxide emitted into the
atmosphere and its absorption by 'sinks' such as the ocean and
biomass.

Two important results arise from the models presented here. Firstly, to
achieve stabilisation of carbon dioxide concentrations at or below 550 ppm by
volume – close to the bench-mark figure of 560 ppm by volume for 'doubled'
carbon dioxide concentration – emissions must be lower than any of the IS92
scenarios, other than IS92c. Secondly, for all the stabilisation scenarios
considered, emissions must ultimately fall well below current levels, in most
cases to less than half current levels. (Over the next 100 years deforestation is
expected to cease – either through better land management or lack of further
forests to exploit – so emissions refer mostly to those from burning fossil
fuels.) Of course, there are uncertainties with these figures that increase with
lengthening time-scales, and a better understanding of the carbon cycle and
the effects of climate feedbacks and carbon dioxide fertilisation may lead to
adjustments. Nevertheless, it is unlikely to change the overall message that
we will be required to reduce carbon dioxide emissions drastically in the
future – and the sooner we start the easier it will be.

Two further factors serve to emphasise this point. Thus far we have
considered anthropogenic carbon dioxide emissions as a global total
without looking at how they shared out between nations. The current mean
(average) *per capita* emission per year is approximately 1.1 tonnes of carbon
(1.1 tC), with developed countries averaging 2.8 tC per person (a range of
1.5 tC to 5.5 tC) and developing countries averaging 0.5 tC. It is possible to
make a rough estimate* of the per capita emissions per year, between 1990
and 2100, that will lead to stabilisation at 550 ppm by volume. It turns out
to be not much below the average 1990 per capita levels, but note the steep
reduction in emissions after this time (see Figure 3.26). But these scenarios
assume, in accordance with the equity principle in Article 3, that

*By taking the accumulated
emissions from this scenario for
1990–2100, approximately 1000
gigatonnes of carbon (1000 GtC),
and the median UN population
estimates for this period, 9 billion
say. The yearly emissions are simply
1000 GtC divided by 110 years,
giving 9 GtC per year. As 'giga'
means $10^9$, the same as a billion, the
per capita emission per year works
out simply at about *one tonne*.

developing countries will be given the chance to grow economically with a possible doubling of *their* per capita emissions over the next fifty years. Developed countries will have to take the lead, cutting back their emissions by at least 50% to move closer to the global mean, i.e. to 1tC per person, a much less palatable message.

Secondly, to be comprehensive, we need to consider the effect of the other greenhouse gases, in particular methane and nitrous oxide. This can be expressed in terms of an equivalent concentration of carbon dioxide. For example, if emissions of methane and nitrous oxide continue on a 'business as usual' basis until 2050, they will add the equivalent of another 100 ppm by volume of carbon dioxide to the levels shown in Figure 3.25. If we wish to stabilise the equivalent carbon dioxide concentration at 550 ppm by volume (close to the doubling figure), carbon dioxide itself can only be allowed to rise to 450 ppm by volume, whereas for stabilisation at 450 ppm by volume, carbon dioxide concentrations cannot exceed *current* levels. Clearly, if there is to be any hope of stabilising greenhouse gases at levels below the 'doubling of carbon dioxide', with all that this implies for climate change, emissions of all the greenhouse gases need to be controlled.

## 5.4   *Options for action*

In this last section we look briefly at some of the immediate actions that countries can take towards meeting the objective of the Climate Change Convention. We take as read the responsibility of developed countries to play a leading part in these actions and to support developing countries, a theme common to the Rio agreements which are explored further in Chapters 5 and 6.

Some actions are already under way. As we saw in Chapter 2, CFCs and the halocarbons known to deplete stratospheric ozone – most of which are powerful greenhouse gases – are controlled by the Montreal Protocol and its subsequent amendments. These agreements are likely to lead to atmospheric concentrations stabilising before the end of the century and then declining slowly over the following fifty years – meeting the stabilisation requirement of the Climate Change Convention. However, the use of CFC substitutes – HCFCs and HFCs, also greenhouse gases – will need to be monitored. Concentrations of HCFCs, for example, are currently rising rapidly, a trend which may continue for some time; HFCs are little used at the moment but, as they do not significantly deplete ozone, they are not covered by the Montreal Protocol.

Action under the Climate Change Convention to reduce or steady greenhouse gas emissions *in the short term* – the agreement by developed countries at Rio to return to 1990 levels by 2000 – has already been noted. Some countries are showing signs of wanting to back away from even this commitment. Others, by encouraging energy conservation measures and switching from coal and oil as fuels to natural gas, which generates 30% to 40% less carbon dioxide for the same amount of energy, are likely to be able to meet it without too much difficulty. What of the longer term? Germany and Denmark have made commitments to reduce their 1990 levels of greenhouse gas emissions by 20% to 25% by the year 2005, without spelling out the details, but most other countries have made no individual commitments beyond the year 2000. This was one of the key points of discussion at the first Conference of Parties in Berlin, in March 1995. While there was a consensus that some cuts would be needed after 2000, no agreement was reached at the meeting on appropriate levels, partly due to

lack of sufficient scientific understanding of key processes. Further studies
were thought necessary to determine how much was needed, and the
discussion is set to resume at the next Conference of Parties in 1997.

The difficulties expressed already by some countries point to one
pragmatic principle that is likely to guide much immediate action: cost-
effectiveness. Some actions to reduce greenhouse gas emissions – many
energy-efficiency and conservation measures, for example – may cost little
or nothing on balance to implement and may even lead to overall savings.
Such actions can be listed in merit order and all those which show a net
gain or no loss should be encouraged. However, incentives may often be
needed to encourage these activities as they will not necessarily take place
of their own accord. There are many ways in which countries can take
immediate and effective action to meet the requirements of the Convention,
at little cost. They can be grouped under three main headings which we
will consider briefly in turn: energy-saving and conservation, improved
land management (including forests), and reduction of methane emissions.

## Energy use

This topic has been covered comprehensively in *Reddish* (1996). Here we
summarise the opportunities for reducing greenhouse gases either by
switching to new forms of supply or by reducing the demand for energy.
In the medium term there are plenty of opportunities for *suppliers* of energy
to reduce greenhouse gas emissions by converting to more efficient plant:
most commercial energy systems will be replaced at least twice by 2100.
At the moment the average efficiency when burning fossil fuels is 30%
compared with the 55% efficiency now possible with *combined cycle gas
turbine generation (CCGT)*, and even higher values available from combined
heat and power systems where the 'wasted' heat energy is used. Switching
to less carbon-intensive fossil fuels has already been mentioned as an
option (for example, if natural gas is substituted for coal in power stations
and for oil in the transportation sector). In the UK power sector a change
from coal to gas is already under way, spurred on by the relatively low

◄   *Indra Prastha coal-
burning power station on
River Yamuna, Delhi, India.
Reliance on fossil fuels,
with increasing demand for
energy in the third world, will
add to the risk of global
warming.*

capital cost of CCGT – the 'dash for gas' – though the change has had high social costs. Its impact on UK greenhouse gas emissions is significant and forms the main basis of the UK government's strategy for meeting the Convention commitment to return to 1990 levels by 2000.

However, using fossil fuels with greater efficiency is not a long-term solution as sooner or later demand will outstrip any efficiency gain. Ultimately, there must be a significant move to more renewable forms of energy. Currently renewables make up just over one-sixth of primary energy production: 10% from biomass, 5% from large-scale hydroelectric projects and only 2% from the newer generation of renewables (solar, wind, geothermal, small-scale hydro, and biomass conversion). The technology of these newer forms is still developing and, although costs continue to fall, their rate of introduction will depend on incentives for use and support for research. Nuclear energy is another option used in several countries, such as France, for electricity production, but because of well-documented safety concerns and high capital costs, it is not an attractive proposition for private enterprise, and further expansion seems likely to be limited.

There is also plenty of scope for reducing energy demand, whether in the home or by industry. It is likely that efficiency gains of 10–30% could be made in most countries at little or no net cost, and of 50% or greater in developing countries, given the support of funding and technology transfer. Industry accounts for one-third of greenhouse gas emissions through energy use and basic industrial processes such as food, chemicals, building materials and iron and steel manufacture. These sectors have improved energy efficiency substantially since the oil price rises of 1974; even so there is potential for further short-term energy savings of about one-quarter. There are still wide variations in energy efficiency of many manufacturing processes in developed nations, a factor of two, for example, for the production of steel. Energy use within buildings – commercial as well as domestic – accounts for another one-third of greenhouse gas emissions. Energy demand from this sector is expected to grow by 2% a year under the business as usual scenarios as more appliances are used and higher standards of heating and lighting are demanded. Again there is large scope for increasing the efficiency of space heating, improving insulation and using low-energy lighting, many of which would provide net savings, and of moving to more efficient appliances. Average new models of cookers, washing-machines and refrigerators currently on sale in the UK, for example, are respectively two, two and a half, and five times less efficient than the best available models worldwide.

The final sector, transportation, accounts for one-quarter of primary energy use and is also the sector with the most rapid growth in greenhouse gas emissions. This sector typifies in many ways the problems faced by policy-makers. Currently, 75% of transport energy usage is in developed countries, but most future growth will come from developing countries. There is plenty of room for efficiency gains through design changes that would not impair performance, ranging from an estimated 30% for heavy goods vehicles, 40% for aircraft to 50% for cars. The sector has been resistant to attempts to manage it in spite of the growing evidence of its disbenefits, but ultimately much of it will have to move to electrical and hydrogen power based on renewable energy – combined with a change in land-use and travel patterns and a move to greater use of public transport systems. Changes may well come in attempts to solve other problems such as congestion, pollution and accidents. They will certainly bring benefits by reducing these and also indirectly by reducing nitrogen oxides and volatile organic carbon compounds, precursors of another major pollutant and greenhouse gas – tropospheric ozone.

## Land management

Better management of the agricultural and forestry sectors can lead to both increasing carbon storage and reduction of emissions of carbon dioxide, methane and nitrous oxide. Options that focus on the more sustainable use of these sectors can produce other positive effects such as conservation of biodiversity, slowing soil degradation and reducing air and water pollution. Current estimates indicate that the total amount of carbon that could be sequestered* in these two sectors over the next fifty years is between 90 and 150 GtC. If we take the IS92a scenario for the same period to provide a comparison, this is equivalent to about 15–25% of all carbon dioxide emissions. About two-thirds of the carbon sequestration can come from more sustainable management of the forestry sector, with tropical forests showing the greatest potential, accounting for 80% of the sub-total. Over the decade of the 1980s, tropical forest cover has been destroyed at a rate of 1% a year (from an estimated 18.9 million $km^2$ in 1980 to 17.2 million $km^2$ in 1990). Much of the change can be achieved by slowing deforestation and promoting natural forest regeneration with increased use of forest biomass as bio-fuels and long-lived wood products to substitute for fossil-fuel products. In the agriculture sector a variety of practices could increase carbon storage by 30–60 GtC: for example, improved care of rangelands, return of crop residues to the soil and restoration of degraded agricultural land. Emissions of greenhouse gases from the agricultural sector now contribute about 20% of the anthropogenic total, and account for 50% of methane and 70% of nitrous oxide emissions.

## Reduction of nitrous oxide and methane emissions

Sources of nitrous oxide in agriculture are mineral fertilisers, legume cropping, animal waste and biomass burning. Anthropogenic emissions could be reduced by 10% through using techniques already available for improving the efficiency of fertiliser and manure use.

After carbon dioxide, methane is the most significant greenhouse gas: stabilising its atmospheric concentration would make a small but important contribution to the Convention objective. Because of its relatively short residence time in the atmosphere (compared with nitrous oxide or carbon dioxide), only a small reduction of emissions – 10% – is needed to stabilise the gas. As we have seen, agriculture is an important source of methane emissions, with enteric fermentation in ruminant animals and rice cultivation being major sources. Although these emissions cannot be prevented, they can be reduced by altering the nutrition of ruminants and irrigation patterns of rice cultivation. Together with altered treatment of animal wastes and reduction of biomass burning, reductions of 10–20% of anthropogenic emissions are possible.

There are three other sources of methane emissions arising from human activity which could be reduced at little or no cost and probably more easily than in the agricultural sector. Firstly, methane emission from biomass burning could be halved if biomass burning was substantially curbed. Secondly, methane could be recovered from landfill sites and other waste material to generate energy. Thirdly, leakage from natural gas pipelines could be reduced significantly at little or no net cost. Together these three actions could provide a further reduction of methane emissions of 15%, which, added to the earlier cuts, give a total of nearly one-third – equivalent to a 10% cut in current carbon dioxide levels.

*Over the long term the carbon sequestered in virgin forests or sustainably managed forests and agriculture will be roughly in balance with that released to the atmosphere by burning or decomposition; they will be $CO_2$ neutral. The process of sequestration described here is achieved by a partial reversal of the biomass destruction that has taken place over centuries in the northern hemisphere and in recent decades in the tropical rainforests. It can only occur as long as carbon fixation from growth exceeds that lost by decomposition or cropping.

## 5.4   Conclusion

In this section we have looked at some of the ways open to countries and industries to reduce greenhouse gas emissions. Most of the options are available at little or no net cost, and could be implemented over reasonably short time-scales. However, because they would require a change of attitude about the use of resources, they are unlikely to occur without pressure from government and environmental organisations. Of course, many developing countries may have more pressing problems to cope with, but in many cases a sustainable approach to development can help rather than hinder them. It is likely, then, that Article 2 of the Climate Convention can be met with little 'pain' in the short term, given the political will.

*Activity 7*

Some experts have said that we cannot hope to control the anthropogenic emissions of greenhouse gases until we stop the growth in world population. Outline some of the pros and cons of this argument.

## 6   Conclusion to Chapters 2 and 3

In the last twenty to thirty years our understanding of the atmosphere and the climate has greatly increased as satellites have provided global collection of data and ever more powerful computers have allowed more sophisticated modelling of processes and scenarios. But increased knowledge has brought with it a radical change of perspective. In the 1970s there was often more concern about the depletion of physical resources than the effects of the waste products from using them, usually seen as local, hazardous pollutants. Certainly few scientists would have placed low-level emissions of *non-toxic, unreactive* chemicals high on the agenda, or suspected that they could have such devastating *global* effects in such short time. Human-induced climate change was a pipe-dream rather than a nightmare: in the 1950s and 1960s much research effort and money was sunk into attempts to 'seed' clouds with fine particles to produce rainfall – with very little success. Yet while these small-scale experiments were going on, we were inadvertently seeding clouds with aerosols on a global scale and producing global climate changes which we are only now beginning to understand.

In Chapters 2 and 3 we have looked at two major global threats arising from the inadvertent release of seemingly harmless gases. We now realise that not only can we alter the atmosphere and climate on a global scale and with unprecedented speed, but we have already done so inadvertently. And in both cases what makes the trace gases harmless – their chemical stability – also makes their threat long-lived, like time-bombs ticking away. The ozone layer – which with luck will be defused in time – will remain damaged for at least fifty years, but global warming is a greater threat burning on a slower fuse and its effects will be with us for much longer.

The response from the international community has also been surprising. In ten short years the Vienna Convention has moved from an unpromising start to successful control of almost all ozone-depleting substances. Many problems remain, but few of those starting down that road would have been confident of the level of international collaboration now achieved. The framework approach of the Vienna Convention has provided the model for the Climate Change Convention and although it may be premature to make a judgement on the programme to date, the problems to be solved are certainly far greater: greenhouse gases are the result of our dependence on fossil fuels (and of poor land management); the long-term solution is to use our land and energy resources in a sustainable fashion. Already we may have altered our climate irrevocably and must learn to live in a world of new responsibilities. The air and water are not infinite, natural resources, but very finite systems under our own erratic management. We must learn to use this stewardship wisely before it is too late.

## References

GRIGG, D. (1996) 'World agriculture: productivity and sustainability', Ch. 2 in Sarre, P. and Blunden, J. (eds) *Environment, Population and Development*, London, Hodder and Stoughton/The Open University (second edition) (Book Two of this series).

REDDISH, A. (1996) 'Energy resources', Ch. 1 in Blunden, J. and Reddish, A. (eds) *Energy, Resources and Environment*, London, Hodder and Stoughton/The Open University (second edition) (Book Three of this series).

SARRE, P. and REDDISH, A. (eds) (1996) *Environment and Society*, London, Hodder and Stoughton/The Open University (second edition) (Book One of this series).

SCHNEIDER, S. (1989) 'The changing climate', *Scientific American*, September.

SILVERTOWN, J. (1996) 'Earth as an environment for life', Ch. 5 in Sarre, P. and Reddish, A. (eds).

## Further reading

The Intergovernmental Panel on Climate Change has provided two scientific assessments with interim supplementary reports. These provide the most authoritative accounts available of the state of knowledge on climate change, with very detailed summaries.

HOUGHTON, J. T., JENKINS, G. J. and EPHRAUMS, J. J. (1990) *Climate Change: the IPCC scientific assessment*, Cambridge, Cambridge University Press.

HOUGHTON, J. T., CALLANDER, B. A. and VARNEY, S. K. (1992) *Climate Change 1992: the supplementary report to the IPCC scientific assessment*, Cambridge, Cambridge University Press.

HOUGHTON, J. T. et al. (1995) *Climate Change 1994: radiative forcing and an evaluation of the IPCC IS92 emission scenarios*, Cambridge, Cambridge University Press/Intergovernmental Panel on Climate Change.

The second IPCC Scientific Assessment, 1995, is published in three volumes (by Cambridge University Press, 1996):
Vol. 1: *The Science of Climate Change*
Vol. 2: *The Impact of Climate Change*
Vol. 3: *Economic and Social Dimensions of Climate Change*.

An authoritative, but readable book has been produced by one of the co-chairs of the three IPCC Scientific Assessment Working Groups, Sir John Houghton:

HOUGHTON, J. T. (1995) *Global Warming: the complete briefing*, Oxford, Lion Press.

A more detailed and lively discussion of the science and the issues is to be found in one of the books for the Open University course S280 *Science Matters*:

WARR, K. K. and SMITH, S. (1993) *Changing Climate*, Milton Keynes, The Open University.

Regular and authoritative updates in the global warming story will appear in popular science journals such as *New Scientist* and *Scientific American*.

# Answers to Activities

## Activity 1

Red light is at the long wavelength (low frequency) end of the visible spectrum. White light contains all the visible wavelengths. 'White hot' objects therefore emit light of shorter average wavelength than cooler 'red hot' ones. Generalising from this, as the temperature of an object decreases, the average wavelength of the radiation it emits will increase.

## Activity 2

*Q1* The amount of radiation incident on the upper surface of the atmosphere is exactly the same as the amount reflected and re-emitted into space from the Earth–atmosphere system as a whole. The radiation budget is balanced.

*Q2* The troposphere is warmed by long-wave radiation emitted from the Earth's surface and successively absorbed and re-radiated as it works upwards. The warmest part of the troposphere is therefore that nearest the ground and temperature decreases with increasing height. In the stratosphere, the warming mechanism involves absorption of incoming solar radiation and temperature therefore increases with increasing height. In effect the two regions are mainly warmed by radiation coming from opposite directions.

*Q3* In addition to the effects of anthropogenic greenhouse gases, variations in the Sun's output may either increase or decrease the radiation reaching the Earth, causing either heating or cooling, but probably not by a significant amount. Volcanic eruptions that put large amounts of sulphate aerosol into the stratosphere can reduce the radiation reaching the Earth significantly, and produce a cooling effect, but the effect is temporary – a few years. You will see in the next section that sulphate aerosols in the lower atmosphere, caused by human activity, can have a similar effect.

Finally, it can be argued that the (very major) role of water vapour serves to amplify the effects of *other* greenhouse gases.

## Activity 3

The sequences are:

(a) Higher temperature, more evaporation, more clouds, more reflection of short-wave radiation, leading to cooler temperatures. Negative feedback tends to oppose the original trend.

(b)  Higher temperature, more ice melts, less reflection of short-wave radiation, leading to higher temperatures. Positive feedback tends to reinforce the original trend.

*Activity 4*

(a)  The projection of carbon dioxide emissions is essentially a problem that needs to be tackled by social scientists. The uncertainties are mainly associated with difficulties in forecasting demographic, economic, political and social factors. They are not of a type which would be significantly reducible.

(b)  The projection of atmospheric concentrations of carbon dioxide – one of the greenhouse gases – is essentially a problem for physical scientists. However, they start with input data for emission rates that are themselves uncertain, as discussed in (a) above. Then there are further uncertainties resulting from deficiencies in the models of the global carbon cycle. These arise partly from a lack of understanding about some of the features of the cycle and partly from the difficulty of constructing models that incorporate all the known biological, chemical and geological elements of the cycle with its many feedback loops. Projections of concentrations of other greenhouse gases are subject to similar constraints. The modelling uncertainties will be reducible by further research, but only in the long term.

(c)  The development of scenarios for climatic futures is also subject to modelling uncertainties associated with an incomplete understanding of the natural processes being modelled, difficulties in parameterisation of some elements and inadequate spatial resolution of the models. Further research and an increase in the storage capacity and speed of computer will reduce these uncertainties, but again only slowly. The complexity of the processes involved make it unlikely that totally accurate models will be built.

*Activity 5*

In carrying out this exercise, you will probably have begun to appreciate for yourself some of the problems that plague scientists working in this field!

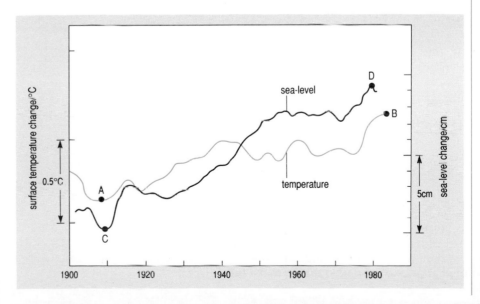

◀   *Figure 3.27*
*Temperature changes and sea-level rises.*

(a)  The temperature plot is a smoothed curve, representing accumulated data over many climatic zones as well as over short-term variations and seasonal cycles in the temperature (compare Figure 3.11). The quoted global temperature rise of 0.5°C seems to correspond to the difference between the lowest and highest points on the graph (A and B respectively in Figure 3.23), with a fairly large uncertainty ($\pm 0.2$°C) to allow for the scattered nature of the individual temperature readings that have been used to construct the graph. On a similar basis, the change in sea-level between the minimum and maximum points on that curve (points C and D respectively) corresponds to an overall rise of about 9cm. This change took place between roughly 1910 and 1980, i.e. over a period of about 70 years. On this basis, one could estimate the sea-level rise as $9 \text{ cm} \times \dfrac{100}{70} \approx 13 \text{ cm}$ per century. During the 1980s scientists were indeed quoting values of the order of $(12 \pm 5)$ cm for the rise in sea-level per century at the then current rate of global warming. As you see, the uncertainties are high ($\pm 40\%$), so if you read other accounts you may find that individual workers quote rather different values.

(b)  There is clearly some correlation between changes in global temperature and changes in sea-level. In particular, both curves have a similar 'dip followed by hump' shape between 1900 and 1920. However, the fine structure of the two curves does not match, and there was a period in the 1940s when the sea-level rose despite a fall in the mean temperature.

(c)  The difficulties of prediction become apparent here: you will have been trying to combine two uncertain quantities, let us take a typical figure, based on Figure 3.23, of a sea-level rise of $(10 \pm 3)$cm for a temperature increase of $(0.5 \pm 0.2)$°C. (Your own figures may differ a little from this, but the principle of the calculation still applies.) One could reason as follows:

Minimum response =

$$\frac{\text{minimum value for sea-level rise}}{\text{maximum corresponding value for temperature change)}}$$

$$= \frac{(10 - 3) \text{ cm}}{(0.5 + 0.2)°\text{C}} = \frac{7}{0.7} \text{ cm/°C}$$

$$= 10 \text{ cm/°C}$$

Maximum response =

$$\frac{\text{maximum value for sea-level rise}}{\text{minimum corresponding value for temperature change}}$$

$$= \frac{(10 + 3) \text{ cm}}{(0.5 - 0.2)°\text{C}} = \frac{13}{0.3} \text{ cm/°C}$$

$$\approx 40 \text{ cm/°C}$$

This could be expressed as a rise of $(25 \pm 15)$ cm/°C, suggesting a sea-level change of between 30 cm and 120 cm for a temperature increase of 3°C. (There are more rigorous statistical methods for combining uncertain quantities, but, with the kind of data we are considering here, the 'commonsense' method outlined above is quite adequate to make the point.)

In performing this calculation, we have made the main assumption that the correlation, such as it is, between sea-level rise and a temperature increase of about half a degree would continue to apply for much larger changes in temperature. We have further assumed that if a temperature increase of $x°C$ causes a sea-level rise of $y$ cm, then an increase of $2x°C$ will result in a rise of $2y$ cm (i.e. what mathematicians call a linear relationship).

## Activity 6

*Q1* Growing trees need carbon to make new wood; they obtain this from carbon dioxide in the air through the process of photosynthesis. An afforestation programme would therefore make some contribution to slowing the build-up of carbon dioxide in the atmosphere. Afforestation cannot provide a permanent solution, because of the problems associated with harvesting the 'crop' once the trees are mature: the wood could not be burnt or otherwise allowed to decay. (There are also snags in relation to the scale of the project: it has been estimated that an area as big as the USA would need to be planted in order to absorb the current anthropogenic emission of carbon dioxide.)

*Q2* It is impossible to attribute a single extreme event to the anthropogenic greenhouse effect, since such events are anyway part of the natural variability of the weather and their occurrence and distribution has always been random. In climatic terms, such effects only assume significance when their frequency and/or severity over a considerable period of time clearly exceeds previous averages.

## Activity 7

The anthropogenic emission of some greenhouse gases is clearly linked to population (compare Figure 3.14). In the absence of alternative technologies an increasing population burns more fossil fuels, so adding to the carbon dioxide emissions. It also requires more land for agriculture and may cut down (and ultimately burn) forests to obtain this. Increased agricultural activity leads to increased emissions of methane and nitrous oxide. A stabilisation of the human population would probably also stabilise (if not actually reduce) the emission rates for greenhouse gases. However, this scenario is unlikely to be attained in less than about 100 years. Measures to control the emissions of CFCs and to reduce the consumption of fossil fuels could be taken on a shorter time-scale, independently of population changes, though the implementation of such a proposition is greatly complicated by differences between industrialised and developing countries in the management of energy resources.

# Chapter 4    Threats to biodiversity

## 1    Introduction

As you read this chapter, look out for answers to the following key questions:

●    What is biodiversity, and why is it important?

●    What values do we attach to biodiversity, and how do we use biological resources?

●    How is biodiversity threatened?

●    How is biodiversity conserved and managed?

●    Can the UN Convention on Biological Diversity ensure the long-term protection of biodiversity?

*Activity 1*

If you studied *Silvertown* (1996a, b), review those two chapters. The present chapter develops several of the issues and ideas first introduced there.

Edward Wilson, the noted Harvard biologist and 'parent of biodiversity studies', has commented that 'humanity has initiated the sixth great extinction spasm, rushing to eternity a large fraction of our fellow species in a single generation', and he argues that, 'every scrap of biological diversity is priceless, to be learned and cherished, and never to be surrendered without a struggle' (Wilson, 1992, p. 32).

The five major documented extinction episodes which have occurred over the last half-billion years, including the most recent Cretaceous extinction 66 million years ago which saw the disappearance of the dinosaurs, were all attributable to natural phenomena such as great volcanic eruptions or climatic changes or possibly asteroid impacts. The sixth one, caused this time entirely by humans, may be the one that 'breaks the crucible of life'. Activities such as the rapid conversion of natural habitats for human use and their fragmentation, pollution of soil, water and atmosphere, intensive agriculture and forestry, over-exploitation of plant and animal species, human-induced global climate change, and the introduction of exotic species are having disastrous effects on natural ecosystems and the species which inhabit them. Fossil evidence suggests that after each of the major extinction spasms, life recovered to a similar level of diversity. However, this recovery took at least 10 million years, perhaps 20 million years for evolution to restore the huge losses after the Cretaceous extinction. These estimates of recovery times should put paid to those who believe that what humans destroy, Nature will redeem. The time-scale required has little meaning for contemporary humanity.

But why should we be concerned with this next potential biological extinction spasm? Many sceptics consider the predicted losses to be exaggerated, citing only a handful of documented extinctions in recent history. However, these are all 'high profile' organisms such as mammals and birds, and most likely reflect our lack of knowledge of many taxa and inadequate monitoring of biodiversity as a whole. In this chapter we will explore some of the scientific issues that are associated with the biodiversity crisis. Gaps in our knowledge of biodiversity need to be plugged and broader development issues need to be tackled (discussed in Chapters 5 and 6). We risk triggering a sixth extinction spasm, leaving the Earth biologically impoverished, if we ignore the impending biodiversity crisis. It is becoming increasingly important that as a species we foster a much greater spirit of stewardship over the world's biological wealth to ensure its long-term conservation (*Sarre and Brown*, 1996).

# 2   What is biodiversity and why is it important?

## 2.1   Introduction

**Biodiversity** suddenly became a 'buzzword' in conservation circles towards the end of the 1980s (Wilson and Peter, 1988). It is simply a contraction of the term 'biological diversity' and is a concept that embraces the whole array of life-forms and their components, from genes, through species to habitats and ecosystems. Biodiversity, then, is the totality of genes, species and ecosystems inhabiting a region, and ultimately encompasses the number, variety and variability of life on Earth.

    **Genetic diversity** refers to the variation of genes within a species. This encompasses genetic variation within populations, which is, for example, very high among Indian rhinos and very low among cheetahs (*Silvertown*, 1996a), and between populations, such as the thousands of traditional rice varieties (*Woodhouse*, 1996). **Species diversity** refers to the different species of plants, animals and micro-organisms in a region. **Ecosystem diversity** encompasses the variety of habitat types (forest, savanna, swamp and so on) found within a region.

    Biodiversity at each of these levels has practical value to humankind. Genetic diversity within crop species is the basis of breeding programmes to maintain resistance to pests and diseases. Species diversity provides us with innumerable products derived directly from plants, animals and micro-organisms, including food, fuel, building materials, medicines, industrial products and cosmetics. At an ecosystem level, a whole range of ecological services is provided, such as the cycling of water, carbon, minerals and other materials. However, the scope of biodiversity is much broader, encompassing aesthetic and philosophical values which are important for the maintenance and development of human societies. For these reasons, biodiversity must be conserved and an essential element in this is the collection and dissemination of knowledge gained from scientific research. The documentation of life on Earth is an urgent priority. The unknown potential of genes, species and ecosystems represents an unending biological frontier of enormous value. However, we are now seeing the disappearance of habitats worldwide and rapidly declining biodiversity, at the same time as an increasing demand for vital biological resources by an expanding human population. If these trends continue, we could soon be living in a biologically impoverished world. As we lose species and habitats we lose potential economic benefits, and we lose the continuity of healthy ecosystems which are essential for clean air and water. We also lose touch with nature and our relationship to it. It is not surprising that the biodiversity crisis is set to become one of the most significant global issues facing us as we enter the next millennium.

Knowledge of the richness of species and their spatial distribution is essential for establishing priorities for conservation planning and management. The biodiversity crisis is highlighting how much we do not yet know about the species that comprise the biosphere (see *Silvertown*, 1996b).

## 2.2   Genetic diversity

Genetic diversity represents the sum of the genetic information contained in the genes of microbes, plants and animals. It ultimately lies in the sequence of the four base-pairs (adenine, cytosine, guanine and thiamine) of the nucleotides that comprise the genetic code (DNA). Each species is a repository of an immense amount of genetic information. The numbers of genes ranges from about 1000 in bacteria to more than 400 000 in many flowering plants. Each species is made up of populations of individuals which are all genetically different. This pool of genetic variation within an inter-breeding population is acted upon by natural selection. Differential survival of individuals results in changes in frequency of genes within this gene pool resulting in population evolution and, with geographic isolation, eventually to speciation. Therefore, the significance of genetic variation is that it is the basis of evolutionary change and, in human times, selective breeding. Conservation priorities at this level are to maintain enough individuals of a species to ensure that individuals are representative of the genetic variability of the species, thus maximising the gene pool. *Silvertown* (1996a) has described the effects of the dramatic genetic bottleneck that the cheetah (*Acinonyx jubatus*) is thought to have gone through and the problems of inbreeding depression on genetic diversity.

## 2.3   Species diversity

Despite two centuries of research by biologists exploring every corner of the Earth, we still do not know to within a factor of ten how many species inhabit the Earth, even though scientists at the end of the great exploration era of the nineteenth century confidently declared that they had found and named most of the species on Earth. More detailed investigations, particularly in the Tropics, are revising the estimated number of species ever upwards. When Linnaeus introduced the binomial system of naming species in 1753, he listed 9000 species of plants and animals. To date, taxonomists have described about 1.75 million species, although published figures vary between 1.4 and 1.8 million species, largely due to differences in the way in which tallies are made. Estimates of the number of undiscovered and undescribed species range from 10 million to 100 million: see Table 4.1.

The problems of applying the biological species concept to micro-organisms hampers all aspects of biodiversity research where the species is used as the standard measure. Much of the variation in estimates of the total number of species on Earth shown in Table 4.1 is a consequence of defining what constitutes a species of micro-organism. The species concept remains a difficult and controversial subject in viral, bacterial and fungal systematics (see Box 4.1).

Some of the figures shown in Table 4.1 are higher than those quoted by *Silvertown* (1996b) and represent a more recent assessment (Heywood and

## Box 4.1   Systematics: the science of biodiversity

*Systematics* is that branch of science dedicated to discovering, organising and interpreting biological diversity. It consists of three main elements:

• *Taxonomy* – the science of discovering, describing and classifying species.

• *Phylogenic analysis* – the search for evolutionary relationships among a group of species. A phylogeny is an evolutionary history.

• *Classification* – the grouping of species, based on taxonomic similarity and evolutionary relationships.

*Table 4.1   Approximate numbers of described species (in thousands) currently recognised and estimates of possible species richness for groups of organisms with more than 20 000 described species and/or estimated to include more than 100 000 species*

| | Described species (d) | Estimated species | | Working figure (w) | Proportion described (d/w) | Accuracy |
|---|---|---|---|---|---|---|
| | | Highest estimate | Lowest estimate | | | |
| Viruses | 4 | 1 000+ | 50 | 400 | 1% | V. poor |
| Bacteria | 4 | 3 000+ | 50 | 1 000 | 0.4% | V. poor |
| Fungi | 72 | 2 700+ | 200 | 1 500 | 5% | Moderate |
| Protozoans | 40 | 200+ | 60 | 200 | 20% | V. poor |
| Algae | 40 | 1 000+ | 150 | 400 | 10% | V. poor |
| Plants | 270 | 500 | 300 | 320 | 84% | Good |
| Nematodes | 25 | 1 000+ | 100 | 400 | 6% | Poor |
| Molluscs | 70 | 200+ | 100 | 200 | 35% | Moderate |
| Crustaceans | 40 | 200+ | 75 | 150 | 27% | Moderate |
| Arachnids | 75 | 1 000 | 300 | 750 | 10% | Moderate |
| Insects | 950 | 100 000 | 2 000 | 8 000 | 12% | Moderate |
| Vertebrates | 45 | 55 | 50 | 50 | 90% | Good |
| Others | 115 | 800 | 200 | 250 | 46% | Moderate |
| Totals | 1 750 | 111 655 | 3 635 | 13 620 | 13% | V. poor |

Notes: Low and high (highly speculative) estimates together with a working figure are quoted. The reliability of *all* estimates is likely to vary greatly. The accuracy of the working figures is quoted as probably accurate within a factor of 2 (Good), within a factor of 5 (Moderate), within a factor of 10 (Poor) or not certain within a factor of 100 (Very poor).

Source: adapted from Heywood, V. and Watson, T. (1995) *Global Biodiversity Assessment*, Cambridge, Cambridge University Press, p. 118.

Watson, 1995). This 'working figure' of about 13.5 million species – although becoming widely adopted – is still fraught with inaccuracies and based on best estimates by specialists in the field. Some of the problems of estimating the number of species have been reviewed by *Silvertown* (1996b). Like other natural resources, the distribution of biodiversity in the world is not uniform. Species richness increases from the Poles to the Tropics. This can be illustrated by considering the statistics in Table 4.2 which show the numbers of species in selected groups of plants and animals in Great Britain and the La Selva Reserve. La Selva is a small area of tropical cloud forest in Costa Rica which has been very intensively studied over the last two decades.

Table 4.2
Numbers of species in selected groups in Great Britain and La Selva, Costa Rica

|  | La Selva Reserve Costa Rica | Great Britain |
|---|---|---|
| Area | 13.7 km² | 233 000 km² |
| Approximate latitude | 10°N | 50–61°N |
| *Number of species* |  |  |
| Flowering plants | 1800 | 1500 |
| Mammals | 104 | 48 |
| Freshwater fish | 42 | 48 |
| Reptiles | 76 | 6 |
| Amphibians | 46 | 6 |
| Breeding birds | 394 | 210 |
| Butterflies | 143 | 59 |

Natural historians have discussed large-scale gradients in species richness for nearly two centuries since Humboldt first commented on the huge disparity in tree species richness between temperate and tropical habitats. It is not latitude *per se* which determines biodiversity gradients, but environmental factors that are linked to latitude. Factors which are thought to be important in determining biodiversity gradients are total precipitation, seasonal variation in precipitation, temperature and irradiation. Glaciation severely reduced the fauna and flora of northern latitudes (*Silvertown*, 1996b).

## 2.4   Ecosystem diversity

We have quite good knowledge of the broad distribution and extent of the world's biomes and their major component ecosystems (*Silvertown*, 1996a). Because individual biomes represent large-scale biogeographical features which share particular physical characteristics and/or species composition, they can be detected and mapped from satellite images and aerial photographs. This has allowed us, for example, to track the rate of deforestation of tropical rain forest on a country-by-country basis or throughout the Tropics as a whole. The most authoritative assessment of the global status of tropical forest remains the UN Food and Agriculture Organisation's (FAO) 1993 report which quoted an annual average loss of 15.4 million ha of tropical forest during 1980–90. This amounts to an annual loss of about 0.8% and a total loss over the decade of an area of tropical forest of almost three times the size of France (Heywood and Watson, 1995). Rates show little sign of slowing.

However, the world's biomes and large-scale ecosystems actually represent a continuum of local ecosystems, habitats and communities whose boundaries are difficult to define. Because of this it is very difficult to estimate rates of change of ecosystems with any great accuracy. This has significant implications for ecosystem management and can account for the widely varying figures quoted for losses of particular ecosystems, for example tropical forests. In addition, we still have much to learn about how ecosystems function and which natural processes and species are vital to the survival and productivity of a particular ecosystem. In practice this means that ecosystem diversity is often evaluated through measures of the diversity of its component species (discussed in the next section).

---

*Activity 2*

Try to build up a picture of how the term 'biodiversity' is used and viewed by the media.

• Review your local newspaper, national newspapers, general and specialist magazines, and investigate the usage on the television and radio. Categorise the way biodiversity is used in these sources. You might also investigate the use of related terms such as biological diversity, species diversity and species richness.

• You could also do a survey amongst friends. What do they understand by the term, and is it a useful term? How do these uses compare with the definitions and usage in this chapter?

• Look at a range of dictionaries in your local library or elsewhere. Does biodiversity appear? How recently do dictionaries need to have been published in order to contain the term biodiversity?

---

## 2.5    *Monitoring biodiversity*

In the past there has been little effort to co-ordinate efforts to estimate the number of species in the world. Charismatic species such as mammals, birds and butterflies have received much attention but so many different methods have been used to record and monitor biodiversity that comparisons have not been possible. There is now an urgent need to co-ordinate these activities internationally and institute more coherent and integrated species assessments based on standardised techniques. Initiatives have begun at both national and international levels to try to address these shortcomings. One example is Systematics Agenda 2000 (SA 2000, 1994), a bold initiative 'to discover, describe and classify the world's species' (see Figure 4.1).

The expected benefits from Systematics Agenda 2000 include:

• an increased number of usable species resources

• an improved database to help conservationists and biological resource managers preserve, as well as utilise, their nations' species diversity

• knowledge to guide the selection of new and improved food crops and medicines

• baseline data for monitoring global climate and ecosystem change, rates of species extinction, ecosystem degradation and the spread of exotic, disease-causing and pest organisms.

The ecological literature is full of debate on how biodiversity should be measured and yet there is still no consensus. Most of this debate has focused on how to measure species diversity (discussed next), with much less effort expended on the measurement of genetic or ecosystem diversity, at least until recently. With the greater availability of modern molecular techniques, such as DNA sequencing using the *polymerase chain reaction technique (PCR)* (see Box 4.2), conservationists are being drawn into the new discipline of molecular ecology. Used by ecologists, this molecular tool-kit can identify relationships among individual organisms, populations and

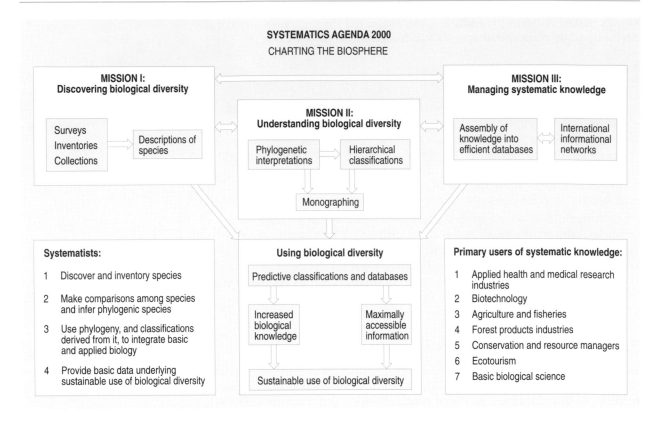

▲   *Figure 4.1*
*The Missions of Systematics Agenda 2000*

species. Given the complexities of defining diversity at the ecosystem level, as described in Section 2.4, any number of ecosystem or community attributes which are components of biodiversity may warrant measurement for specific objectives. Consequently there is no authoritative measure of ecosystem diversity. Weight can be given to *trophic diversity*: for example, a

---

*Box 4.2    The polymerase chain reaction*

The polymerase chain reaction (PCR) depends on a heat-resistant enzyme discovered by Thomas Brock in the bacterium *Thermus aquaticus* which occurs naturally in hot springs in Yellowstone National Park. Discovered by Kary Mullis, who shared the 1993 Nobel Prize for chemistry for the discovery, PCR is now a basic tool in molecular biology and used routinely in diagnostic medicine as well as much biotechnology. The reaction allows the multiplication of DNA millions of times in a few hours. Thus individual genes can be multiplied and easily sequenced.

PCR is now being applied to the problem of the species concept in micro-organisms and in an attempt to document microbial diversity, and in particular bacterial diversity. Most research has concentrated on the gene for an RNA component of ribosomes, known as 16S rRNA. This gene is present in all organisms and evolves so slowly that it is identical in every member of a given species, and also of some very closely related species. Researchers have applied this technique to investigating bacterial diversity in a wide range of environments, including freshwater pools, the open ocean, soil and hot springs. Results from these habitats indicate that estimates based on PCR show a bacterial diversity several orders of magnitude greater than that indicated by conventional culturing techniques.

Microbiologists now have the tools to probe the microbial world for the first time. However, molecular prospecting is time-consuming and expensive, and still with some technical difficulties. As yet, no single habitat has been fully surveyed, but we are at a new frontier.

Source: Lovejoy, 1995; Holmes, 1996

hypothetical ecosystem which consisted of several species of plants and herbivores would be considered less diverse than another ecosystem containing the same total number of species, but which included carnivores and omnivores in addition to plants and herbivores.

## Measurement of species diversity

Species diversity, as an ecological measure, is a function of the number and abundance of species. It is often reduced to species richness, the number of species in a given area, but technically it is a measure of the richness and evenness of species in an ecological community. An ecological community is defined as being composed of individuals of different species; a community with high species diversity is one that has about equal numbers of individuals (high evenness) of many different species (high richness). It is not the total number of individuals that is important in determining species diversity, but rather the relative proportions of individuals of each species and the number of species. There are many different types of diversity index, each of which aims to summarise the diversity of a community in a single value. One widely used index is *Simpson's diversity index* which has the attractive property that the maximum theoretical value (highest diversity) is very close to one (although it can never equal one) and the minimum theoretical value is zero. The calculation of Simpson's index for any community is based on the number or the biomass of individuals of each species in that community.

## Biodiversity hot-spots

A principal goal of conservation activity is to ensure the long-term survival of as many species as possible. The high rate at which natural habitats are being lost means that we need information on where species occur, and in particular those areas of high species diversity, *now*. Scientists are discovering various short-cuts to provide the more urgent information needed for biodiversity conservation without classifying every species first. The value of these approaches is that they identify key areas for conservation priority. In species-rich countries the only practical way to conserve biodiversity is to protect the natural vegetation of selected areas, rather than taking measures for individual species, one at a time. However, campaigning organisations use certain animals as flagship species to raise funds for the biodiversity of key areas.

Several approaches have been adopted. One approach, described by *Silvertown* (1996b) is the concept of *megadiversity countries* which merit special international attention because of their high species richness. However, it is not necessarily the total number of species which is the most important criterion for determining priority areas for conservation. Another approach is to identify areas with the greatest numbers of *endemic species*. Endemics are species of restricted range; the only place in the world that they may be found is on a single mountain top, a small island or in a small area of forest. At the global level, areas of high endemism are of high conservation priority because if unique species are lost, they are lost forever. Such areas were labelled **hot-spots** by Norman Myers in 1988. His two reviews identified 18 hot-spots which collectively are the exclusive home of 49 955 endemic plant species, almost 20% of the world's plant species. However, these hot-spots occupy just 746 400 km², about 0.5% of the world's land surface (Myers, 1988, 1990), and comprise a diverse group of forests and Mediterranean-type heathland (Figure 4.2).

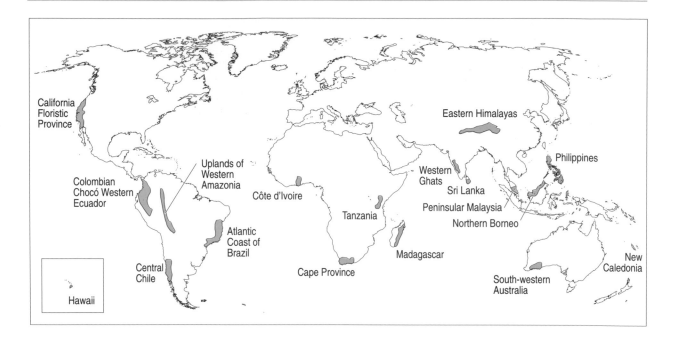

With so much of the world's attention directed towards the Tropics, and tropical rain forests in particular, insufficient attention is directed toward threatened temperate species. In this respect it is worth highlighting the Cape Floristic Province of South Africa. Occupying 89 000 km² at the southern tip of Africa is a specialised heathland called *fynbos*. It is characterised by one of the world's most unusual and diverse floras, comprising 8600 species of plants. Of these, 6300 (73%) exist nowhere else in the world. One third of the *fynbos* has been lost to agriculture, development and invasion by exotic plant species in recent years; the remainder is being rapidly fragmented and degraded. At least 26 species are known to be extinct and another 1500 are rare and threatened – equivalent to the total flora of Great Britain! This puts the Cape Province flora amongst the most threatened in the world. (See Plate 10.)

More recently, the Centres of Plant Diversity Project has identified 234 sites worldwide which are species-rich, contain large numbers of endemic species and need to be conserved (WWF and IUCN, 1994).

These types of study have their limitations. Because there is a high number of endemic species in one taxon, it does not follow that the same will be true of other taxa. However, there are some good correlations. In 1992 The International Council for Bird Protection (ICBP) completed its Biodiversity Project which identified 221 Endemic Bird Areas (EBAs) worldwide. More than 25% of the world's bird species are restricted to these EBAs which occupy only 5% of the world's land surface. This study also showed that the majority of EBAs were also important for plants and other animals such as amphibians and reptiles (Bibby *et al.*, 1992).

What is important is that these approaches, uneven as they may be, highlight the conservation challenge that we are facing and help to identify priorities. The establishment of the Rapid Assessment Programme (RAP) created by the Washington-based NGO Conservation International has taken this approach into the field. Their goal is rapidly to investigate poorly known ecosystems that might be local hot-spots, in order to make recommendations for action. Areas targeted are limited in extent, being

▲  *Figure 4.2*
*Myers' forest and heathland hot-spot areas.*

*Note: The 18 hot-spots identified here are forests and Mediterranean heathland well known enough to be included with certainty, but the map, based on preliminary study, is far from complete. Other forest types not shown are endangered, as well as a large number of lakes, river systems and coral reefs. The broader areas depicted, such as the coastal forests of Brazil and the Philippines, actually consist of many smaller hot-spots scattered across local mountain ridges, valleys and islands.*

isolated mountain-tops or single valleys, for example. The RAP team is composed of leading field biologists, including host country scientists, who examine an area's biological diversity and degree of endemism, the uniqueness of its environments and the degree of risk of extinction of its most threatened species. This is usually based on well-known groups such as flowering plants, mammals, birds and butterflies. A high species richness across these taxa would be sufficient to recommend protection and further investigation of other, lesser-known taxa. However, the team has no control over how its recommendations are treated by the host government.

# 3   Values and uses of biodiversity

The importance of biodiversity to humanity has been recognised in art, literature and religion since the beginnings of recorded history. These cultural, ethical, social and economic values of biodiversity have been core elements in the development of human societies worldwide. Given that these values are so fundamental to human activities, it is not surprising that the principles of conservation are recognised and practised by most cultures. However, the process of development has tended to stress the material benefits of biodiversity, and in order to compete for attention in a modern world, policies regarding biodiversity need first to demonstrate a monetary value to a country's social and economic development. Only then will government decision-makers take heed of underlying aesthetic and ethical values and the moral imperative of conserving biodiversity for future generations.

Biodiversity has a range of both direct and indirect uses for humankind. The value of biodiversity is difficult to define and often impossible to estimate, unless it can be assigned a direct monetary value. The value of biodiversity is often thought of more broadly in terms of its contribution to human welfare. In this context, biodiversity is of value to humankind in a variety of practical (use values) and non-practical (non-use values) ways, as illustrated in Table 4.3.

Table 4.3   Classification of biodiversity values

| Value type | Examples |
| --- | --- |
| A. Use values | |
| (i)  Direct | |
| Consumptive | Home-grown food; forest fruits |
| Productive | Plant breeding; natural medicines |
| Non-consumptive | Ecotourism |
| (ii)  Indirect | Ecological processes |
| (iii)  Option | Future value of drugs |
| B. Non-use values | Existence value of wildlife |

Source: based on Flint, M. (1991) *Biological Diversity and Developing Countries: issues and options*, London, ODA, p. 17.

## 3.1   Use values: biodiversity as a resource

Use values represent the present, and potential, practical value of elements of biodiversity to humankind. They can be sub-divided into direct, indirect and option values.

### Direct use values

**Direct use values** are those concerned with the enjoyment or satisfaction received directly from biodiversity. They are easy to measure and so a price can be attached to them. In this way they contribute directly to the local and world economy. The main difference between consumptive and productive values is that the former are consumed directly without being traded. So home-grown garden crops, wild-collected fruits and forest-collected fuelwood would represent *consumptive* use values, whereas the commercial value of traded timber, cereals and fish would represent *productive* use values. Consumptive use values seldom appear in national income accounts, but there is nothing to stop a financial value being assigned to them. For example, firewood and dung collected from the wild provides over 90% of the total primary energy needs in Nepal, Tanzania and Malawi, and more than 80% in many other developing countries, and this contribution to the economy should be directly recognised. The value of products that are commercially harvested and traded in formal markets are reflected in national income accounts and can have a major impact on national economies.

Biodiversity confers direct use value in at least three major areas of our day-to-day lives: agriculture, medicine and industry. Arguably the most important direct use of biodiversity is as food. All of our food comes from other organisms. However, less than twenty species of the thousands of edible plants known to exist produce most of the world's food; three species – wheat, maize and rice – account for 54% of the calorific consumption in developing countries. Their domestication and importance in world agriculture is described by *Grigg* (1996). In addition to these twenty major food crops, some 5000–10 000 plant species provide an important source of food and materials, including timber and fuelwood, rattans, forage and pasture crops, rootcrops, beverage crops, spices, fruits and fibres. Many of these are currently only of local importance, but have wider potential. Wilson (1992) describes a range of edible plant species which could be developed as new foods with export potential, thus providing much needed income to developing countries. These included a potential superstar, the winged bean (*Psophocarpus tetragonolobus*) from New Guinea. Described as a one-species supermarket, the entire plant is palatable: the leaves are used as a spinach-like vegetable and the young pods as green beans; the young seeds are like peas, and the tubers have the versatility of potatoes and contain more protein; the mature seeds resemble soybeans and can be cooked, ground into flour or liquefied into a caffeine-free drink that tastes like coffee. Being a legume, it possesses nitrogen-fixing bacteria in root nodules, enabling it to grow without much need of fertilisers, and to be used as a green manure. It has a phenomenal growth rate, reaching a length of 4 metres in a few weeks. With a small amount of genetic improvement through selective breeding, the winged bean has great potential for raising the standard of living in many of the poorest tropical countries. However, there are inherent dangers in promoting single cash crops as the key to raising living standards through exports. These include ecological problems of monocultures, economic problems of gluts and the reduction in biodiversity.

Much less is known about the diversity of these lesser used species compared with the major food crops, and very little is being done to conserve them, either *in situ* or *ex situ*.

---

### Box 4.3   The value of new species

The discovery in the late 1970s of a new species of wild corn has yielded multi-million dollar rewards. The new species, named *Zea diploperennis*, was discovered in the Sierra de Manantlán montane cloud forest of south-western Mexico. This plant, the most primitive known relative of modern maize (*Zea mays*), was found to be resistant to the seven major viral diseases that infect cultivated maize. The viral resistance of *Z. diploperennis* has been transferred to *Z. mays* and four new commercial lines have been produced with an estimated annual value of US$4.4 billion. At the time of its discovery, this species was surviving in only three tiny habitat patches totalling about four hectares. Its habitat was under the imminent threat of destruction from loggers and squatter cultivation. *Z. diploperennis* is a perennial, unlike other forms of maize which are annuals; it grows at elevations of 2500–3300m, and is adapted to climates that are cooler and damper than habitats where maize is currently grown. These characteristics offer great potential to expand the cultivation range of maize and reduce some of the associated costs, such as ploughing and re-seeding. Given that maize is one of the three most important crops worldwide, with an estimated annual value of US$60 billion, this find is of enormous significance. It offers an economic justification for the inventory work discussed in Section 2.

Source: Iltis, 1988

---

Wild species have been a vitally important source of medicines for at least two thousand years since the Greek physician Hippocrates prescribed willow bark (*Salix* spp.) as an analgesic. The precursor of aspirin was first isolated from the leaves and bark of white willow (*Salix alba*). Indigenous tropical forest peoples have been using many plant species for medicines for similar lengths of time. Medicinal drugs derived from natural sources make up an important global contribution to health care. About 80% of people in less developed countries rely on traditional medicines for primary health care and this shows no signs of decline despite the availability of western medicine. Approximately 119 chemicals extracted from about 90 species are used in medicines throughout the world, and many cannot be manufactured synthetically. For example, the cardiac stimulant digitoxin, the most widely used cardiotonic in western medicine, is extracted directly from dried *Digitalis* spp. (foxgloves). Synthetic vincristine, used to treat childhood leukaemia, is only 20% as efficacious as the natural product derived from *Catharanthus roseus* (Madagascar periwinkle).

Table 4.4 shows a selected list of some of the plant-based medicines that we rely on for our health. It is estimated that 25% of prescriptions in the USA have active ingredients with plant origins. Sales of these prescriptions in 1990 were estimated to have cost US$ 4.5 billion.

Two recent discoveries highlight the importance of biodiversity and systematic understanding:

- *Taxol*, initially isolated from the bark of the Pacific yew (*Taxus brevifolia*), is a powerful drug used in the treatment of ovarian and breast cancer. It takes the bark of three trees to provide sufficient taxol to treat one cancer patient. Removal of the bark kills the tree. A random search for other plants which contain a similar chemical would undoubtedly have taken years, and still may have been unsuccessful. However, because the taxonomy of its close relatives was well known, a selective search soon showed that leaves of the European yew (*Taxus baccata*) were rich in taxol which could be extracted without killing the plant. Clippings from *T. baccata* trees in the UK are now sent to a French firm for taxol extraction.

*Table 4.4    Some pharmaceuticals derived from plants and fungi*

| Drug | Use | Plant source | Plant's native range |
| --- | --- | --- | --- |
| Bromelain | Controls tissue inflammation | Pineapple (*Ananas comosus*) | Tropical America |
| Caffeine | Stimulant, central nervous system | Tea (*Camellia sinensis*) | South and East Asia |
| Cocaine | Local anaesthetic | Coca (*Erythroxylon coca*) | East Andes |
| Codeine, Morphine | Analgesics | Opium poppy (*Papaver somniferum*) | Western Mediterranean |
| Digitoxin | Cardiac stimulant | Foxgloves (*Digitalis* spp.) | Europe, Mediterranean |
| Diosgenin | Source of female contraceptive | Wild yams (*Dioscorea* spp.) | Tropics |
| L-Dopa | Parkinson's disease suppressant | Velvet bean (*Mucuna deeringiana*) | Tropics |
| Gossypol | Male contraceptive | Cotton (*Gossypium* spp.) | Warm temperate and Tropics |
| Monocrotaline | Anti cancer (topical) | *Crotalaria sessiliflora* | Tropics and subtropics |
| Penicillin | General antibiotic | Penicillium fungi (esp. *Penicillium chrysogenum*) | |
| Quinine | Anti malarial | Yellow cinchona (*Cinchona ledgeriana*) | Andes to Costa Rica |
| Reserpine | Reduces high blood pressure | Indian snakeroot (*Rauvolfia serpentina*) | Tropics |
| Scopolamine | Sedative | Thornapple (*Datura metel*) | South and North America, but widely naturalised |
| D-tubocurarine | Active component of curare; surgical muscle relaxant | *Chondrodendron* and *Strychnos* species | Tropics |
| Vinblastine, Vincristine | Anti cancer especially childhood leukaemia | Madagascar periwinkle (*Catharanthus roseus*) | Madagascar |

Source: data from Wilson, E. O. (1992) *The Diversity of Life*, London, Allen Lane, pp. 286–7; range data from Mabberley, D. J. (1990) *The Plant Book*, Cambridge, Cambridge University Press.

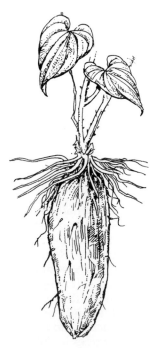

•    The discovery of an insignificant forest vine from Korup National Park in Cameroon, West Africa, has caused a great deal of excitement. An extract from the vine, *Ancistrolcladus korupensis*, has yielded a compound, *michellamine B*, which proved so successful in protecting human cells from the HIV virus *in vitro* that it is now undergoing extensive tests as a potential treatment for AIDS.

The World Health Organisation has listed over 21 000 plant names
(including synonyms) that have reported medicinal uses around the world.
Very few of these have been scientifically tested. Only about 5000 vascular
plants have been thoroughly screened and investigated as potential sources
of new drugs. This represents less than 5% of the approximately 270 000
species of vascular plants so far described by taxonomists. Most of these are
temperate species and the biochemical potential of tropical plants has yet to
be realised. The US National Cancer Institute has identified over 1400
tropical forest plants with the potential to fight cancer. Prospecting for
nature's biochemical riches (*bioprospecting*) is a rapidly growing activity.

The recent rise in ecotourism provides a good example of a non-
consumptive use of biodiversity. Natural ecosystems protected as national
parks can generate significant levels of income from wildlife tourism. It has
been estimated that elephant viewing in Kenya generates US$25 million per
year, far more than the elephant's potential value as dead trophies.
However, more important for developing countries is how much of the
money generated by the tourism industry remains in the local economy and
how it is distributed in the local community (Cater, 1995). If this money is
repatriated by foreign companies or spent on tourism-related imports
rather than going to local communities or the national park system, then
there will be little economic incentive for investment in the recurring costs
of conservation activities. The World Bank has estimated that developing
countries lose 55% of gross tourism revenues in this way. The equitable
distribution of money is an issue of much wider concern and is applicable
to earnings from all biological resources.

© 1984 FarWorks, Inc/Distributed by Universal Press Syndicate

"What? . . . They turned it into a WASTEbasket?"

## Indirect values of biodiversity

Many natural and semi-natural ecosystems, some of which may be of high biological diversity, are of considerable benefit to humankind more for the services provided than the goods; these are termed **indirect use values**. These life-support functions include:

• the role of forests in watershed regulation, in stabilisation of soils in erosion-prone areas and in carbon storage

• the role of mangroves in coastal zone stabilisation and as nursery areas for fisheries species

• the role of coral reefs in supporting important subsistence fisheries

• regulation of climate, at both macro- and micro-climatic levels, including influences on temperature, precipitation and air turbulence

• maintenance of the gaseous quality of the atmosphere

• waste assimilation functions, such as the retention or detoxification of pollution

• control of the hydrological cycle, including reduction of the probability of serious floods and droughts.


In most of these cases it is the whole ecosystem that is more important than the actual nature of species diversity. However, since the ecosystem is the reflection of its component species diversity, then these services are directly linked to biodiversity and the conservation of biodiversity is a prerequisite for maintenance of these vital services. How much of the world's biodiversity is redundant is a question that ecologists have not yet managed to answer. We do not yet know whether a 'weedy' world from which most of its species diversity had been lost would maintain those ecological services which maintain the Earth as a hospitable place for humanity to live.

Several hypotheses have been developed to describe the observed or the expected consequences of the loss or addition of species to an ecosystem. These hypotheses are based partly on observations following a change in species composition, and partly on predictions from ecological modelling and theory. We need to establish how much alike different species are to one another with respect to their roles in community or ecosystem processes. Some species may be 'keystones', whilst others may be of lesser importance, but for most ecosystems we simply do not yet know.


Q   What do you think a 'keystone species' is?

A   A **keystone species** is a species which has a large impact on its community or ecosystem. It can significantly affect ecosystem function. This impact is greater than would be expected from its relative abundance.

Q   Do keystone species share common biological traits or characters?

A   None, other than their critical importance to the ecosystem. Keystone species have been reported from a wide range of ecosystems, at various trophic levels, and from many taxa. Their pivotal importance is often not appreciated until their relative abundance starts to decline, or they are lost from the system.

---

*Box 4.4    Sea otters as keystone predators in Alaskan kelp forests*

Alaskan kelps (large brown seaweeds) are heavily grazed by sea urchins, which themselves are a staple food for sea otters. In the absence of otters, the numbers of sea urchins increase and the kelp is over-grazed. In this way a lush kelp forest can be converted to a 'sea urchin barren', a community composed primarily of encrusting coralline algae and sea urchins. These urchin barrens can be completely devoid of kelps and other large seaweeds. The diverse assemblage of macroscopic benthic and pelagic invertebrates, fishes and marine mammals associated with the species-rich kelp forests is also absent from these urchin barrens. Kelp forests also provide a range of ecosystem services which are lost when forests are converted to urchin barrens. Dissolved and particulate matter derived from kelp forests provide food for filter-feeding clams, mussels and barnacles in adjacent intertidal and subtidal communities. Kelp forests can also provide important nursery areas for many open-water fishes and often protect shorelines from the full force of waves during winter storms.

The otter is thus a *keystone species* because it exerts a disproportionately large influence over the whole kelp forest community, as well as indirect influences on adjacent ecosystems.

Source: Heywood and Watson, 1995

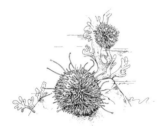

The redundancy hypothesis suggests that species overlap with one another in functional properties to such an extent that the loss of one species has a negligible effect on the ecosystem. If this were true, then it may be that many species are redundant from the viewpoint of ecosystem services. The uncertainties of the relationships between species diversity and ecosystem services led to the formulation of the 'rivet hypothesis' (Ehrlich and Ehrlich, 1981). In the same way that it would be foolish to remove rivets at random from the wing of a plane, this hypothesis suggests that, given the complexity of ecosystems and our lack of detailed knowledge of their functioning, it is foolish to allow species to be removed at random from ecosystems. There are more rivets in the wing of an aircraft than are necessary to maintain structural integrity, but removing too many will lead to structural failure. Moreover, at some point the removal of a single rivet can cause the rest to pop out and the wing to fail. In the same way, there may be some redundancy of species in maintaining ecosystem services, but there may be an extinction threshold and at present we have no idea at what point of species loss this may be reached. Similarly, the removal of a keystone species could result in a cascade of extinctions and the loss of ecosystem function. It remains a major challenge to ecologists to improve their understanding of the relationship between biodiversity and ecosystem functioning, to determine what proportion of the world's species diversity is redundant, if any, and so determine the overall level of biodiversity which is required for the delivery of ecosystem services.

## *Option values of biodiversity: 'The precautionary principle'*

At present a relatively small proportion of the world's biodiversity is actively exploited by humankind. The rest may be important in terms of values which are unused or unknown at present, but which could enhance the material well-being of humankind in the future if they were to be discovered and exploited: these are termed **option values**. These may be either in the form of new drugs, foods and other materials or in ecosystem services. As a result, society may be willing to pay to retain the option of having future access to this biodiversity.

## 3.2   *Non-use values of biodiversity: 'existence value'*

Resource-based arguments for the maintenance of biodiversity focus on relatively few species. The fundamental justification for striving to maintain all species cannot be linked to any resource-based arguments and is usually based on ethics and aesthetics. These argue that there are intrinsic values unassociated with actual or potential use, which reflect the satisfaction that people receive from simply knowing something exists. Many people find satisfaction in knowing that there are whales in the oceans. They are prepared to support this financially by contributing to conservation organisations to protect endangered species, even though they do not

---

*Box 4.5   An ethical basis for conserving biodiversity developed by IUCN's Working Group on Ethics and Conservation*

• The world is an interdependent whole made up of natural and human communities. The well-being and health of any one part depends upon the well-being and health of the other parts.

• Humanity is part of nature, and humans are subject to the same immutable laws as all other species on the planet. All life depends on the uninterrupted functioning of natural systems that ensure the supply of energy and nutrients, so ecological responsibility among all people is necessary for the survival, security, equity and dignity of the world's communities. Human culture must be built upon a profound respect for nature, a sense of being at one with nature and a recognition that human affairs must proceed in harmony and balance with nature.

• The ecological limits within which we must work are not limits to human endeavour; instead, they give direction and guidance as to how human affairs can sustain environmental stability and diversity.

• All species have an inherent right to exist. The ecological processes that support the integrity of the biosphere and its diverse species, landscapes and habitats are to be maintained. Similarly, the full range of human culture adaptations to local environments is to be enabled to prosper.

• Sustainability is the basic principle of all social and economic development. Personal and social values should be chosen to accentuate the richness of the flora, fauna and human experience. This moral foundation will enable the many practical values of nature – for food, health, science, technology, industry and recreation – to be equitably distributed and sustained for future generations.

• The well-being of future generations is a social responsibility of the present generation. Therefore, the present generation should limit its consumption of non-renewable resources to the level that is necessary to meet the basic needs of society, and to ensure that renewable resources are nurtured for their sustainable productivity.

• All persons must be empowered to exercise responsibility for their own lives and for the life of the Earth. They must therefore have full access to educational opportunities, political enfranchisement, and sustaining livelihoods.

• Diversity in ethical and cultural outlooks toward nature and human life is to be encouraged by promoting relationships that respect and enhance the diversity of life, irrespective of the political, economic or religious ideology dominant in a society.

Source: McNeely *et al.*, 1990, p. 26

expect to visit or use the resource they are helping to conserve. Our changing attitudes to Nature are discussed by *Sarre and Brown* (1996).

Many governments have already made an ethical commitment to nature by signing the World Charter for Nature, adopted by the United Nations in 1982, which expresses absolute support for the principles of conserving biodiversity. However, it is a non-binding law and seems to have been forgotten by all those involved; it has not led to any real protection of biodiversity. Drawing on the principles of the World Charter for Nature and the World Conservation Strategy, IUCN's Working Group on Ethics and Conservation produced an ethical basis for conservation (see Box 4.5). Again, this seems to have been largely ignored. Additional justification is required to change the way in which governments take decisions, which will usually require an economic argument, giving further support to the need to combine economic arguments with ethical justifications.

### Activity 3

One of the major, on-going debates in the conservation of the African elephant is whether to lift the ban on the trade in ivory. Elephants are currently listed on CITES Appendix I and a total ban on trade in ivory was imposed in 1989, after a decade of heavy poaching. Propositions have been put forward to downgrade elephants to Appendix II and allow commercial ivory trading to recommence. Scientific and economic arguments have been put forward in order to justify this action. Some of these arguments are listed below.

(a)  For a relaxation of the ban

*Scientific*

- Elephant populations are now thriving.
- Habitat destruction problems with increasing population size.

*Economic*

- Present stockpile of ivory cannot be sold.
- Profits from the sale of ivory could be used to support conservation actions.
- Insufficient aid available to support alternative forms of conservation.
- Increased requirement for land.
- Trading would encourage local communities to protect park and elephants.

(b)  Against a relaxation of the ban

*Scientific*

- Uncertain accuracy of population estimations.
- Unsure of population's ability to recover.
- Difficulty of distinguishing between legal and illegal ivory.

*Economic*

- Relaxation of ban 'too risky' – black market prices soared at hint of relaxation.
- Ecotourism is the key to saving African mammals.

Consider the 'ethical principles' listed in Box 4.5.

● Do any of the scientific or economic arguments consider this ethical dimension?

● Try to apply some of the principles listed in Box 4.5 to add a third category for and against the relaxation of the ban.

● Consider to what extent your feelings in this are linked to the charisma of elephants. Would these differ when considering less attractive organisms?

# 4    Biodiversity under threat

Biological diversity is under threat in a number of ways. Natural changes in the environment can eliminate once-successful species or reduce their numbers, eventually to mere remnant populations. Humankind disrupts ecosystems and destroys species, both deliberately and accidentally. This section will explore some of the key issues and causes.

## 4.1   Loss of biodiversity

The last section clearly demonstrated that biodiversity has real economic, social and cultural values in today's world. Why, then, is it being lost at such a rapid rate and how do we know that it is being lost?

Estimates from the fossil record suggest that on average a species persists between one and ten million years before going extinct due to such factors as climate change, natural catastrophe or genetic mutation. Background extinction rates have been conservatively estimated at one mammal species every 400 years and one bird species every 200 years. However, since 1600, 484 species extinctions have been documented with 58 mammal and 115 bird extinctions, most of them on islands: 63% of total, 59% of mammals and 90% of birds (WCMC, 1992). These, however, are only the recorded extinctions and almost certainly underestimate the real number. As Table 4.1 in Section 2.3 above showed, we know so little about life on Earth that many extinctions must go unrecorded, especially in the Tropics. The widespread habitat and population destruction, especially in the species-rich Tropics, has only occurred in the last few decades. The full effects of this will take several generations to appear.

Edward Wilson has estimated that we are now pushing 27 000 species a year into extinction (Wilson, 1992, p. 268), and Robert May (1990, p. 304) has suggested that about half of all terrestrial species are likely to become extinct over the next 50 to 100 years. Given that about half of the extant species appeared over the last 50 to 100 million years, if these estimates are true then the rate of extinction and rate of speciation are out of balance by a factor of a million! (*Silvertown* (1996b) describes some of the uncertainties associated with calculating rates of extinction.)

## IUCN Red List Categories

During the 1960s Sir Peter Scott developed the concept of the Red Data Book (RDB) as an attempt to categorise species at risk according to the severity of the threats facing them. IUCN–The World Conservation Union established five categories of threat – Extinct, Endangered, Vulnerable, Rare and Indeterminate – and RDBs were compiled by IUCN on a global basis.

*Q*   What is a 'threatened species'?

*A*   A *threatened species* is one thought to be at significant risk of extinction in the wild in the foreseeable future.

*Q*   How do we know that species are threatened with extinction?

*A*   Data on species and changes in their distribution are collated globally. The World Conservation Monitoring Centre (WCMC), in Cambridge, UK, is the major repository of data on threatened species.

More recently Red Lists have appeared which merely list globally threatened species without the accompanying distribution data included in RDBs. WCMC compiles and updates these Red Lists in association with the IUCN Species Survival Commission (SSC) network of Specialist Groups. SSC Specialist Groups comprise world experts in a particular group of plants or animals, including orchids, cacti, cats, crocodiles and alligators, and elephants. The 1995 Global Biodiversity Assessment estimated that the minimum number of species threatened with extinction includes 5 366 animals and 26 106 plants (Table 4.5). However, these estimates cover only those species known and examined by taxonomists.

These categories of threat have been criticised in recent years, particularly on the grounds that they can only be applied to species where there are full data on the decline of a species across its entire range. After a five-year consultation period, IUCN adopted a new, expanded set of Red List Categories in 1994. They are based on a more quantitative approach and include estimates of population decline and possible time to extinction. The revised system has ten categories: see Figure 4.3.

**WORLD CONSERVATION
MONITORING CENTRE**

*Table 4.5   Numbers of species considered at threat by the World Conservation Monitoring Centre*

|  | Category of Threat | | | | | Approx. % threatened [Total as % of described species] |
|---|---|---|---|---|---|---|
|  | Endangered | Vulnerable | Rare | Indeterminate | Total | |
| Plants | 3632 | 5687 | 11 452 | 5302 | 26 106 | 10 |
| Mammals | 177 | 199 | 89 | 68 | 533 | 12 |
| Birds | 188 | 241 | 257 | 176 | 862 | 9 |
| Reptiles | 47 | 88 | 79 | 43 | 257 | 6 |
| Amphibians | 32 | 32 | 55 | 14 | 133 | 3 |
| Fishes | 158 | 226 | 246 | 304 | 934 | 4 |
| Invertebrates | 582 | 702 | 422 | 941 | 2 647 | 0.2 |

Source: based on figures from Heywood, V. H. and Watson, R. T. (eds) (1995) *Global Diversity Assessment*, Cambridge, Cambridge University Press, p. 234

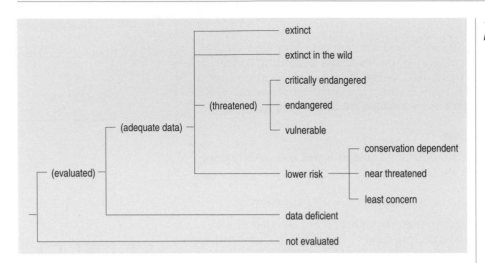

Five quantitative criteria have been developed for the critically endangered, endangered and vulnerable categories. These are population and habitat decline, habitat area, population fragmentation or isolation, population size, and population viability analysis. Different quantitative thresholds have been used for each category. For example, the population size criterion to qualify as critically endangered is less than 50 mature individuals; for endangered it is less than 250; and for vulnerable less than 1000. The revised categories have been tested on a range of species and are now being field-tested. It is too early to evaluate their effectiveness, but the use of more objective, scientifically-based criteria is an important step in the monitoring and classification of threatened species.

## 4.2   Causes of biodiversity loss

The current losses in biodiversity have both direct and indirect causes. The direct causes are relatively well-known and are chiefly: habitat loss and fragmentation; invasion by introduced species; over-exploitation of plant and animal species; pollution; industrial agriculture and forestry; and global climate change. However, these are not the root of the problem which lies more in the way humankind has exploited the environment, and the costs and benefits of both exploitation and conservation. In the long term the loss in biodiversity can only be stemmed if we address these indirect causes, ensuring that the exploiters pay the full costs of their exploitation and that the conservers earn more of the benefits of their actions.

### Direct causes of loss of biological diversity

Analysis of the known animal extinctions, where the cause was known, has shown that the major factors involved in these extinctions were: introduced animals (39%), habitat destruction (36%) and over-exploitation (23%) (WCMC, 1992). Although this analysis was based on island animal species, and there are difficulties extrapolating to plants and continental species, it is generally agreed that habitat loss, introduced (exotic) species and over-exploitation are the major threats to biodiversity worldwide.

Over one hundred species of vertebrates, invertebrates and plants are thought to have become extinct in the UK this century, largely due to human activity. These include the mouse-eared bat (*Myotis myotis*), last recorded in 1990, due to excessive disturbance at, and destruction of, nursery sites; horned dung beetle (*Copris lunaris*), last recorded in 1955, due to ploughing up of pastures on chalky and sandy soils; viper's bugloss moth (*Hadena irregularis*), last recorded in 1979, due to loss of Breckland heath to agriculture and development; hairy spurge (*Euphorbia villosa*), last recorded in 1924, due to the cessation of woodland coppicing at its only site; summer lady's tresses (*Spiranthes aestivalis*), last recorded in 1959, due to drainage of the bogs where this orchid grew.

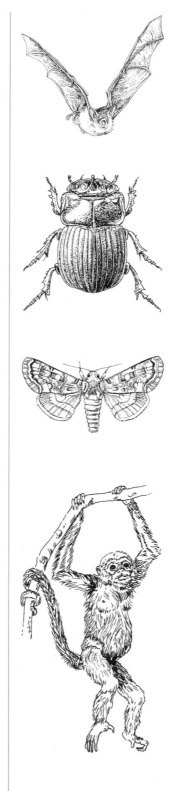

### (a)   Habitat loss and fragmentation

Virtually any form of human activity results in some modification of the natural environment. Recent estimates have suggested that more than half the habitable surface of the planet has already been significantly altered by human activity (Heywood and Watson, 1995). The area of natural ecosystems undisturbed by humankind has reduced dramatically over the past few decades as human population and resource consumption have grown inexorably. This inevitably means a reduction in population sizes of other species, with the resulting loss in genetic diversity. This can lead to increased vulnerability to disease and random population changes and in extreme cases to extinction. This may result from the habitat being made unsuitable for the species (for example, through clear-felling of forests or severe pollution of rivers), or through the habitat becoming fragmented. Habitat fragmentation divides previously contiguous populations of species into small sub-populations. If these become too small, then chance processes lead to increased probabilities of extinction within a relatively short time. *Silvertown* (1996b) discusses habitat loss in Britain and some of its consequences.

Quoted global extinction rates are derived from extrapolations of measured and predicted rates of habitat loss, and estimates of species richness in different habitats. These two estimates are interpreted in the light of the principle derived from island biogeography (MacArthur and Wilson, 1967) which suggests that the size of an area and of its species complement tend to have a predictable relationship: fewer species are able to persist in a number of small habitat fragments than in the original unfragmented habitat, and this can result in the extinction of species. MacArthur and Wilson suggested that the species number halved for each reduction in area by 90%. For example, the Atlantic coastal forest of Brazil, a Myers' hot-spot (shown in Figure 4.2), is home to numerous endemic plant and animal species, including the famous golden lion tamarin and woolly spider monkey. The forest once formed a continuous belt over 2000 km long from Bahia in the north to Rio Grande do Sul in the south. Primary forest has been reduced to less than 10% of its original extent and what remains is in scattered fragments, very few larger than 1000 ha. Since most of the remaining fragments individually comprise less than 0.1% of the original forested area, their biodiversity could be reduced to $(0.5 \times 0.5 \times 0.5 = 0.125)$ one-eighth that of the intact forest, based on this model.

Concerns about the deleterious effects of habitat fragmentation have led to the idea of providing wildlife corridors to link up separate habitats. The Greenways programme in the USA is particularly well developed. Greenways are corridors of protected open space that are managed for conservation as well as recreation. They follow natural land or water features, such as ridges or rivers, or human landscape features like

*Forest in the Mata Atlantica, Brazil.*

abandoned railroads or canals. They link natural reserves, parks, and cultural and historic sites. Greenways protect environmentally sensitive land and wildlife, and provide people with access to outdoor recreation, thus providing ecological, social and economic benefits. Important wildlife corridors in the UK are hedgerows, railway-lines and canal banks. However, such corridors can also provide escape routes for introduced species (see below). For example, Oxford ragwort (*Senecio squalidus*), introduced to the UK from southern Europe, escaped from Oxford Botanic Gardens along railway-lines, has become naturalised and has also hybridised with the native groundsel (*Senecio vulgaris*).

### (b)   Introduced species

Introduced species are responsible for many recorded species extinctions, especially on oceanic islands. When a species is introduced into a new environment, it is often freed of its natural competitors and pathogens and invades the natural habitat. This can lead to a population explosion which can displace local species unable to survive the alien invasion. In Hawaii, eighty-six introduced plant species are seriously threatening its unique native biodiversity. These include guava (*Psidium cattleianum*) which now dominates large tracts of wet evergreen forest on Hawaii, as well as the Indian Ocean island of Mauritius, to the detriment of the rich endemic native flora. In the UK, *Rhododendron ponticum* now forms a thick understorey in many woodlands, preventing natural regeneration of native trees. It also spreads on open hillsides and is a serious problem in, for example, the Snowdonia National Park. (See Plate 8.)

The mongoose was introduced from India to many Caribbean and Hawaiian Islands to control heavy infestations of roof rats in sugar-cane fields in the nineteenth century. However, the rat was nocturnal and arboreal, whilst the mongoose was mostly diurnal and terrestrial. Instead of destroying the rats, the mongooses quickly exterminated many defenceless native birds. The effects of the introduction of the Nile perch on the endemic cichlid fauna of Lake Victoria are well documented (see, for example, *Silvertown*, 1996b).

The eradication of invasive species is extremely difficult, and efforts to remove alien pest species with pesticides or by physical means are expensive, time-consuming and probably not sustainable on a large scale. These problems have encouraged the development of **integrated pest management (IPM)**, which involves a reduction in the dependence on pesticides and increased reliance on biological control. (See Box 4.6.) The use of IPM in rice cultivation in Asia is discussed by *Woodhouse* (1996). The predators, parasites and diseases of pests used in biological control are a large component of the world's biodiversity. Knowledge of the distribution and ecology of invasive species in their native habitat and of their natural enemies is an important first step in identifying potential biological control agents. This adds another powerful justification to the importance of species inventory work discussed in Section 2.

"And now Edgar's gone. ... Something's going on around here."

---

*Box 4.6   Using biodiversity to protect biodiversity*

The endemic gumwood, *Commidendrum robustum,* is the national tree of the South Atlantic island of St Helena, a UK Dependent Territory. The last remaining stands of gumwoods on St Helena – two small forest patches comprising about 2000 trees – were threatened by an introduced scale insect, *Orthezia insignis.* Populations of this sap-sucking insect, which has no natural enemies on St Helena, were leading to the death of these gumwoods. This could have lead to the extinction not only of *C. robustum,* but also several endemic insect herbivores which rely on the gumwoods for their continued survival.

*O. insignis* originated in South America, but is widespread in many tropical countries. It was unintentionally introduced on to St Helena in the 1970s or 1980s, probably on plant material brought in from South Africa. *O. insignis* has a wide host range and several other endemic trees on St Helena could also be under threat. Biological control was the only long-term solution. Since *Orthezia* is such a widespread pest, a search for natural enemies in South America had already identified a specialised predatory coccinellid beetle, *Hyperaspis pantherina,* and a successful biological control programme was developed. The release of this predatory beetle on St Helena in 1993 and 1994 has dramatically reduced the *Orthezia*

population and the gumwoods are recovering. Predator and pest populations are being monitored and they seem to survive in low numbers without any apparent adverse effects on the native ecosystems.

The continuing search for natural enemies of potential pest species depends on the conservation of native habitats and their biodiversity.

*High Knoll, St Helena*

Source: Simon Fowler, International Institute of Biological Control, and Michael Maunder, Royal Botanic Gardens, Kew (personal communication).

---

*(c)   Over-exploitation of plant and animal species*

The over-harvesting of species for food is probably the most obvious type of human destruction of biological resources. Numerous forest, fisheries and wildlife resources have been over-exploited, sometimes to the point of extinction. Once individuals are taken at a rate faster than can be sustained by natural reproductive capacity, the species is on a cataclysmic decline. Well-known cases include flightless birds such as the dodo (1665) and the great auk (1844) which were hunted to extinction (see *Silvertown*, 1996b, for other examples).

*(d)   Industrial agriculture and forestry*

During the twentieth century there has been a radical shift away from diversity of crops and livestock varieties towards monocultural agriculture (see *Grigg*, 1996; *Taylor*, 1996). Mixed forests have been converted to single-species plantations. Modern plant-breeding programmes have led to the planting of fewer varieties of crops that respond better to fertilisers, pesticides and water with the consequent reduction in genetic diversity. In Bangladesh, for example, more than 60% of rice varieties derive from a single material plant, and 80% of vegetable varieties grown in the USA a century ago have been lost. The reliance on fewer varieties has highlighted the need to conserve the genetic diversity of crop plants and their wild relatives. This has been achieved for many important crop plants by various international centres of agricultural research, now linked together as the Consultative Group on International Agricultural Research (CGIAR).

*Woodhouse* (1996) described the work of the International Rice Research Institute, one of the thirteen-strong CGIAR network, in building a collection of the existing rice varieties worldwide. The Rome-based International Plant Genetic Resources Institute (IPGRI) co-ordinates a network of national and regional centres for genetic conservation and concentrates on the conservation and utilisation of genetic resources of agricultural crops and forest tree species. The International Centre for Research in Agroforestry (ICRAF), based in Nairobi, is concerned with the role of agroforestry and in particular the incorporation of multi-purpose tree species in sustainable farming systems. The most recently established, the Centre for International Forestry Research (CIFOR), is based in Bogor, Indonesia, and focuses on the conservation of forest genetic resources. They work on both natural ecosystems and their management, as well as on plantations.

### (e) Chemical pollution

We have known for a long time that pollutants can have disastrous effects on local populations of sensitive species. The publication in 1962 of Rachel Carson's *Silent Spring* alerted the public to the insidious and widespread effects of pollution. Carson pointed out that pesticides were causing declines in many non-target organisms and warned that they may even drive some species into extinction. Many of the environmental impacts of polluting industries (mining, manufacturing and energy) were discussed in *Blunden and Reddish* (eds, 1996).

Chemical pollution is a complex problem expressed in many forms. Atmospheric pollution, particularly as oxides of sulphur and nitrogen, has caused devastating direct effects on vegetation around the world when deposited as acid rain (*Reddish and Rand*, 1996). Excessive use of agricultural chemicals has led to the contamination of watercourses resulting in ecological imbalances in wetlands, estuaries and shallow seas. Persistent pesticides, like DDT, accumulate in food chains and are particularly harmful to top carnivores like raptors (birds of prey). Their direct effect was to cause thinning of the egg-shells so that broods were destroyed. However, species can recover if the pollution source is removed, as have raptor populations in the UK since DDT and its relatives were banned in the early 1970s. However, barn owl populations in the UK have fallen by 10% since the introduction of new rodenticides. The use of illegal pesticides to control crayfish along the boundaries of Spain's Costa Donana National Park in 1985 killed 30 000 birds. Release of toxic chemicals and heavy metals from industrial sources has direct impacts on land, freshwater and inshore sea ecosystems.

### (f) Global climate change

Current computer models predict a doubling of atmospheric carbon dioxide by the middle of the twenty-first century which will result in an increase in the Earth's average surface temperature of 1.5–4°C (see Chapter 3 for details). This rate of warming is unprecedented, being much faster than rates experienced after the last Ice Age. In northern temperate zones a 1°C rise in temperature could push a species' southern range limit about 120 km northwards, or 150m vertically up a mountain: see Figure 4.4. The ability of species to migrate both far enough and fast enough to keep pace with this predicted warming will be critical for their survival, especially in a modern landscape such as the UK where many species are restricted to isolated fragments of semi-natural habitats. A number of species of Arctic and boreal species are at their southern limit in the UK and warming may cause them to become extinct in the UK. Victims of local UK extinction could be the dotterel (*Eudromias morinellus*) which nests in

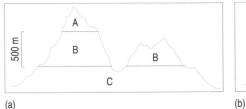
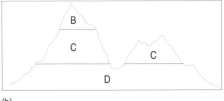

(a)                                                          (b)

▲  *Figure 4.4    (a) Present altitudinal distribution of three species, A, B and C.*
*(b) Species distribution after a 500m shift  in altitude in response to a 3°C rise in*
*temperature. Species A becomes locally extinct; species B shifts upward, and the*
*total area it occupies decreases; species C becomes fragmented and restricted to a*
*smaller area, while species D successfully colonises the lowest altitude habitats.*

northern Scotland and the yellow mountain saxifrage (*Saxifraga hirculus*).
Wetlands in eastern and northern Britain and upland blanket bogs could
become drier, favouring colonisation by grasses and trees. Conditions for
species at their northern limit in southern Britain such as the Dartford
warbler (*Sylvia undata*) may become more favourable. In addition, the
predicted rise in sea-level could completely submerge many of the world's
island and low-lying coastal areas resulting in the loss of entire ecosystems.
The Royal Society for the Protection of Birds (RSPB) has estimated that
1–2% of wintering waders are at risk from rising sea-levels and that some
nesting waterfowl – including its emblem species, the avocet (*Recurvirostra
avosetta*) – could become extinct in the UK.

   Regional effects such as El Niño, hurricane activity and monsoon
patterns also seem to be affected by humankind's activities and these, too,
may have significant effects on biodiversity in the future.

   Global climate change and the predicted effects on species distribution
will have a significant effect on the widespread policy of habitat and species
protection, based on protected areas (discussed in Section 5.1). In a near-
future altered landscape, will our Parks and Protected Areas be in the right
place to protect the habitat, and so the biodiversity, which they were
established to protect? Figure 4.5 illustrates a possible scenario.

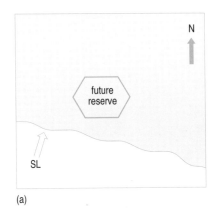
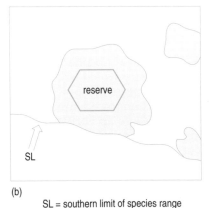
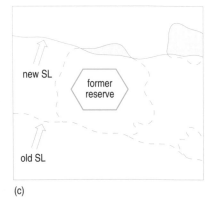

(a)                              (b)                              (c)

SL = southern limit of species range

▲  *Figure 4.5    How climatic warming may turn biological reserves into former reserves.*
*Shading indicates in (a) species distribution before human habitation; in (b) fragmented species distribution after*
*human habitation and in (c) species distribution after warming.*

◄ *Plate 1*
*The classic Apollo 17 view of the Earth (taken in December 1972), showing Africa and Antarctica.*

▼ *Plate 2*
*A satellite remote-sensing image of the world ocean, over a period of 18 months. The image has been computer-processed, so the colours are not natural, to show plankton concentrations: yellow and orange areas have the most plankton, purple areas the least. Black indicates an area with not enough data for the image.*

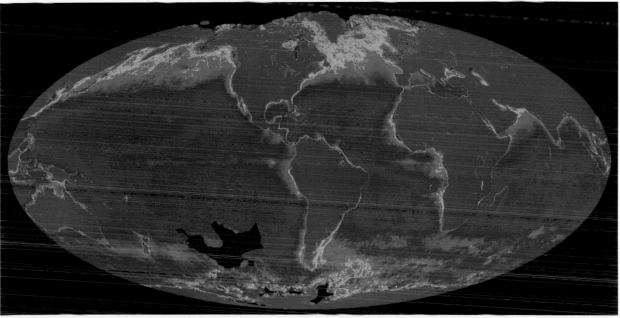

(a) Observed surface air temperature, 1971–90 minus 1881–1920

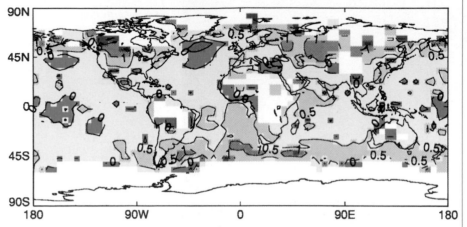

(b) Simulated (greenhouse gas only)

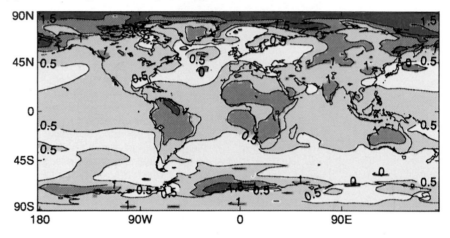

(c) Simulated (greenhouse gas plus aerosol)

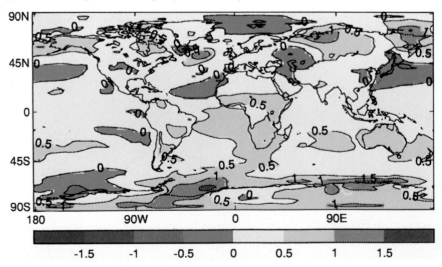

-1.5    -1    -0.5    0    0.5    1    1.5

◀ *Plate 3*
*Global climate change over the last century, reconstructed using a coupled ocean–atmosphere GCM from the Hadley Centre in the UK Meteorological Office which includes the cooling effects of sulphate aerosols. Plate 3a shows the observed change in global surface air temperature (1979 to 1990 minus 1881 to 1920). The temperature changes simulated by the model are shown with greenhouse gases (Plate 3b) and with greenhouse gases and aerosols (Plate 3c).*

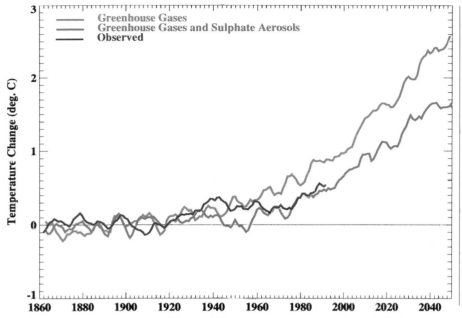

◄ Plate 4
Observed (1860 to present day) and predicted (1860–2050) surface air temperature changes (in °C), for greenhouse gases and for greenhouse gases and sulphate aerosols combined, based on the coupled global circulation model developed at the Hadley Centre in the British Meteorological Office.

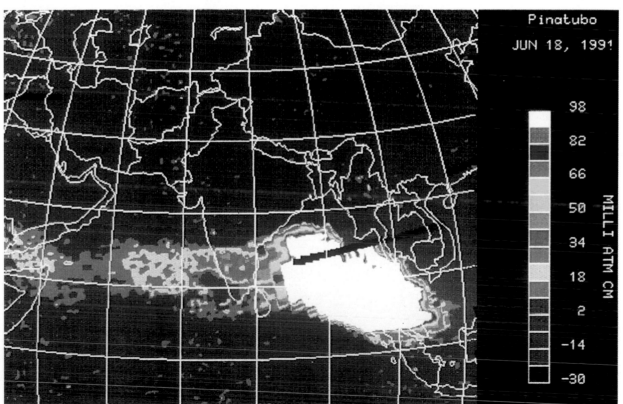

▲ Plate 5
A false-colour map showing the initial spread of sulphur dioxide, released by the eruption of Mount Pinatubo volcano in the Philippines, June 1991. The main concentration of gas, shown as white, lies between the Philippines and the Indian sub-continent, but smaller amounts have spread within a few days as far as the Arabian Sea. The image was prepared from data recorded on 18 June by the Total Ozone Mapping Spectrometer (TOMS) on the Nimbus-7 satellite.

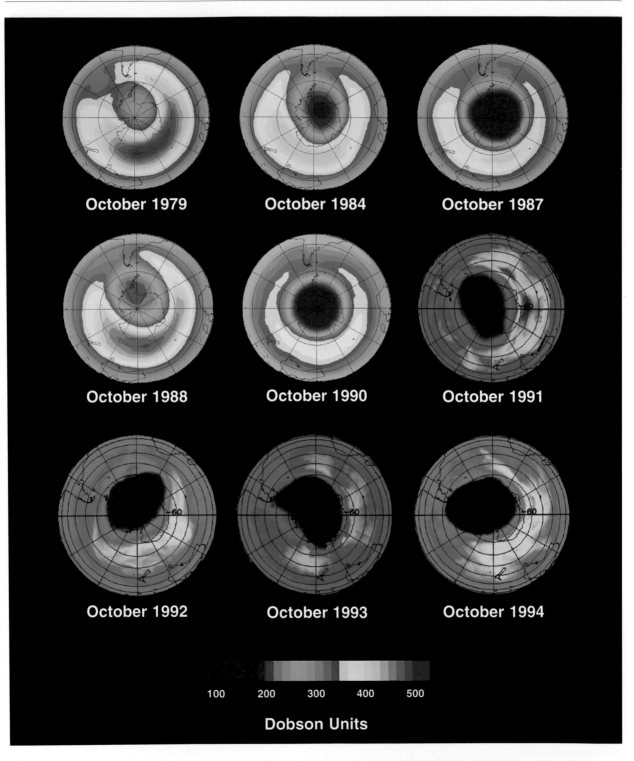

October 1979 October 1984 October 1987

October 1988 October 1990 October 1991

October 1992 October 1993 October 1994

100 200 300 400 500

**Dobson Units**

▲ *Plate 6 'Maps' of the ozone column (monthly mean total ozone) over the southern hemisphere for 1979, 1984, 1987, 1988 and 1990–94, based on the TOMS data from the Nimbus-7 satellite (1979–91) and the Meteor-3 satellite (1992–94). In the ozone 'hole' (purple) over Antarctica, the ozone abundance is roughly half that seen at the start of the series. This broader view conveys dramatically the increase of ozone depletion apparent in the ground-based data from Halley Bay (Figure 2.15).*

▲ Plate 7
Polar stratospheric clouds
seen from the NASA DC-8
aircraft at nearly 39 000 feet
in the polar regions north of
Stavanger, Norway, during
the Airborne Arctic
Stratospheric Expedition in
January/February of 1989.

◄ Plate 8
Rhododendron ponticum in
flower at Staple Plain,
Quantock Hills, Somerset,
England.

◄ *Plate 9*
*Lady's Slipper Orchid*
*(Cypripedium calceolus).*

► *Plate 11*
*Biosphere Reserve, Cape*
*Tribulation, Queensland,*
*Australia, designated under*
*the UNESCO Man and the*
*Biosphere (MAB)*
*programme.*

▼ *Plate 10*
*Fynbos, Storms River Valley, Cape*
*Province, South Africa.*

► *Plate 12*
*'Coalbrookedale, by night',*
*Philip James de*
*Louthebourg, 1801.*

◄ Plate 13
Smokestack effluent from a lignite (brown coal) power station in Bitterfeld, Germany. Air pollution from the combustion of lignite is a serious health hazard in eastern Germany and other former Soviet bloc countries.

▲ Plate 14
This 'gasifier' is a good example of self-build intermediate technology at work in Dhanawa village, India.

## *Indirect causes of loss of biological diversity*

The factors considered in the previous section are the processes that directly cause loss of biodiversity. These processes, however, are underpinned by more fundamental causes. As we gradually awaken to the global damage that is being caused by unsustainable development and exploitation of biodiversity, we must search for solutions to the real problems that have caused the biodiversity crisis. These lie in a re-evaluation of the way we live as a species and our fundamental relationship with the other species with which we share this planet. In the same way that biodiversity is an essential resource for sustainable development, finding sustainable ways to live is essential if biological diversity is to be conserved.

The Global Biodiversity Strategy published in 1992 by the World Resources Institute (WRI), IUCN–The World Conservation Union and the United Nations Environment Programme (UNEP) identified six fundamental causes of biodiversity loss (WRI, IUCN and UNEP, 1992):

- the unsustainably high rate of human population growth and natural resource consumption
- the steadily narrowing spectrum of traded products from agriculture, forestry and fisheries
- economic systems and policies that fail to value the environment and its resources
- inequity in the ownership, management and flow of benefits from both the use and conservation of biological resources
- deficiencies in knowledge and its application
- legal and institutional systems that promote unsustainable exploitation.

These are all social issues – economic, political and cultural – linked amongst other causes to the problems of a population explosion out of control. Science can only take us so far. Through identifying, monitoring and investigating biodiversity, scientists can provide politicians and decision-makers with the stark reality of the consequences of the loss of biodiversity. Political pressure must be exerted to ensure that this is taken seriously. These political dimensions are discussed in the following chapters of this book.

# 5   *Conservation and management of biodiversity*

Traditionally, two main types of management approach have been adopted for the conservation of biodiversity:

- protection of individual species
- protection of the habitat.

Efforts directed at the species and population level are both time- and resource-intensive and as such can only ever be directed towards a very small percentage of threatened species. These have inevitably been the 'charismatic megavertebrates' including whales, elephants and pandas, and showy plants such as orchids. Measures include legal protection for individual species, the development of management plans for the protection of these species, supported by *ex situ* **conservation** in zoos, botanic gardens and gene banks.

Q    What are the fundamental differences between *ex situ* and *in situ* conservation measures?

A    *In situ* **conservation** is the protection of species within their natural habitat where evolutionary processes can continue. *Ex situ* conservation constitutes measures to protect species away from their natural habitats and provide an insurance policy against extinction of species or genetic diversity in nature. Conservation biologists now appreciate that these are complementary approaches and that integrated conservation strategies should take both types of activity into account.

Habitat conservation has mainly taken the form of the identification of protected areas designated as national parks or nature reserves. Historically, human activity has been excluded or very strictly controlled in these protected areas. The protected area concept is being re-evaluated in an attempt to better accommodate humans whilst safeguarding biodiversity. It is becoming increasingly evident that wildlife conservation will not be possible without the active collaboration of the local population.

Whilst these approaches will continue to be important, conservation must become more innovative and wide-ranging if biodiversity is to be safeguarded against the increasing threats that it faces. The effective application of conservation technology requires additional resources including finance, personnel and political commitment as well as cross-sectoral support. However, individual species can still serve as symbols for fund-raising.

## 5.1   Legislation and protected areas

Most national governments have established legal mechanisms for protecting and regulating the use of habitats that are important for conserving biodiversity and biological resources. These include national legislation to establish National Parks and other protected areas as well as local laws to protect particular forests or wetlands. In 1985 IUCN published a system of eight categories of protected areas, each designed to achieve a particular series of management objectives. In 1994 these were revised and the six new categories range from Strict Nature Reserves and National Parks to Managed Resource Protected Areas. In the former, no form of human exploitation is permitted but scientific and educational activities are allowed, and in the latter sustainable utilisation of certain biological resources is allowed; thus biodiversity is conserved whilst sustainable benefits are provided to local communities. The UK's interpretation of National Parks, where human exploitation is allowed, is described by *Sarre* (1996) and by *Blunden and Curry* (1996).

Just over 6% of the world's land area is now protected under the aegis of the IUCN system, but with wide variation between regions. North America has set aside nearly 12% of its land in parks and reserves, Central America and Europe just over 9%, South America, Africa and Asia 5–6% each, whilst the countries of the former Soviet Union have protected just over 1%. The current patchwork of protected areas has raised concerns about the size of protected areas and ecosystem representation, as well as

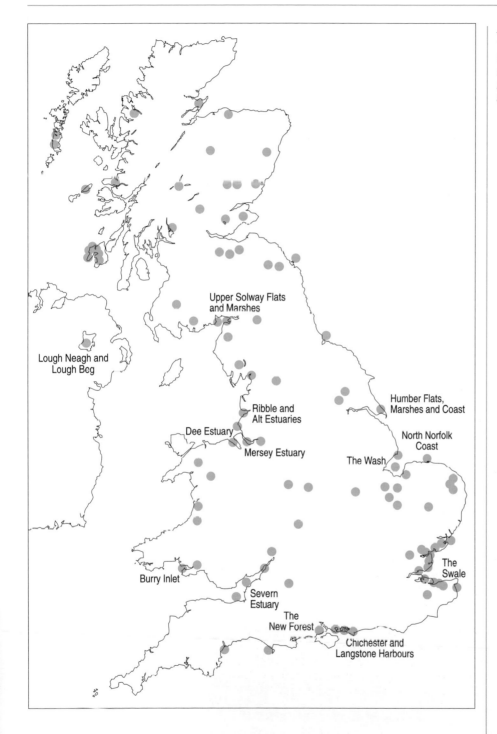

◀  *Figure 4.6*
*The Ramsar Sites in the*
*United Kingdom, as at*
*February 1996. Those*
*named are sites over*
*5000 ha in extent.*

concerns about lack of community participation and management
effectiveness. Many protected areas around the world are threatened by
farmers, poachers, illegal logging and mining, and the lack of enforcement
of the very laws established for their protection. There is also concern that
many of these protected areas may become natural habitat islands
surrounded by a sea of increasingly disturbed and human-modified land.

These islands may also be too small to support viable populations of the large animals, especially predators, which need a large home range.

This habitat approach is supplemented at an international level by a number of binding Conventions. The Ramsar Convention on Wetlands of International Importance, drawn up in 1971, is the only global nature conservation convention designed to cover a particular broad habitat type, that of inland, coastal and marine wetlands. These so called Ramsar Sites are of particular importance to migratory waterfowl. By January 1996, 92 parties had ratified the Convention and 775 sites have been designated covering over 52 million hectares. In the UK 96 sites have been designated ranging in size from 14 ha (Llyn Idwal in Wales) to the 63 124 ha of the Wash in England: see Figure 4.6.

The Convention concerning the Protection of the World Cultural and Natural Heritage (Paris, 1972) provides for the protection of unique natural and cultural areas. The Convention has over 100 State Parties and has established a network of 95 World Heritage Sites which include Queensland Rainforests (Australia), Manu National Park (Peru), Serengeti National Park (Tanzania) and Great Smokies National Park (USA).

A new management approach which seeks to reduce human impact on protected areas involves the development of *Integrated Conservation-Development Projects (ICDPs)*. These ICDPs promote the management of protected areas within the context of the management of the surrounding landscape, often by establishing **buffer zones** around the protected area.

*Q*   What are buffer zones?

*A*   *Buffer zones* are regions adjacent to protected areas which provide local communities with sustainable, income-generating opportunities. These may include selective logging, hunting or firewood collection.

*Biosphere Reserves* are well-known examples, consisting of a core protected area, often an existing park, surrounded by buffer zones where limited human activity is allowed. These transitional areas extend the total area

◀   *The Isle of Rhum, a designated Biosphere Reserve.*

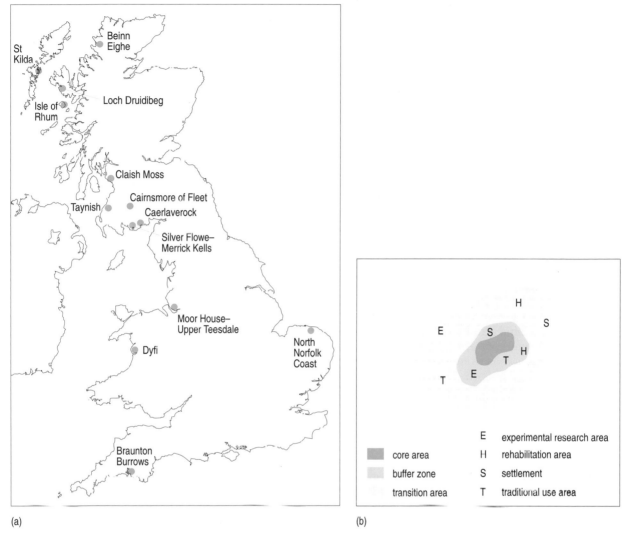

▲   *Figure 4.7   (a) Designated Biosphere Reserves in the UK. (b) A model biosphere reserve.*

that can be managed for biodiversity beyond the core area, whilst offering local people economic opportunities through sustainable utilisation of a variety of biological resources. In this way they have conservation and development as equal management objectives. The establishment of Biosphere Reserves is part of an international scientific programme, the UNESCO Man and the Biosphere (MAB) Programme. (See Plate 11.) Biosphere Reserves differ from Ramsar Sites and World Heritage Sites in making provision for people who are considered as an important component for the management of Biosphere Reserves. Some 324 sites considered as Biosphere Reserves have been designated, forming an international network in 82 countries covering nearly 162 million hectares. Thirteen areas in the UK, totalling 44 258 hectares, have been designated as Biosphere Reserves. These include Moor House National Nature Reserve in Upper Teesdale (described by *Silvertown*, 1996a), the Isle of Rhum and the sand dune systems of Braunton Burrows in Devon: see Figure 4.7.

ICDPs are often multi-sectoral collaborations, involving governmental agencies, private groups and non-governmental organisations (NGOs) in partnership in environmental management. Private groups and NGOs are often in a better position to raise money to purchase land for protection and to support conservation in existing protected areas.

One funding mechanism that has met with some success is the *Debt for Nature Swap*. This mechanism is most useful in countries whose debts are heavily discounted, such as in Latin America, where debt burdens are very high and economies stagnating. The debt-swap mechanism involves the purchase of a developing countries' secondary debt in the secondary debt market. This is usually purchased by an international conservation organisation, such as WWF or Conservation International, but may be by an individual or another government. Such secondary debt is sold by existing holders at a discount which reflects the markets' judgement on the probability of repayment: the less likely the repayment, the greater the discount. The debt-holding is then presented to the debtor country in exchange for local currency which is invested in a conservation project in that country, usually managed by a local NGO. The money may be used to purchase land or to train conservation managers, or in other ways that directly improve in-country conservation. The first Debt for Nature swap was negotiated with Bolivia by Conservation International in 1987. Since then it has been used for effective conservation in many countries including Madagascar, Zambia and Ecuador. Costa Rica, one of the most effective debt-swappers, has converted over US\$ 80 million of debt into funds for the protection of hundreds of millions of hectares of tropical forests.

Despite their problems, these examples indicate that protected areas still play a critical role in conserving the world's biodiversity.

## 5.2   *Conservation of individual species*

Establishing protected areas does not always ensure adequate protection of threatened species. Poachers do not respect protected areas, regardless of penalties. The excessive harvesting of wildlife species for commercial gain is a major threat to biodiversity. Poaching of animals, often just for one product, such as elephant ivory, and the illegal collection of plants, notably orchids, are pushing some species towards extinction. In an effort to curb poaching and to regulate trade in endangered species, the Convention on International Trade in Endangered Species of Wild Fauna and Flora (CITES) was negotiated in 1973. It came into force in 1975 and, by January 1996, 130 states were party to what has been described as 'perhaps the most successful of all international treaties concerned with the conservation of wildlife'.

CITES controls and monitors international trade in wild specimens and derived products of threatened species. The Convention operates by a licensing system. At the heart of the Convention are three Appendices, or species lists. Appendix I lists those species of animals and plants in which trade in wild specimens is prohibited. International trade is allowed, subject to licensing, in artificially propagated or captive-bred specimens of Appendix I plant and animal species. The most well-known Appendix I animal is the African elephant (*Loxodonta africana*). In addition all apes, the giant panda, cheetahs and tigers are on Appendix I. Appendix II includes species whose survival in the wild is not yet threatened, but may become so. Trade is allowed in Appendix II species, subject to licensing. All orchids and cacti are on Appendix II with the most endangered species listed on

Appendix I. This reflects the years of over-collection in the wild. The UK's most famous orchid, the Lady's Slipper Orchid (*Cypripedium calceolus*), has been reduced by over-collecting to a single individual plant which now has a full-time warden looking after it. (See Plate 9.) It is the subject of a species recovery and re-introduction programme co-ordinated by English Nature and the Royal Botanic Gardens, Kew. Appendix III acts as a support mechanism to domestic legislation where countries ask other Parties to monitor trade in particular species which are not listed on Appendix I or II. *Swietenia macrophylla* (mahogany) is currently listed on Appendix III so that trade in this valuable timber species can be monitored.

WCMC maintains a database that includes all records of international trade in species listed on CITES Appendices involving a Party State. It also monitors trade on some unlisted species with a view to possible inclusion. The TRAFFIC (Trade Records Analysis of Flora and Fauna in Commerce) Network is an international NGO, funded mostly by WWF, whose international headquarters are based at WCMC in Cambridge, UK; it monitors trade in and utilisation of wild plants and animals and now has seventeen offices worldwide.

## 5.3    Ex situ conservation

We have already seen that the most effective and efficient mechanism for the conservation of biodiversity is habitat protection (Section 5.1). However, off-site (*ex situ*) activities can be critical components of a comprehensive conservation programme. The techniques and facilities available include:

- gene banks such as seed banks, field banks and sperm and ova banks
- *in vitro* plant tissue and microbial culture collections
- captive breeding of animals and artificial propagation of plants, with possible re-introduction into the wild
- collections of living organisms in botanic gardens, zoos and aquaria for research, and public education and awareness.

### Botanic gardens

About 1500 botanic gardens and arboreta have been established around the world, of which about 800 are thought to be currently involved in active plant conservation. However, there are still relatively few in tropical countries, where plant diversity is greatest. Greatly increased international support is needed for tropical botanic gardens to enable them to participate fully in the international effort to conserve biodiversity.

Botanic Gardens Conservation International (BGCI) co-ordinates the activities of botanic gardens at an international level. The Secretariat disseminates information to promote and co-ordinate the *ex situ* conservation of threatened wild plant species. It also provides technical guidance, data and support for botanic gardens in about 100 countries and assists and promotes the development of botanic gardens and their land conservation programmes. BGCI currently has a worldwide membership of 317 botanic gardens. Botanic gardens are playing an increasing role in conservation and maintenance of genetic diversity. This role is especially important in newly created tropical botanic gardens. For example, the Conservatoire et Jardin Botanique de Mascarin on Reunion Island, founded in 1987, contains over 60% of the island's threatened plant species in

**Botanic Gardens Conservation International**

cultivation. Active programmes in many botanic gardens include: documenting the local flora; maintaining threatened species in cultivation; propagating species for possible reintroduction and habitat restoration projects; maintaining herbaria of named specimens with abundance, distribution and economic use data; and maintaining and monitoring nature reserves either within or associated with the garden.

The genetic diversity maintained in most *ex situ* collections in botanic gardens is inadequate for conservation purposes. Living collections rarely constitute representative samples of the genetic variation of the species. Potentially important species are often represented by a single genotype. Efforts are now being made to re-evaluate the genetic management of collections to maximise genetic diversity.

## Zoos and aquaria

As with botanic gardens, the conservation role of zoos has been increasing. Currently there are approximately 700 000 individuals of 3000 species of mammals, birds, reptiles and amphibians in about 800 professional zoos around the world. About 5800 species of fish, mostly collected from the wild, are held in aquaria. Many zoos now have captive breeding programmes for endangered species, but populations of many rare and endangered species are too small to maintain their genetic diversity. It has been estimated that a breeding population of 150 is needed to maintain 95% of its genetic diversity over 50 generations. To help prevent further inbreeding and genetic depression, the International Species Inventory System maintains a breeding database. There are now stud books for 104 species of birds and mammals, 68 of which are threatened in the wild. The major zoos involved in captive-breeding programmes form a collaborative network to share breeding-stock, frozen sperm and captive-breeding techniques.

# 6   Trends in biodiversity protection: the political response

The Convention on Biological Diversity, signed by 155 states plus the EU at UNCED, reflects the growing international concern with the worldwide loss of biodiversity. The challenge facing the post-UNCED world is how to translate this international political commitment into effective action at all levels to solve the biodiversity crisis.

## 6.1   From Stockholm to Rio: the context for the Convention on Biological Diversity

The 1972 UN Conference on the Human Environment in Stockholm brought together developing and industrialised nations to try to establish the right of the human family to a healthy and productive environment. The Stockholm Conference had several important outcomes. At an international level, the United Nations Environment Programme (UNEP) was created to promote awareness and action on environmental issues within the UN. It also promoted the development of national environmental policies which resulted in the creation of environmental agencies and ministries in many countries. The decade that followed saw a series of meetings and initiatives, two of which are of particular note.

## The World Conservation Strategy

The World Conservation Strategy was launched in 1980 by UNEP, IUCN and WWF with three global objectives: to maintain essential ecological processes, to preserve genetic diversity, and to ensure the sustainable utilisation of species and ecosystems. This recognised, for the first time, that conservation and development were 'two sides of the same coin' and, with this, a new agenda was established, one that was taken forward by the World Commission on Environment and Development in 1983. Chaired by Gro Harlem Brundtland, the Prime Minister of Norway, the Commission was asked to formulate 'a global agenda for change'. The message of the Commission's report, *Our Common Future* (the Brundtland Report), was that questions of environment and questions of development are inseparable, so that governments and their people must take responsibility not just for environmental damage, but the policies that cause the damage. Some of these policies threaten the survival of plant and animal species, including the human race. The overall recommendation was that human activities needed to be redirected towards a pathway of sustainable development.

The need for an International Convention on Biological Diversity was recognised in the mid-1980s as the weight of scientific evidence began to document the increasing loss of global biodiversity, particularly in the Tropics. In 1987 UNEP called upon governments to examine the possibility of establishing an international legal instrument on the conservation and sustainable use of biodiversity. With UN resolution 44/228 in December 1989, the UN General Assembly agreed to convene the United Nations Conference on Environment and Development (UNCED), or Earth Summit, which was held in Rio de Janeiro in Brazil in June 1992. The Earth Summit gathered together the largest number of Heads of State in a single forum to address a common global issue – the global environment. *Reddish* (1996) discusses some of the broader issues during this process.

## The Inter-governmental Negotiating Committee for a Convention on Biological Diversity

Negotiations soon identified the central battlefield – arguments between countries over the exchange of biological resources in return for access to technologies, particularly biotechnology. This largely reflected North–South differences. The developing nations of the South, where biodiversity tends to be greater, insisted on sovereign rights over their own genetic resources, as well as access to benefits derived from exploiting these resources. Many developed countries of the North viewed biodiversity as a common heritage of humankind, and biotechnology to be the concern of industry, not governments, and so found these demands unacceptable. The final draft of the Biodiversity Convention prepared in Nairobi in May 1992 reflected compromise on both sides.

The road towards the conservation of global biodiversity and sustainable development which started with the Brundtland Report and culminated with UNCED was helped along the way by the publication of *Caring for the Earth* (IUCN, UNEP and WWF, 1991) which emphasised that biodiversity must be conserved as a matter of principle, as a matter of survival and as a matter of economic benefit, and by the *Global Biodiversity Strategy* (WRI, IUCN and UNEP, 1992). The Global Biodiversity Strategy called on all nations and peoples to initiate and sustain a Decade of Action to conserve the world's biodiversity for the benefit of present and future generations, and also highlighted the need for a Convention on Biological Diversity.

## 6.2   The United Nations Convention on Biological Diversity

Much was achieved at the Earth Summit (see Chapter 6). The major success in Rio for the world's animal and plant species was the signing of the UN Convention on Biological Diversity by 155 states plus the EU. Several more states, including the USA, signed later. The Convention came into force on 29 December 1993, and can be hailed as a landmark in the environment and development field. It is the first time that biodiversity, as an issue, has been comprehensively addressed, and the first time that genetic diversity was specifically covered in a globally binding treaty. At the heart of the Convention is conservation, sustainable use and sharing of benefits. Each country which is a party to the Convention has responsibility for the conservation and sustainable use of its own biological diversity and in addition has the responsibility of controlling its own activities which may threaten biodiversity, regardless of where their effects occur. As of January 1996, 138 countries have ratified the Convention (the UK ratified the Convention on 3 June 1994). Regrettably 38 of the original signatories, including the USA, have still not ratified the Convention.

The objectives of the Convention, set out in Article 1, are:

* the conservation of biological diversity
* the sustainable use of its components
* the fair and equitable sharing of the benefits arising from the use of genetic resources, including appropriate access to genetic resources and appropriate transfer of relevant technologies.

The Convention addresses four main issues:

* National sovereignty and the common concern of humankind
* Conservation and sustainable use
* Access
* Funding.

*National sovereignty and the common concern of humankind*
The original idea that biodiversity should be treated as the 'common heritage' of humankind was rejected very quickly with the recognition that most components of biodiversity are located within national boundaries and are national assets. However, this emphasis on sovereign rights over biological resources is balanced by the duties that derive from sovereignty itself and from the fact that conservation of biological diversity is a 'common concern' to the entire international community.

*Conservation and sustainable use*

The Convention contains a series of far-reaching obligations related to the conservation of biological diversity and the sustainable use of its various components. Emphasis is placed on *in situ* conservation as the fundamental form of protection of biodiversity by calling for the establishment of systems of Protected Areas for the protection of natural habitats and the maintenance of viable populations of species (see Section 5.1).

The need for the restoration of degraded land is recognised as well as the need for recovery programmes for threatened species. As a complement to *in situ* conservation, the Convention recognises the need for *ex situ* conservation measures. This defines a new role and direction for zoos and botanic gardens (see Section 5.3).

Q    What do you think would be the limitations of this view of *in situ* conservation?

A    Much more than just Protected Areas needs to be considered. Unless we have everyone living in cities, i.e. artificial ecosystems, surrounded by Protected Areas, there will always be large areas of managed, semi-natural ecosystems which are exploited, but need sensitive treatment.

Q    What do you think would be the limitations of *ex situ* conservation?

A    All species cannot be protected in zoos and botanic gardens; the gene pools will be small for most species; re-introduction of species will be impossible if the habitat has been destroyed.

*Access*

Under the Convention, biodiversity (genetic resources) is no longer freely available to all-comers. Instead, as a national asset, it can only be exploited on the basis of agreements between signatory countries. Wild, cultivated and improved species are all considered to have an economic value. These agreements form the basis of partnerships between the developing countries, which possess the biodiversity, and the developed countries, which possess the biotechnology and the finance. Three types of access are recognised: access to genetic resources; access to and transfer of relevant technology, including biotechnology; and access for the developing countries to benefits ultimately gained from the use of the genetic resources in the developed country. (See Box 4.7.)

*Funding*

A flow of resources from North to South is needed to achieve the goals of the Convention. The Global Biodiversity Strategy estimated the worldwide costs for conserving biodiversity to be approximately US$17 billion per year. Compared with the estimated US$1 *trillion* annual world military expenditure, this is a relatively modest amount (WRI, IUCN and WWF, 1992). Despite considering a number of innovative methods, agreement finally came down on a classical financial mechanism whereby new funds are paid in by developed country Parties for the exclusive use of developing country Parties. The Global Environment Facility (GEF), set up in 1990 and jointly operated by the World Bank, the United Nations Development Programme (UNDP) and the United Nations Environment Programme (UNEP), has become the interim financial mechanism under the Convention.

## Box 4.7   INBio – Costa Rica's national biodiversity institute: a success story

The Instituto Nacional de Biodiversidad de Costa Rica (INBio), established in 1989, is a non-profit institution dedicated to the conservation of Costa Rican biodiversity, and the brainchild of molecular biologist, Rodrigo Gamez and ecologist Daniel Janzen. INBio's major activities are focused on biodiversity inventory, bioprospecting, information-gathering and dissemination. Most of their activities are conducted within Costa Rica's National System of Conservation Areas, a network of government-owned parks and protected areas that covers 25% of the country.

Their first aim is an All Taxa Biodiversity Inventory (ATBI) for Costa Rica within a decade. This is thought to amount to 500 000 species of animals, plants and micro-organisms. The inventory is conducted largely by paid 'parataxonomists' – carefully trained locals who are stationed at biodiversity offices located in the parks. They spend much of their time in the field collecting specimens and recording a range of ecological data. The fieldwork is co-ordinated by INBio inventory managers, local generalist taxonomists who liaise between parataxonomists and international specialists who identify difficult material and describe new species. Plants and insects are the initial focus and information on identities, distributions and natural history is being computerised. The local expertise in inventory and information management is being shared with other tropical countries through on-site workshops.

INBio has actively sought out partners who may want to use this biodiversity information and has signed two successive two-year bioprospecting contracts with the US pharmaceutical conglomerate Merck & Co. Merck paid INBio US$1.3 million for each contract in return for samples of plants, animals and micro-organisms which can be screened for active chemicals for use in the pharmaceutical and medical industries. Expansion into other areas is planned, including organisms as sources for genes for biotechnology, biological control and pesticides. 10% of these funds and 50% of any royalties go to the Costa Rican Ministry of Natural Resources, Energy and Mines for general conservation work, whilst the rest is used for biodiversity development at INBio and in the Conservation Areas.

This was the first joint venture between a multinational biotechnology company and a biodiversity-rich country, in a venture which may become the model for similar partnerships in the spirit of the Biodiversity Convention.

*Tropical forest, Costa Rica*

Source: Heywood and Watson, 1995

Other obligations under the Convention include the need to develop national strategies, plans or programmes for the conservation and sustainable use of biodiversity. (See Box 4.8.)

**Biodiversity Assessments** to identify and monitor the components of biodiversity and **Environmental Impact Assessments (EIAs)** to evaluate the impact on biodiversity of any development projects are further obligations under the Convention. Information exchange, technical and

---

### Box 4.8   The UK National Biodiversity Action Plan

In January 1994 the UK government became one of the first countries to produce a Biodiversity Strategy and Action Plan in accordance with Article 6a of the Convention on Biological Diversity when it published *Biodiversity: the UK Action Plan*. The overall goal of the UK Action Plan is to 'conserve and enhance biological diversity within the UK and to contribute to the conservation of global biodiversity through all appropriate mechanisms'. The major objectives are to enhance:

- the overall populations and natural ranges of native species and the quality and range of wildlife habitats and ecosystems

- internationally important and threatened species, habitats and ecosystems

- species, habitats and natural and managed ecosystems that are characteristic of local areas

- the biodiversity of natural and semi-natural habitats where this has been diminished over recent decades.

In order to achieve these objectives, a programme of activities was described in the Action Plan, the key components of which were:

- to develop costed targets for the most threatened and declining species and habitats

- to improve the accessibility and co-ordination of biological datasets

- to increase public awareness and involvement by targeting key sectors

- to recognise the importance of local biodiversity action plans which complement national action plans.

A Biodiversity Steering Group was established to report on how these four key components of the Action Plan could be achieved. The Steering Committee's report (1995) identified a 'long list' of 1250 species of concern, of which 400 were short-listed because they are either globally threatened or are rapidly declining in the UK. Costed action plans were prepared for the 116 most threatened species, from mosses to orchids and insects to mammals. In addition, 15 key habitats were identified which are the primary habitat for the greatest number of species of conservation concern, broadleaved woodland being the most important. (The UK Biodiversity Action Plan classified the UK into 37 habitat types.) Costed action plans were prepared for the 15 key habitats. Additional funds need to be made available to implement these recommendations. Indicative annual costs were prepared for 1997, 2000 and 2010 and are estimated at £3.8m, £2.9m and £2.4m respectively, for the species action plans, and £12.9m, £24.5m and £37.2m for the habitat action plans.

Other recommendations include the establishment of a United Kingdom Biodiversity Database, the first phase of which will be to bring together disparate existing datasets. The development of Local Biodiversity Action Plans should form an integral feature of the local Agenda 21 process. The profile of environmental education needs to be raised and 'champions' identified who can act as key players in each sector to convey the importance of biodiversity conservation. For example, if children identify with TV soap stars then it is they whom we need to persuade to put across key messages. The Steering Committee's report was submitted to the Government on 13 December 1995. The Environment Minister, John Gummer, was enthusiastic about the report, announced that English Nature is planning a new grants programme to support the species action plans and pledged a full response to the report during the Spring of 1996.

Sources: Anon, 1994, 1995a, 1995b

---

scientific co-operation, research and training, public education and awareness are all important considerations under the Convention.

Several Parties, including representatives from the European Commission (now European Union), thought the Convention to be too timid, and to generalise the activities each national government should undertake, rather than setting out a framework for truly international action for the conservation of species. Certainly some of the language used is weak, with obligations commonly qualified with 'as far as possible' or 'as appropriate'. As a result of the protracted arguments over access to genetic diversity and technology transfer, the Convention has concentrated too heavily on these issues, at the expense of examining methods to conserve biodiversity.

The UN Convention on Biological Diversity has provided a broad framework and the challenge to solve the biodiversity crisis. The Conference of the Parties, who will meet regularly to review the Convention and its implementation, will need to find ways to strengthen the legal framework and ensure success. The issue is too important to allow failure.

*Activity 4*

Woodlands once covered most of the British Isles, but now account for around 10% of the land area (*Silvertown*, 1996a). One of the goals of the UK Biodiversity Action Plan (Box 4.8) is to 'conserve and enhance biological diversity within the UK'.

• Based upon your own experience and reading (as well as material provided in this series), list some of the ways in which biodiversity in British woodlands could be enhanced.

• Contact some local agencies such as the County Naturalist (Conservation or Wildlife) Trust, the Forestry Commission and your local council to see what woodland conservation activities are going on and what the feeling of these organisations are towards the UK Biodiversity Action Plan and conservation of local biodiversity.

• Try and find out how much discussion goes on to elicit the views of the local people, and whether these ideas are incorporated when new activities or initiatives are being planned.

• How far should local people fund these initiatives, or should the responsibility lie with the government?

## 6.3   How does the Convention on Biological Diversity address loss of biodiversity?

An essential contribution towards sustainable management of the biosphere is to name and describe its component species and this is now encapsulated in Article 7 of the Convention. This article is about the ordering and use of information on biological diversity and biological resources as described in Section 2. It requires Parties to:

• identify the components of biodiversity important for conservation and sustainable use

• monitor the components of biological diversity

• identify and monitor processes and categories of activities having or likely to have significant adverse impacts on the conservation and sustainable use of biological diversity

• maintain and organise these data derived from the identification and monitoring activities.

The significance of this comes in Article 8, *In-Situ* Conservation, which provides the main set of Convention obligations to conserve biodiversity. *In-Situ* Conservation is recognised by the Convention as the primary approach for biodiversity conservation. The importance of protected areas as a major conservation strategy is encapsulated in Article 8. It also highlights the importance of including people as part of the solution to protecting biodiversity, rather than excluding them as the main problem. Major requirements under Article 8 are to:

- establish a system of protected areas
- develop guidelines for the selection, establishment and management of protected areas
- regulate and manage biological resources to ensure their conservation and sustainable use
- promote the protection of ecosystems, natural habitats and the maintenance of viable populations of species in natural surroundings
- promote environmentally sound development in areas adjacent to protected areas (Biosphere Reserves are given as a possible model)
- respect, preserve and maintain the knowledge, innovations and practices of indigenous and local peoples
- regulate and manage detrimental processes and activities identified under Article 7.

In Section 4.2 we identified both direct processes – including deforestation, drainage of wetlands, unsustainable agriculture and pollution – and indirect processes – including population growth and the international economic climate – as having adverse effects on biodiversity. Articles 7 and 8 provide the legal framework to tackle them. Equally important is to ensure that data derived from the identification and monitoring process (Section 2) are passed on in a comprehensible way to decision-makers for management actions. The need to do this is highlighted in the commentary of Article 12(b), to promote research which contributes to conservation and sustainable use. The challenge has been thrown down!

# 7   Summary and conclusions

In this chapter we have discussed five fundamental questions which are at the heart of the biodiversity crisis that we face today. What conclusions can we draw from this discussion and what is the long-term prognosis for biodiversity?

The UN Convention on Biological Diversity defined biodiversity as 'the variability of all living organisms from all sources', and we have seen that this variability is manifest at three levels : genetic, species and ecosystem. Biodiversity started as a scientific issue, documenting the Earth's species and relating genetic diversity to the long-term survival and continuing evolution of plants, animals and micro-organisms. However, biodiversity is increasingly recognised as having wider relevance to all those concerned with the human environment and global welfare. Although scientific endeavours must continue since we know so little about the Earth's biodiversity, the importance of biodiversity and its uses and values as biological resources are increasingly central to development issues and concepts of fair and equitable distribution of these resources.

Approximately 1.75 million species have so far been scientifically named and described. However, this is thought to represent a small fraction of the species inhabiting the Earth. Microbial diversity, in particular, is little understood and the tropics in general are under-investigated. More resources need to be put into discovering and naming species before it is too late.

Biodiversity has always been used by humans, as a source of food, shelter and medicine. However, its value goes much deeper than such practical uses. Plants and animals have been central elements in the religious and cultural development of societies. The vital importance of biodiversity in regulating ecological processes has long been recognised. Yet despite these wide-ranging uses and the intimacy of our relationship with the natural world, biodiversity is under severe threat from a wide variety of activities, mostly human-driven. Some estimates suggest that human activities are pushing 10 000 species a year towards extinction. Approximately 10% of plants and 12% of mammals are thought to be threatened with extinction by activities such as habitat loss and fragmentation, pollution and over-exploitation.

Urgent measures are needed to protect and conserve biodiversity. Efforts directed at protecting habitats as well as conserving individual species need to be co-ordinated and also need to be integrated into wider development issues. The interdependence of conservation and development, first recognised in the Brundtland Report, has finally been enshrined in legally binding legislation: the Convention on Biological Diversity. Some have described the rate of progress in the implementation of the Convention as glacial. However, the Convention on Biological Diversity remains one of the most innovative international instruments developed in recent years. It is comprehensive in addressing biodiversity in terms of genes, species and ecosystems. The Convention recognises the importance of biodiversity for humanity both in terms of its intrinsic value and because humans depend on its maintenance to achieve sustainable development. It is also the first treaty to address equity issues, including recognising the role of indigenous and local communities in conserving biodiversity. Conserving biodiversity and using biological resources sustainably will benefit all of society. There is a crucial need to re-establish contact between people and nature at first hand. Therefore, action at the local and national level to promote the objectives of the Convention on Biological Diversity is needed immediately. International consensus will be impossible to attain unless all sectors of society come together to contribute to the common goal of achieving the conservation of biodiversity, the sustainable use of resources and the equitable sharing of benefits derived from biodiversity. Unless we manage this, we are damning future generations to life in a world of increasing biological impoverishment.

## References

ANON (1994) *Biodiversity: The UK Action Plan*, London, HMSO.

ANON (1995a) *Biodiversity: the UK Steering Group Report, Volume 1: Meeting the Rio Challenge*, London, HMSO.

ANON (1995b) *Biodiversity: the UK Steering Group Report, Volume 2: Action Plans*, London, HMSO.

BIBBY, C. J. *et al.* (1992) *Putting Biodiversity on the Map: priority areas for global conservation*, Cambridge, International Council for Bird Preservation.

BLUNDEN, J. with CURRY, N. (1996) 'Competing demands in the countryside: a UK case study', Ch. 5 in Sarre, P. and Blunden, J. (eds).

BLUNDEN, J. and REDDISH, A. (1996) (eds) *Energy, Resources and Environment*, London, Hodder and Stoughton/The Open University (second edition) (Book Three of this series).

CARSON, R. (1962) *Silent Spring*, Boston, Houghton Mifflin Company.

CATER, F. (1995) 'Consuming spaces: global tourism', in Allen, J. and Hamnett, C. (eds) *A Shrinking World? Global unevenness and inequality*, London, Oxford University Press/The Open University.

EHRLICH, P. R. and EHRLICH, A. H. (1981) *Extinction: the causes and consequences of the disappearance of species*, New York, Random House.

FLINT, M. (1991) *Biological Diversity and Developing Countries: issues and options*, London, Overseas Development Administration.

GRIGG, D. (1996) 'World agriculture: productivity and sustainability', Ch. 2 in Sarre, P. and Blunden, J. (eds).

HEYWOOD, V. H. and WATSON, R. T. (eds) (1995) *Global Biodiversity Assessment*, Cambridge, Cambridge University Press.

HOLMES, B. (1996) 'Life unlimited', *New Scientist*, Vol. 149, No. 2016, pp. 26–9.

ILTIS, H. H. (1988) 'Serendipity in the exploration of biodiversity: what good are weedy tomatoes?', pp. 98–105 in Wilson, E. O. and Peter, F. M. (eds).

IUCN (1994) *IUCN Red List Categories*, IUCN Species Survival Commission, Gland, Switzerland, IUCN Publications.

IUCN–THE WORLD CONSERVATION UNION, UNITED NATIONS ENVIRONMENT PROGRAMME AND WORLD WIDE FUND FOR NATURE (1991) *Caring for the Earth*, Gland, Switzerland, IUCN.

LOVEJOY, T. E. (1995) 'The quantification of biodiversity: an esoteric question or a vital component of sustainable development?', pp. 81–7 in Hawksworth, D. L. (ed.) *Biodiversity: measurement and estimation*, London, Chapman and Hall.

MABBERLEY, D. J. (1990) *The Plant Book*, Cambridge, Cambridge University Press.

MACARTHUR, R. H. and WILSON, E. O. (1967) *The Theory of Island Biogeography*, Princeton, NJ, Princeton University Press.

McNEELY, J. A. *et al.* (1990) *Conserving the World's Biological Diversity*, Gland, Switzerland, IUCN.

MAY, R. M. (1990) 'How many species?', *Philosophical Transactions of the Royal Society of London, Series B*, Vol. 330, No. 1257, pp. 293–304.

MYERS, N. (1988) 'Threatened biota: hot spots in tropical forests', *Environmentalist*, Vol. 8, No. 3, pp. 187–208.

MYERS, N. (1990) 'The biodiversity challenge: expanded hot spot analysis', *Environmentalist*, Vol. 10, No. 4, pp. 243–56.

PETERS, R. L. (1988) 'The effect of global climatic change on natural communities', pp. 450–61 in Wilson, E. O. and Peter, F. M. (eds).

PETERS, R. L. and DARLING, J. D. (1985) 'The greenhouse effect and nature reserves', *BioScience*, Vol. 35, No. 11, pp. 707–17.

REDDISH, A. (1996) 'Environment and development', Ch. 4 in Sarre, P. and Reddish, A. (eds).

REDDISH, A. and RAND, M. (1996) 'The environmental effects of present energy policies', Ch. 2 in Blunden, J. and Reddish, A. (eds).

SA 2000 (1994) *Systematics Agenda 2000: charting the biosphere*, Technical Report, SA 2000.

SARRE, P. (1996) 'Environmental issues in Cumbria', Ch. 1 in Sarre, P. and Reddish, A. (eds).

SARRE, P. and BLUNDEN, J. (eds) *Environment, Population and Development.*, London, Hodder and Stoughton/The Open University (second edition) (Book Two of this series).

SARRE, P. and BROWN, S. (1996) 'Changing attitudes to Nature', Ch. 3 in Sarre, P. and Reddish, A. (eds).

SARRE, P. and REDDISH, A. (eds) (1996) *Environment and Society*, London, Hodder and Stoughton/The Open University (second edition) (Book One of this series).

SILVERTOWN, J. (1996a) 'Ecosystems and populations', Ch. 7 in Sarre, P. and Reddish, A. (eds).

SILVERTOWN, J. (1996b) 'Inhabitants of the biosphere', Ch. 6 in Sarre, P. and Reddish, A. (eds).

TAYLOR, A. (1996) 'World trade and its environmental effects', Ch. 6 in Sarre, P. and Blunden, J. (eds).

WILSON, E. O. (1992) *The Diversity of Life*, Cambridge, MA, The Bellknap Press of the Harvard University Press (London, Allen Lane, Penguin Press).

WILSON, E. O. and PETER, F. M. (eds) (1988) *BioDiversity*, Washington, DC, National Academic Press.

WOODHOUSE, P. (1996) 'Farming a wetland ecosystem: rice cultivation in Asia', Ch. 3 in Sarre, P. and Blunden, J. (eds).

WORLD COMMISSION ON ENVIRONMENT AND DEVELOPMENT (1987) *Our Common Future*, Oxford, Oxford University Press (the Brundtland Report).

WORLD CONSERVATION MONITORING CENTRE (1992) *Global Biodiversity: status of the Earth's living resources*, London, Chapman and Hall.

WORLD RESOURCES INSTITUTE, IUCN–THE WORLD CONSERVATION UNION, and UNITED NATIONS ENVIRONMENT PROGRAMME (1992) *Global Biodiversity Strategy*, Washington, DC, WRI.

WORLD WIDE FUND FOR NATURE and IUCN–THE WORLD CONSERVATION UNION (1994) *Centres of Plant Diversity: a guide and strategy for their conservation*, 3 volumes, Cambridge, IUCN Publications.

# Answers to Activities

## Activity 1

Since informed debate on the biodiversity crisis needs to be based on sound biological and ecological data, much of the contents of *Silvertown* (1996a, b), comprise useful background material for this chapter. Of direct relevance is the section on an inventory of the biosphere, which discusses taxonomic classification, global species number and distribution, and which species are under threat and why (*Silvertown*, 1996a). In *Silvertown* (1996b) the examples given for Amazonian tropical rainforest ecosystems, harvesting from wild populations and genetic diversity all complement discussions in this chapter.

## Activity 3

(a) The scientific and economic arguments are for the most part based on the human value of ivory and as such find a justification in direct use values of biological resources. Little, if any, account is taken of the intrinsic value and rights of the elephant as a species (non-use values).

(b) It is difficult to justify a relaxation of the ban on the basis of non-use values. The only 'ethical' arguments for relaxing the ban would take the perspective of the local human community rather than the elephant's viewpoint. These might include:

• the provision of a sustainable living for the local community who co-exist with the elephant community

- the contribution that sustainable exploitation can make to empowering local communities to control their own lives, and be less dependent upon outside influences, such as international aid and loans.

Arguments against a relaxation of the ban would concentrate on the rights of the elephant over and above those of the local community, and certainly those of the international community.

These might include:

- our moral responsibility to future generations so that they can view elephants in the wild

- the right of all species to exist, unmolested by humans

- we should not exploit animals purely for human profit – ivory is a 'luxury' item.

## Activity 4

All woodlands have some value for wildlife, but ancient semi-natural woodlands are the most valuable and diverse, and are of particular importance since they cannot be replaced. New plantings since the First World War have largely been of non-native conifers, of much less value for biodiversity, although the planting of native broadleaved species has expanded significantly in recent years. New woodlands established close to existing ancient, semi-natural woodland, and woodlands that are managed to mimic nature by developing a varied spatial and structural pattern, have the greatest potential benefit for wildlife conservation.

Possible activities to enhance woodland biodiversity include:

- identify, conserve and restore ancient and semi-natural woodlands

- encourage the sensitive management of existing forests and woodlands, by, for example:
    - using traditional management techniques such as coppicing
    - the creation of woodland edge habitats
    - restructuring of even-aged plantations to introduce diversity of species and a range of ages
    - retaining old and dead trees on site to promote decomposition activities and provide habitat for animals, e.g. woodpeckers (standing trees) and dormice
    - removal of invasive species, such as *Rhododendron ponticum*
    - improved management of rides and riparian (stream banks) zones
    - incorporate open spaces into woodland

- create new woodland (e.g. community woodlands), especially in areas of low current wildlife value

- promote the use of local genotypes through careful selection of seed sources, where native broadleaved tree species are being planted.

The techniques described above are some that will enhance biodiversity of woodlands. Discuss these with people that you are able to talk to at your local conservation and woodland management agencies. Many of these techniques are more labour-intensive and costly than those of less value to biodiversity, such as plantations. Is money a major obstacle? Many people are willing to contribute towards biodiversity-sensitive management. Try to gauge potential response in your area.

## Chapter 5 Sustainable development

## 1 Introduction

This chapter follows four others on particular global issues that have been the subjects of intense international debate in recent decades, not least at the Rio UNCED conference in 1992. It differs from the previous ones, and to some extent embraces them, in the exceptionally wide-ranging nature of the questions raised – ultimately, no less than the whole course of human activity and its impact on the planet. For that reason it cannot be said that as much progress has been made in defining the nature of sustainable development, nor establishing policies to promote it, as for the four previous issues. The Rio conference set out primarily to address the recommendations on sustainable development that had been evolved in *Our Common Future* – the Brundtland Report of 1987 – but concluded with much fuzzier agreements on policy than many environmentalists had hoped. At the same time the continuing negotiations on Climate Change and Biodiversity, with firmer scientific analysis and clearer objectives, were able to converge at Rio in well-defined and binding Conventions, as we have seen. Nevertheless, progress on this larger issue has not been negligible, and the processes started then are still going on, with consequences that will take decades to emerge.

To trace the progress of these ideas, this chapter is divided into five main sections, each considering one of the key questions (see margin). Consideration of these questions will then lead into a more general discussion, in the next chapter, of the development of international environmental policy, and a final epilogue speculating on possible futures.

As you read this chapter, look out for answers to the following key questions:

• What do we understand by global development and its environmental effects?

• How did the concept of *sustainable* development evolve?

• How was this expressed in the 'Agenda 21' programme debated at Rio?

• How have the international politics of sustainable development evolved since Rio?

• In particular, how are the institutions set up there – the Commission on Sustainable Development (CSD), the Global Environment Facility (GEF), and Local Agenda 21 programmes – progressing?

## 2 Global development and its environmental effects

In a broad sense, 'develpment' might be said to describe all those changes in human technology and society which have brought our species, over a million years or so, from local bands of hunter–gathering primates to our present global society, with its complex settlements, institutions and cultural diversity. You no doubt already have considerable insight into the various stages of this process, particularly if you have read the three previous books of this series.

*Activity 1*

Review what you know about technical and social change, and its environmental effects, on three scales: the last 10 000, 1000 and 100 years.

More narrowly, terms like 'developed' and 'developing' countries, 'underdevelopment' and 'uneven development' have come to refer particularly to the process of **industrialisation**, its social implications and the present inequalities in the world associated with the way it has spread over the last few hundred years. Whether this will be the right way to consider 'development' for the future remains to be seen.

If we live in the United Kingdom, we are particularly well placed to consider this process – though it may be difficult to stand back from it, as we are so accustomed to it. The special conditions in eighteenth-century England saw the beginning of what we know as the Industrial Revolution, initiating global changes which still continue. To understand these conditions, we would need to go back somewhat further, to the development of the 'capitalist' financial system over a rather longer period. Historians differ in their account of this, but one view places its origins at least in the fifteenth century (Braudel, 1979). A complex development started with the day-to-day operations of trade and markets – in food, clothing, energy, metals, craft goods of all kinds – with money replacing barter. Then came the financial institutions – accounting systems, banks, commodity and stock exchanges – evolved to support increasingly wide-ranging trading operations, and finally the larger-scale political processes influencing and influenced by these financial activities. The evolution of capitalism has progressively affected finance and power relations at local, national and international levels.

By the eighteenth century the financial centre of Europe had moved from Venice and Amsterdam to London, with growing British internal and seafaring trade. The resulting wealth, and relative freedom from the local effects of bad harvests, provided both capital and labour to exploit technological change – in particular the invention of the steam engine, and a new view of the way in which energy could be used in factories, forming the system of industrial capitalism. Here profit is sought not so much from supporting the enterprises of merchants as from those of manufacturers.

The environmental pollution and exploitation of workers in the explosive growth of the factory system in the nineteenth century has been well documented and led to criticisms of various kinds. On the one hand we recognise the divergent attitudes to Nature characterised in *Sarre and Brown* (1996), with Stewardship forgotten in the ruthless Imperialism of some industrialists, the more benignly intended Utilitarianism of others, and the Romantic reaction against both. (See also Plate 12.)

*Activity 2*

If you are familiar with them, review your understanding of these four attitudes to nineteenth-century industrial developments, and to the global environmental problems of today – Stewardship, Imperialism, Utilitarianism and Romanticism. If you are not, turn to the Answer given at the end of the chapter.

At the same time, opposition also took a more overtly political form, particularly in the criticisms of the relationship between owners of capital and workers associated with Marx and Engels, which provided the basis of Communism. The milder socialism of William Morris, with its respect for craft skills and distaste for mass production, and the 'grass-roots' anarchism of Peter Kropotkin continue to resonate with present-day green politics. (See Box 5.1 and Box 5.2.)

Meanwhile, however, the industrialisation process continued to spread throughout the world, in spite of the recurrent crises which seem to characterise it. Economic historians looking at production, price and wage statistics claim to recognise various regular patterns in their fluctuations. A particular cycle of about fifty years is associated with the name of Kondratieff, and has been interpreted as an inherent tendency for investment to follow success, leading to over-production, market collapse and loss of employment, only to stimulate a new wave of invention and new markets. In the 1840s a crisis in the cotton industry and the political revolt of the Chartists was followed by a new expansion associated with the invention of railways. A huge growth in railway mileage in Britain led to an even greater one, much of it British built, in Europe, America and the rest of the world, opening up new resources and markets to the global economy. The British colonial empire continued to expand at this time; a further economic depression in Europe and America in the 1880s stimulated imperial expansion from all the industrial powers, with rival colonial empires in competition. By the end of the nineteenth century there was a clear distinction between the wealthy industrial 'core' of Europe and North America and the dependent 'periphery' of Africa, South America and much

▲  *Barrow-in Furness Haematite Works, c. 1875. The industrialisation process required great capital investment.*

Box 5.1

Karl Marx

Friedrich Engels

In *The Conditions of the Working Class in England* the German philosopher Friedrich Engels painted a vivid picture of urban and industrial squalor in Manchester in the 1840s, a time of explosive industrial expansion. In 1848 he and Karl Marx issued the *Communist Manifesto* as a tract for revolutionary change. They saw poverty as the central problem, resulting from inevitable class conflict between capitalists, who owned the means of production, and the workers they employed. Industrialisation in itself they saw as a desirable means of wealth creation; what mattered was the more equitable distribution of its fruits by a change of ownership to the workers themselves – with underlying problems about how their interests would be fairly represented by a State supposedly acting in their name. They were given to Imperialist images of the mastery or conquest of Nature, though they did have some recognition of the human place in Nature too; the larger-scale environmental consequences of industrially-based population growth, resource depletion and pollution were, of course, not so evident then.

While their writings had a pervasive influence on more moderate forms of socialism in the established industrial societies over the next century, their followers only acquired full political control in Russia after the First World War, extending to Eastern Europe and China after the Second. The disastrous environmental consequences of the resulting centralised system of control are seen by many as a damning indictment of Marxist thinking, though it will continue to be argued as to what extent the practice of the post-revolutionary USSR was an inevitable outcome of the theories of the 1840s. Contemporary 'neo-Marxists' continue to contribute to present debates about the resolution of environment and development concerns on the global scale.

of Asia. In this process of **colonialisation**, local economies in the periphery were distorted to provide food and raw materials for the core, and expanding markets for manufactured goods. Although this colonial system was restructured after the 1914–18 Great War between rival European powers, and was destroyed after the wider 1939–45 conflict, it left inequalities in the level of industrialisation, wealth, access to resources and terms of trade which continue to underlie present stresses between the countries of the world (see, for example, *Taylor*, 1996).

The criticisms of capitalism voiced in the nineteenth century by Marx and Engels centre on the inevitable class conflict between workers and the owners of capital; it was argued that this could only be resolved by nationalisation of the means of production and central planning of the economy. Though influential in moderating some of the worst excesses of capitalism, this argument was resisted in the older industrial economies, but in 1919 a group committed to its implementation gained power in Russia, still at an earlier stage of industrialisation. Their narrow and ruthless interpretation of Marxism led to one form of totalitarian regime.

Box 5.2

*William Morris*

*One of Morris's designs: 'Acorn', a design for wallpaper from 1880.*

Later in the nineteenth century two further critics of industrialisation, with different emphases, were William Morris (1834–96) and Peter Kropotkin (1842–1926). The English 'socialist' William Morris, rich by inheritance, was influenced in his Oxford University days by the works of Thomas Carlyle and John Ruskin. His hatred of the Imperialist or Utilitarian values of capitalism led him from a Romantic yearning for pre-industrial way of life to revolutionary convictions culminating in the founding of the Socialist League in 1885. Direct experience as craftsman, designer, poet and owner of craft workshops led him to a view of the joy of creative labour, and a resistance to consumerism, at odds with both capitalist and communist images of industrial production. The international as well as local dimensions of this conflict are recognised in a typical insight:

> To further the spread of education in art, Englishmen in India are .... actively destroying the very sources of that education – jewellery, metal work, pottery, calico-printing, brocade weaving, carpet-making – all the famous and historical arts of the great peninsula have been ... thrust aside for the advantage of any paltry scrap of so-called commerce.

(*The Art of the People*, 1879)

The Russian 'anarchist' Peter Kropotkin, largely writing in exile in London, was influenced by the ideas in the socialist movement stemming from the concept of ecology first formulated in the 1860s by the German biologist, Haeckel. (See *Silvertown* [1996] for a more detailed account of the term 'ecology' in its scientific sense.) Its political implications arise when human society is seen as part of the global ecosystem rather than outside it, and human needs are provided for by the 'grassroots' interactions of people in their personal and working relationships rather than by states and industrial corporations. His *Fields, Factories and Workshops Tomorrow* (1899) is seen by some modern environmentalists as one of the essential texts of this political 'ecological' tradition. His ideas on the decentralisation of both agriculture and industry in small-scale self-sufficient communities, and of the importance of co-operation rather than competition in evolution and society, continue to be echoed in more recent thinking.

While neither Morris's nor Kropotkin's ideas formed a powerful political force in the twentieth century to counter capitalism as Marxism did, the newer attitudes developing since the collapse of (the USSR version of) communism and growing global environmental concern recognise their value.

Stresses in the older industrial system between the wars encouraged the growth of other forms of authoritarian government in Italy, Germany and Japan, with their varied doctrines of national power, military expansion and racial purity. They were opposed by the alliance of Britain, the USA, the USSR and a developing China against them in 1939–45.

Soon after this conflict, communism became dominant in China too. Its uneasy alliance with the USSR formed the rival ideology of the 'Second

▲   *The construction of the railways in the nineteenth century brought vast new areas into the global economy: (top) laying Union Pacific Railway in North America, 1869; (below) the opening of the Madras railway in India, 1856.*

World' in the 'Cold War' with the 'First World' countries in the West. With their 'mixed' or 'welfare' economies, European countries like the UK had established forms of capitalism differing considerably from that of the nineteenth century, some deriving from the ideas for state intervention of the economist John Maynard Keynes in the 1930s. Arguments within them about the balance between 'free' markets, private ownership, nationalised industries and state welfare provision continue, of course. The Cold War terminology of First and Second Worlds led to the 'Third World' as a loose term for those developing countries as yet uncommitted to either camp, though it has come to be used more freely for the less industrialised

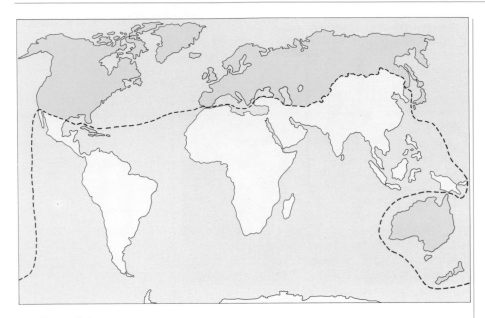

▲  *Figure 5.1*
*The Brandt Commission's 'North–South' divide.*

countries, regardless of these implications of capitalist/communist rivalry.
Throughout the twentieth century continuing technological changes have
influenced and been influenced by these social and political changes. Along
with the development of increasingly destructive weapons (chemical,
nuclear and biological), innovations in transport (cars, aircraft and space
flight) and electronics (radio, television and computers) continued to make
the world seem a smaller and more vulnerable place, reinforced by the
images from space of the 1960s (see Plate 1).

The collapse of the communist system in Russia and the adjacent
European countries in the late 1980s, along with some restoration of
markets in China, makes the 'three worlds' terminology no longer
appropriate. An even simpler classification into rich industrialised 'North'
and poor developing 'South', largely reflecting the earlier colonial
framework, was adopted by the Brandt Commission of 1980–83 which
preceded Brundtland and Rio (see Figure 5.1). This is still evident in the
divisions between the OECD or 'G7' group of industrialised countries and
the 'G77' (now more like 130) developing countries in international debate,
but all these groupings oversimplify more complex variation. The
apparently dominant 'capitalism' takes different forms even in the
established industrial economies: in his discussion of British economic
problems, for example, Hutton (1995) contrasted the British system
unfavourably with other European, US and Japanese forms in innumerable
details of class structure, industrial, financial, labour and welfare policies.
He continued to argue for the reforms advocated by Keynes (but took no
account of the 'environmental economics' beginning to arise from more
recent concerns). At the same time, the rapidly industrialising countries of
South East Asia now have their own momentum: the oil-rich countries of
the Middle East are peculiarly influential; the countries of the former Soviet
bloc are now 'economies in transition'; China and India in different ways
are becoming industrial powers. Large **transnational corporations** operate
across state boundaries, and the international financial system moves

capital rapidly without particular reference to national traditions or needs. United Nations institutions and the World Bank have struggled to express capitalist principles in large-scale development projects and economic 'restructuring' with questionable success.

However, it can be argued that 'capitalism' never was a 'designed' system in the sense that communism was, but is just a description of what has in fact evolved, in all its diversity, from the markets that seem to characterise *Homo sapiens*. Our species is ecologically distinct from others in that we do not live off the same land as we inhabit – we are the 'trading apes' whose 'territory' is the whole Earth, whether we visit it or not. This is not to say that the present form of this evolution from simple markets is either inevitable or desirable. The expansion of 'capitalism' from the financing of trade to that of manufacture has continued to evolve into the complex financial services of today, with profit sought from more abstract manipulation of stocks, insurance, 'futures', 'derivatives' and money itself in the global financial markets. Their electronic operation, at speeds much greater than that of the flow of the underlying goods, or than that of government decision-making, challenges the ability of society to control them; they are increasingly divorced from their social or environmental consequences. The national governments of industrialised countries have been influenced by, but have also sought to control, the capitalist economic processes within them, variously interpreted by economic theories from Adam Smith to Marx and Keynes. In a 'globalised' economic system, international institutions have some way to go in establishing the balance between efficiency and equity that may have been sought in various ways within different nation-states, but that they no longer have the power to reach independently. Can a new set of socio-economic principles be developed in response?

In any case, the political conflicts sketched above have been mainly about the struggle for fairness or power between different human groups. What they have not been about, until very recently, is the total impact of all this human activity, whether fair or not, on the rest of the planetary system – the environmental effects that are the primary concern of this series of books. It is evident that no existing political system can claim to be free from environmental effects, of varying severity. The earlier skills of human hunters eliminated many large mammals; settled agriculture and urbanisation removed forests and transformed landscapes. But it is particularly the effects of industrialisation over the last two centuries that we associate with severe environmental stress. In this respect both capitalist and communist systems have proved to be damaging in their own ways. The gross local pollution of the last century may have been overcome in the West by improved environmental legislation; the collapse of communism has revealed appalling levels of industrial pollution and unexpected effects of large-scale environmental manipulation in many parts of the former Soviet Union and Eastern Europe. (See Plate 13.) But at the same time capitalist financial pressures continue to contribute to the loss of tropical forests, and increasing burdens of toxic and hazardous waste disposal, some of which is now a large international activity. Improved control of local effects still leaves the longer-range effects of acid rain and greenhouse gas emissions. Wealthy societies consume disproportionate amounts of finite resources; poor populations struggle with fuelwood shortages, while subsistence agriculture on unsuitable terrain and low productivity soils leads to desertification.

There are varying points of view about the way to solve these problems. Some would argue for improved ecological awareness, or for more enlightened technology. Others would insist that more drastic sociopolitical change is needed to improve distribution and eliminate poverty. Some combination of all of these may well be needed. It is this growing recognition of the interlinked nature of 'development', as at present understood, and environmental change that has led to the 'sustainable development' doctrine of recent years.

## 3   *The concept of sustainable development*

What does 'sustainable development' mean? To the extent that 'sustainable' implies continuity, and 'development' implies change, there is even perhaps an intrinsic contradiction in the term, though, as we shall see, the words can be interpreted more creatively than that. At the least it is a convenient shorthand for the diverse ideas about directions for human society which have come together in the late twentieth century. They might imply anything between a radical reconstruction of industrial capitalist society and relatively minor adjustments of technology to overcome particular environmental problems. There is no single point at which the concept can be said to have started life, but the gathering ideas of the 1970s and '80s were certainly crystallised in this form, and brought on to the political agenda, by the Brundtland Report, *Our Common Future*, in 1987.

By the 1970s the inherited stresses of industrial capitalism from the nineteenth century, the two World Wars and the Cold War between the resulting rival ideologies, continuing fears about nuclear conflict, the space exploration programme and its reminder of the unity and fragility of the Earth system – and associated technological changes in systems analysis, communications and computing – led to growing questioning, from various directions, of the direction of global development. The British radical environmentalist, Edward Goldsmith in 1972 asserted in the first editorial of his new journal *The Ecologist* that, 'That principal defect of the industrial way of life with its ethos of expansion is that it is not sustainable.' This may not be the first use of 'sustainable' in this way, but it is a clear statement of a position which led him to *A Blueprint for Survival* and a series of subsequent proposals for drastic social change, in rather authoritarian form. E. F. Schumacher's *Small is Beautiful* (1973) gave thoughtful expression to related ideas echoing the 'anarchism' of the late nineteenth century, which continue to inform 'green' politics.

From a very different direction, the Club of Rome, an international 'non-political' group of industrialists, bankers, economists and scientists, commissioned a group of systems analysts at MIT in the United State to develop a computer model of the world economy which appeared in 1972 as *Limits to Growth* (Meadows *et al.*, 1972), stimulating much debate and criticism from all sides. They quoted the then UN Secretary General, U Thant:

> . . . the members of the UN have perhaps ten years left in which to subordinate their ancient quarrels and launch a global partnership to curb the arms race, to improve the human environment, to defuse the population explosion, and to supply the required momentum to development efforts . . .

Well, we are still here. The same authors' *Beyond the Limits* (1992) assesses progress, expresses continuing concern but also optimism, as long as 'development' is not identified with 'growth'.

Of course, there were critics of the 'eco-doomsters', like John Maddox in *The Doomsday Syndrome* (also 1972), whose optimism at the time about agricultural developments and nuclear power, for example, might now be questioned. He was confident that improved technology would solve all the perceived problems. This tension between 'ecocentric' and 'technocentric' views was identified in O'Riordan's *Environmentalism* (1981); he, too, has continued in later publications to discuss how the transition to 'sustainability' might take place.

A different form of radicalism, with a concern for the welfare of the planet over very long time-scales, rather than that of humans over our lifetime, was also beginning to emerge in the 1970s from James Lovelock, as an outgrowth of his work on possible life on Mars in the space programme. This came to public notice in his *Gaia: a new look at life on Earth* (1982) and *The Ages of Gaia* (1989), which have captured many people's imagination, if viewed sceptically by scientific orthodoxy. Nevertheless, a major ten-year scientific collaboration, the International Geosphere–Biosphere Programme: A Study of Global Change (IGBP), was launched in 1990 to 'describe and understand the interactive physical, chemical and biological processes that regulate the total Earth system' which, without mentioning 'Gaia', does not seem too far from Lovelock's aims.

Returning to 1972, this also saw the first appearance of these global concerns on the international political stage in the UN Conference on the Human Environment in Stockholm. A number of useful principles were established:

•   harm from pollution should be minimised, but regulations should also take into account economic and social goals

•   the 'sovereign right' of a state not to have its territory polluted as a result of activities in another state takes precedence over the sovereign right of a state to do what it wants within its own boundaries

•   the international community should determine limits on the use and abuse of 'global commons', such as the high seas and ocean resources, the atmosphere, outer space and Antarctica.

Concerns were also expressed about climate, forests and population, all contributing to an evolving process of international debate, though at this stage it had a stronger emphasis on the environment concerns of the North than the development concerns of the South. The setting up of the UN Environment Programme (UNEP) followed, and the subsequent agreements on the Law of the Sea, Trans-Boundary Air Pollution, and Ozone Depletion. But developing countries' concerns about inequality, poverty, debt and development found expression in the World Conservation Strategy (1980) and the reports of the Brandt Commission in 1980, *North–South : a programme for survival*, and 1983, *Common Crisis: North–South cooperation for world recovery*. These formed the background to the major study started in 1983 which culminated in the 'Brundtland Report' of the World Commission in Environment and Development, *Our Common Future*, in 1987. This firmly identified 'sustainable development' as a global objective. (See Box 5.3.)

Although, like its predecessors, this study had its origins in an initiative from the developed world, a new element was the representation on the commission of twenty-two nations, more than half of which were from developing countries. It re-emphasised the problems of poverty and human

*Box 5.3*

The United Nations World Commission on Environment and Development chaired by Gro Harlem Brundtland, the Prime Minister of Norway, presented its report in 1987. Its published title is *Our Common Future,* but it is most commonly referred to as the Brundtland Report.

The Brundtland definition of sustainable development is:

Development that meets the needs of the present without compromising the ability of future generations to meet their own needs. It contains within it two key concepts:

- the concept of needs, in particular the essential needs of the world's poor, to which over-riding priority should be given;

- the idea of limitations imposed by the state of technology and social organisations on the environment's ability to meet present and future needs.

*Gro Harlem Brundtland*

need, challenging the consumerism of the developed world and the inequalities fostered by the current economic system. In particular it stressed that global environmental concerns must be reconciled with development aspirations: 'It is both futile and indeed an insult to the poor to tell them that they must remain in poverty to protect the environment.'

The result of their exhaustive hearings was a study generally regarded as a major analysis of the world's interlocking crises, seen as environmental, developmental and energy-related. Problems are recognised with population growth, inequitable resource use and the need for technology with low environmental effects to be developed and

disseminated, rather than the polluting high-resource use methods of earlier industrialisation (that is, the three terms in the 'PRQ' formula discussed in *Reddish*, 1996). (See Plate 14.) The desperate problem of debt repayments from the developing world, with a net outflow of capital to the developed world, was emphasised. The recognition that 'ecology and economy are becoming ever more interwoven – locally, regionally, globally' clearly casts doubt on orthodox economic theory which has ignored environmental effects, though attempts are now being made to develop an 'environmental economics' to redress this. The Report's resulting 'strategic imperatives' for sustainable development are summarised in Box 5.4.

Note that although the main objectives stress sustainable development, this is interpreted as 'growth', at least for developing countries; note also that there is no mention of climate change, which leapt into prominence around the same time.

To the ecocentric and technocentric perspectives that the Report seeks to reconcile must be added the view of neo-Marxists like Michael Redclift (1987). He continues to see 'structural inequalities' in the global economic system as a major constraint on the achievement of equity. While more

---

**Box 5.4   *Strategic imperatives***

1 *Reviving growth* Sustainable development must address the issue of poverty, as poverty increases pressure on the environment. The very logic of sustainability, of sustainable development, implies an internal stimulus to growth in the developing countries.

2 *Changing the quality of growth* This means making 'growth' less materialistic, less energy-intensive and more equitable in its impact. Economic and social development have to be 'mutually reinforcing'.

3 *Meeting essential human needs* More food – not just to feed people but to attack undernourishment; basic housing, fresh water supply, health, energy. The need for energy cannot be universally met unless energy consumption patterns change.

4 *Ensuring a sustainable level of population* The challenge is to tackle the highest rates of population growth, especially in Africa. This is linked to improving the quality of life, raising incomes and working for a more equitable distribution of population between rural and urban areas.

5 *Conserving and enhancing the resource base* There is a moral as well as an economic argument here. Included are agricultural resources (land, soil, water), forestry, fisheries and energy. There is a pressing need for efficient, low-waste technologies.

6 *Reorienting technology and managing risk* The implications of 1–5 above are for the reorientation of technology in two principal ways: the capacity for innovation needs to be greatly enhanced in developing countries; and technological development generally must pay greater attention to environmental factors. This is closely linked to the issue of 'risk management', wherein environmental impact has to be effectively minimised.

7 *Merging environment and economics in decision-making* Economics and ecology should not be seen in opposition but as interlocking. Sustainable development requires the unification of economics and ecology in international relations.

To achieve 1–7 the Report sees the following as essential requirements:

- A *political system* that secures effective citizen participation in decision-making.

- An *economic system* that is able to generate surpluses and technical knowledge on a self-reliant and sustained basis.

- A *social system* that provides for solutions for the tensions arising from disharmonious development.

- A *production system* that respects the obligation to preserve the ecological base for development.

- A *technological system* that can search continuously for new solutions.

- An *international system* that fosters sustainable patterns of trade and finance.

- An *administrative system* that is flexible and has the capacity for self-correction.

modern versions of the old criticism of industrial capitalism continue to have force in the present world of global capital movements, the environmental impact of an expanding industrial society was little considered by earlier Marxism and must form part of any new synthesis about socio-economic behaviour.

The Rio conference – the United Nations Conference on Environment and Development – in 1992 was set up as a twenty-year follow-up to Stockholm – with 'environment' and 'development' in the title to reflect changed perceptions about international relations. It had the particular objective of implementing the ideas developed in the Brundtland Report, though the Climate Change and Biodiversity negotiations also taking place were timed to lead to the corresponding Conventions at Rio. The result of the Sustainable Development discussions, deals and horse-trading was Agenda 21, a much more diffuse document than either of the Conventions.

# 4   Agenda 21

## 4.1   Outline

The vast document negotiated in a series of preparatory meetings and finally agreed at Rio as 'Agenda 21', covering over 500 pages and 40 chapters, is intended as a programme of action for achieving sustainable development in the twenty-first century. It was complementary to the Rio 'Declaration on Environment and Development' which the organisers had hoped would be be a ringing 'Earth Charter' briefly expressing a new global ethic. What in fact emerged from this was only a rather laborious set of twenty-seven 'Principles' which expressed many of the conflicts and compromises also found in Agenda 21, so need not be discussed separately.

Agenda 21 as a whole is a remarkable, but also in many ways an unsatisfactory, document. It does represent an agreed statement from most of the 178 national governments represented at UNCED, with all their history of conflict and inequality. However, it is not legally binding, which gives it a much weaker status than the Conventions discussed in the four previous chapters: the 'umbrella' convention on the Law of the Sea, and the 'framework' conventions on the protection of the ozone layer agreed in Vienna, and on climate change and on biodiversity agreed in Rio. Nevertheless, its wording was debated scrupulously, and at times bitterly; it does have something to say about a huge range of relevant and controversial issues, and it will be used as a guidebook and touchstone of local and national policy for some years to come.

Before attempting to assess its merits, deficiencies and ultimate influence, it is first necessary to summarise the subjects covered, which are very numerous. The chapter headings are listed in a series of boxes, followed in each case by some comment and linkage to the earlier parts of this book and its three companion volumes. A fuller account is to be found in *The Earth Summit Agreements: a guide and assessment* (Grubb *et al.*, 1993).

---

> ## Box 5.5    The structure of Agenda 21
>
> The first 'preamble' chapter asserts that, 'Humanity stands at a defining moment in history.' It recognises disparities within and between nations, the worsening of poverty, hunger, ill-health, illiteracy and the ecosystems on which we depend, but is confident that a 'global partnership for sustainable development' will overcome these problems.
>
> The remaining 39 chapters are divided into four sections:
>
> I    Social and economic dimensions (Chapters 2–8)
>
> II    Conservation and management of resources (Chapters 9–22)
>
> III    Strengthening the role of major groups (Chapters 23–32)
>
> IV    Means of implementation (Chapters 33–40)
>
> Section II comprises almost half the total length.
>
> The more detailed contents of these four sections will be dealt with in turn in the following boxes and discussion.

The preamble chapter includes two significant general points:

1    'Its successful implementation is first and foremost the responsibility of Governments', so recognising the status quo of existing sovereign states with all their conflicts and instabilities. However, these shall be 'supported and supplemented' both by larger groupings, such as the UN, and other international and regional organisations, and by more local participation of the public, non-governmental organisations and other groups. Much of what follows grapples with mechanisms for achieving this. While Sections I, II and IV may reflect the more usual concerns of international bodies and negotiations, the inclusion of Section III on 'major groups' perhaps reflects a stronger role for local consultation than has been seen previously.

2    The objectives will require a substantial flow of 'new and additional financial resources' to developing countries – a fundamental issue about the world economic order that was much debated at Rio, and remains contentious.

What the four major sections cover will become clearer as their separate chapters are outlined. Not surprisingly perhaps, the approach is firmly **anthropocentric**, recognising human welfare as a primary concern, and that the 'environment' should be protected more as a support to human life rather than for its intrinsic value. Thus Section I is concerned with the *human* social and economic systems that seek to provide for *human* needs. Section II is the nearest to showing concern for planetary stability, but emphasises environmental protection as a means of ensuring that the required 'resources' continue to be available to *humans*. Section III concerns wider consultation with *human* groups who may have had less influence on decision-making in the past, and Section IV larger-scale *human* organisations.

## 4.2   *Social and economic dimensions*

---

---

The 'international co-operation' chapter heading implies that the nature of 'sustainable development' is clear, and supports existing ideas about more efficient and equitable trading systems, mechanisms for financial help and debt reduction, and better administration without corruption. The beneficial effects of trade liberalisation are at least open to debate: see, for example, Taylor (1996). There is no mention of transnational corporations. The chapter is, of course, concerned about improving equality between human groups rather than the total human effect on the planet – which may well be increased if the measures discussed are successful.

Chapter 3, on poverty, emphasises the achievement of 'sustainable livelihoods', and is more 'bottom-up' in its reference to local community involvement than was often the case previously, the result of successful NGO lobbying.

The next two chapters are concerned with resource use and population – respectively the 'R' and 'P' of the 'PRQ' formula for environmental impact discussed in *Reddish* (1996). The former represents an important success of the developing countries in getting the unsustainable consumption patterns of the developed countries recognised (the subject of much of *Blunden and Reddish* [eds, 1996]), and the latter the strongest expression that could be negotiated on population policy (see *Findlay*, 1996).

Two comparatively uncontroversial chapters follow. Chapter 6, on health, emphasises the role of the existing World Health Organisation and the need to consider environmental pollution in integrated environment and development policies. Chapter 7, a larger chapter on settlements, gives detailed recommendations on shelter, management, infrastructure, energy, transport, construction, and disaster planning, providing a basis for Local Agenda 21 plans to be developed and carried out by local authorities.

The rather dry-sounding final chapter, on the need for integration of economic, social and environmental issues in development decision-making, is potentially important. It is near to being a global endorsement of the 'polluter pays' principle: economic instruments are seen as complementary to legal ones in the achievement of sustainability. This recognises the need to develop, and to standardise at an international level, satisfactory economic accounting measures which include environmental effects – a welcome move away from existing standard measures like Gross National Product (GNP)* which neglect environmental degradation. However, a specific Scandinavian proposal on formal environmental accounting for transnational corporations was not adopted.

*See Box 5.10.

# 4.3   *Conservation and management of resources for development*

---

---

Chapter 9 refers to existing Conventions on ozone, climate change and transboundary air pollution (see *Reddish and Rand*, 1996, and *Blunden*, 1996; also Chapters 2 and 3 of this volume). It proved to be very controversial in its proposals about policy implications, with compromise wordings about lifestyle, consumption, energy efficiency, and new and renewable sources of energy (*Olivier et al.*, 1996).

Chapter 10, on land use, stresses analysis and information and avoids the highly political issues of land ownership and distribution. The chapter on deforestation was only able to cross-refer to the 'Forest Principles' also agreed at Rio, not the wider-ranging Forest Convention that the developed countries had hoped for. Of the two following chapters on managing fragile ecosystems, that on desertification and drought was much discussed, culminating in a response to African pressure for a Desertification Convention to be negotiated soon after Rio, while that on mountain ecosystems was readily accepted. Chapter 14, a very long chapter on agriculture, reflects conflicting concerns about intensive agriculture and environmental degradation versus hunger, oversupply, trade and price support (see much of *Sarre and Blunden*, eds, 1996).

The chapter on biological diversity conservation effectively reiterates the provisions of the Biodiversity Convention separately agreed (see Chapter 4 of this volume), while the discussion of biotechnology in the following chapter goes beyond the Convention in its recommendations on food production, health, solutions to environmental problems and safety.

Another long and complex chapter on the oceans, seas and coastal waters extends the provisions of UNCLOS (see Chapter 1 of this volume), whilst that on freshwater draws on the conclusions of the 1992 Dublin conference about integrated water management, safe drinking water and sanitation, resource assessment, and urban and rural supply.

Four aspects of waste management are covered in the final chapters – on toxic chemicals, hazardous wastes, solid wastes and sewage, and radioactive waste (some of these issues were introduced in *Blunden and Reddish*, eds, 1996, especially Chapters 4–7).

Together the chapters of this section cover a formidable range of environmental concerns, and provide a wide array of separate management proposals. What is perhaps missing is any sense of the Earth as an integrated single system, whether in the natural links between physical and biological processes (such as are being studied in the International Geosphere–Biosphere Programme), in the total impact of an increasingly globalised human society (such as was investigated in the 'Limits to Growth' and later models) or in the more controversial Gaia hypothesis.

## 4.4    *Strengthening the role of major groups*

---

*Box 5.8    Section III: Strengthening the role of major groups*

Chapter 23    Preamble to Section III

Chapter 24    Global action for women towards sustainable and equitable development

Chapter 25    Children and youth in sustainable development

Chapter 26    Recognising and strengthening the role of indigenous people and their communities

Chapter 27    Strengthening the role of non-governmental organisations: partners for sustainable development

Chapter 28    Local authorities' initiatives in support of Agenda 21

Chapter 29    Strengthening the role of workers and their trade unions

Chapter 30    Strengthening the role of business and industry

Chapter 31    Scientific and technological community

Chapter 32    Strengthening the role of farmers.

---

This section has a rather different flavour to the others, with their emphasis on governmental and international institutions. Pressure from NGOs was influential in ensuring its inclusion, and its presence does oblige governments to be more open in their planning and information disclosure than might otherwise have been the case.

The preamble emphasises the importance of widespread consultation in all decision-making on environment and development issues. It represents something of a triumph that the principle of participatory democracy seems to have been acknowledged even by those countries which do not practise it. The groups discussed in the succeeding chapters are evidently of very different significance in effecting change, though perhaps not in the effects of change on them; the need for some of them to be included at all says something about existing institutional arrangements.

Women comprise half the population and – from African farmers to Gro Brundtland – clearly have major effects on both environment and development; nevertheless they are poorly presented in the hierarchical government structures of the 'developed' world. This chapter looked forward to the 1995 Beijing conference and a review in the year 2000 of the 1979 Convention on the Elimination of All Forms of Discrimination against Women. Children and youth are similarly members of all human societies, and it should go without saying that their welfare, education and future experience are interwoven with our environment concerns and development decisions.

In the case of 'indigenous peoples', Chapter 26 turns to a threatened minority of the human population, ruthlessly pushed aside by earlier 'development'; this chapter belatedly acknowledges their rights, the value of their cultures and their relationship with the local environment.

With 'NGOs' the groups concerned are extremely diverse, from small local societies to powerful and highly organised international movements; Chapter 27 acknowledges their right to information, consultation and contribution to decision-making.

In 'local authorities' we recognise part of formal governmental structures worldwide, though of varying power and influence. This chapter specifically proposed consultation and planning to arrive at a 'Local Agenda 21' consensus by 1996.

'Workers and trade unions' are categories with powerful resonances in the industrial societies of the North, as are 'business and industry', with traditions of confrontation and, formerly, little regard for environmental questions, particularly on the global scale. How the proposals for consultation and largely voluntary initiatives (like the International Chamber of Commerce 'Business Charter for Sustainable Development' or the chemical industry's 'Responsible Care' scheme) will fare in a world of transnational corporations, lightning movements of capital and extreme diversity of industrial culture, remains to be seen. The absence of proposals for more formal regulation of international business was seen by many environmentalists as a major gap in the Rio outcome, though this no doubt reflected what was thought to be realistically possible at present.

The 'scientific and technological community' evidently has a major impact on the course of development and its environmental effects, and is urged to contribute more to open debate and decision-making – the linkage of science and technology may be unfortunate, as implying greater identity between them than many practising scientists, engineers and entrepreneurs would accept – but a greater transparency in the influence of 'expert' groups can only be desirable.

Finally, the chapter on 'farmers' emphasises that the development of sustainable agriculture will require decentralisation of decision-making, and consultation with rural people working in farming (also fishing and forestry), and their organisations.

## 4.5   Means of implementation

---

Box 5.9   Section IV: Means of implementation

| | | | |
|---|---|---|---|
| Chapter 33 | Financial resources and mechanisms | Chapter 37 | National mechanisms and international co-operation for capacity-building |
| Chapter 34 | Transfer of environmentally sound technology, co-operation and capacity-building | | |
| | | Chapter 38 | International institutional arrangements |
| Chapter 35 | Science for sustainable development | Chapter 39 | International legal instruments and mechanisms |
| Chapter 36 | Promoting education, public awareness and training | Chapter 40 | Information for decision-making. |

---

This final section includes many of the most contentious and disputed areas. The first two chapters, on financial resources and technology transfer, were key issues for the developing countries.

The costs of carrying out the various chapters of Agenda 21 had been estimated by the UN Secretariat at about $600 billion per year, of which about $125 billion per year would need to be provided to developing countries in grants or concessional loans. This last figure is more than twice the current Official Development Assistance (ODA) from developed to developing countries, though close to the existing official UN 'target' for the richer countries to transfer 0.7% of their GDP. (In 1994 this proportion was only exceeded by the Scandinavian countries and the Netherlands; the UK figure was about 0.3% and the US below 0.2%.) In any case, such aid is often more than compensated by a flow of capital in the reverse direction, as debt repayments. There was bitter conflict over this target, the timetable for achieving it and the mechanisms for administering it. It was provisionally agreed that the recently established Global Environment Facility (GEF), should be used, but only on condition that it was restructured to be outside the control of the World Bank, as it was at that time. Subsequent progress with this restructuring is discussed in Section 6 below; the funds so far available to GEF, however, fall far short of the estimates above. Compromise wording was agreed for the Agenda 21 text, but the lack of significant additional financial commitments at the time of Rio was seen by many as its principal failure. Perhaps it was some achievement that one view of needs is spelt out in some detail; the required re-ordering of national economic priorities was hardly likely to take place instantly, if at all. The transfer of 'environmentally sound' technology also ran into difficulties about terms and conditions, though again the need is identified.

In contrast, the chapters on science and education were uncontentious. That on science evidently overlaps one of the 'major groups' chapters, Chapter 31; neither mentions the Secretary General's proposal for an 'Earth Council' of independent experts, though supporting independent advisory groups in general.

*Activity 3*

If you had to assemble an Earth Council to observe and advise about human welfare and impact on the planet, how would you like to see it constituted?

The next two chapters are important complementary ones, on the need for national and international institutional arrangements respectively. On the national scale a timetable for review, reporting and international co-operation on 'capacity-building' was agreed. This abstract-sounding term recurs frequently in Agenda 21 and refers to the ways in which individual countries will plan for sustainable development and monitor environmental impact by institutions of all kinds, from government and NGOs to business, universities and research centres; comparison with the methods of other countries will be helpful. Internationally, the role of existing UN agencies was reviewed, and perhaps most important, a new *Commission on Sustainable Development (CSD)* was set up, with the twin tasks of co-ordinating international efforts, and reviewing national plans for sustainable development. There was resistance to the obligatory submission of national plans, but voluntary submission was agreed, recognising that it might strengthen arguments for more international help. The setting up of the CSD was one of the most concrete outcomes of Agenda 21. It initiated a

---

Many governments regard Gross Domestic Product (GDP) or Gross National Product (GNP) as valuable indicators of national wealth, with their growth as a goal of policy.

*Gross Domestic Product* is the total value in money terms of all the production in a country in a year. It is measured in three different ways (which should all give the same total): the total value of all the goods and services produced; the total expenditure on them; and the total income received from producing them. (Thus activities like unpaid work – production without money changing hands – or gifts – money changing hands without production – are excluded.)

*Gross National Product* is GDP plus rents, interest, profits and dividends flowing in from abroad, minus similar payments flowing out to people in other countries. Thus GNP measures the total income of the *inhabitants* of a country, and so depends on their location within the country; GDP, on the other hand, depends on where the *economic activity* is located. GNP can thus be either larger or smaller than GDP depending on the balance between investments abroad and incoming foreign investment.

It is clear that such indicators do not consider environment and sustainability in at least two ways: the consumption of finite resources is not accounted for; activities producing waste and pollution *and* those concerned with reducing its effects are added together, so the elimination of such activities would actually reduce GDP twice over.

A number of attempts to provide alternative indicators are being made by 'environmental economists' (such as Pearce and Daly). The Agenda 21 requirement for sustainable development indicators to be developed is beginning to materialise – for example, in a UK Department of the Environment report in March 1966 which showed many current policies, particularly in transport, to be unfavourable. The international effects of such a change in economic assessment of policy – if it becomes general – may be profound.

---

process which has continued after Rio in the way discussed in more detail in Section 6 below.

The final chapters, on legal instruments and information for decision-making, again included important provisions about the need for international environmental standards and national indicators of sustainable development – which, if they can evolve to replace GNP/GDP as measures of national prosperity, may have a major long-term influence.

## 4.6   *Discussion*

This unavoidably laborious outline of Agenda 21 gives some idea of the contents of an enormous document which, for all its deficiencies, is the most complete recent statement we have of a world agreement about sustainable development. It is a first step, but it is disappointing in many ways. First of all, the laboriously constructed language of international compromise destroys any sense of the wonder and challenge both of the natural world and of the rich inventiveness of human culture, covering it all with a grey film of dry abstractions. In addition, the North–South divide dominated the discussions, dividing countries into 'developed' and 'developing' unquestioningly, with economic inequalities and controversy over aid taking precedence over the sense that the 'E' of UNCED might require radical reappraisal of the consumption patterns of the *North* and a different 'D' and financial framework in the future for both North and South – a polarisation of views that fell some way short of a 'global partnership for sustainable development'. A generally anthropocentric view of human welfare, rather than an ecocentric view of planetary welfare, was perhaps to be expected, but a dependence on the decisions of national governments, co-ordinated by institutions like the UN, obviously restrains

the extent to which sustainability might be shown to depend on a
reconstruction of these very institutions themselves. There is missing an
integrated view of the Earth as a system; instead there is a set of separate,
perhaps contradictory, prescriptions – at worst what has been described as
a 'rag-bag' of issues, though this is perhaps less than fair to the undoubted
significance of most of them in some degree. Most of the chapters are 'wish
lists' of hopes, containing many 'shoulds' without any indication of how
they might be achieved. As different groups within and between nations
contributed to each of the chapters, they are repetitive in their cross-
references to each other, and in the reiterated pleas for more research, data-
gathering and dissemination, monitoring and participation – all valuable
and necessary, but numbing in their frequent reassertion. However, it can
be argued that the approach is much more 'bottom-up' in its insistence on
local and NGO involvement – and in its recognition that much is not
known – than many earlier international statements, and it does succeed in
intertwining environmental and developmental concerns.

What was achieved? Evidently, a very substantial handbook of relevant
topics, which will continue to be referred to in local, national and
international debate. But what will change as a result? Those seeking more
immediate results can easily be critical, at levels both below and above the
actual recommendations. On the one hand, even the topics which *are*
covered seem to be in constant danger of being forgotten or sidelined in the
day-to-day operations of people, institutions and governments pursuing
their own well-established patterns of survival and competition.
Nevertheless, sustainable development is now on the agenda, and cannot
any longer be entirely forgotten. More serious are the criticisms of what is
absent from the document – perhaps most importantly, an integrated
ecocentric view of the natural order, possible alternatives to the dominance
of sovereign states, and proposals for the control of international finance
and transnational corporations.

Agenda 21 is in no sense a final statement of objectives to which everyone
can be expected to subscribe for the next decade or more. Rather it is one
stage in a continuing process discussed further below. Its lengthy provisions
do, however, provide a snapshot of the tangled global issues on which some
sort of agreement could be negotiated at a particular moment in 1992.

*Activity 4*

Look back over the measures in Agenda 21 and distinguish those
generating controversy from those accepted easily. Can you discern a
pattern?

# 5   The politics of sustainable development since Rio

## 5.1   Introduction

The implementation of the comprehensive 'programme of action for
sustainable development' outlined in Agenda 21 must take place mostly at

the national and local level. Nevertheless, Agenda 21 was the product of international politics, and the international dimension has remained important in the follow-up process. These final sections of Chapter 5 examine some of the key international processes in the implementation and further development of Agenda 21 since the Rio conference in 1992.

## 5.2 Some characteristics of international politics

First, some comments are in order about the character of international politics and of its participants. States have a privileged position in international politics. They have sovereign authority to legislate within their territories, and cannot legally be forced to obey an international law to which they do not subscribe. Thus, whereas states and their central governments do not generally control the economic, social and environmental activities of concern, they must play an important role in any international responses to unsustainable practices.

Nevertheless, international politics consists of much more than inter-state diplomacy. This is true for all issue areas, but it is particularly true of the international processes that are relevant to implementation of Agenda 21. Local and national economic, social and environmental practices are powerfully shaped by global and regional processes. International institutions and international conferences can stimulate states into improving their environmental performance by:

* promoting co-operation
* increasing awareness and knowledge of problems and opportunities
* generating political pressures and concerns
* facilitating technology and resource transfers, and
* helping weak or poor states to increase their capacity to implement desirable programmes.

Some international organisations are directly involved in local projects. Moreover, a few *supranational* organisations, such as the EU, have the authority to regulate activities inside member-states.

Non-state actors are also very important. Links and coalitions between non-state actors (such as NGOs, businesses, and local communities) in different countries can empower such actors in their dealings with governments or international organisations, and can directly affect their environmental activities. Some key actors, such as environmental organisations (WWF, Greenpeace), scientists' organisations, and multinational corporations (Shell, IBM, IKEA) are *transnational*: that is, they are non-state actors, operating across state borders.

As in other areas of international politics, the outcomes of international efforts to promote sustainable development in each of the areas covered by Agenda 21 characteristically depend on the power relations between the main actors involved, their patterns of interests and values, and the learning processes and levels of knowledge associated with the issue. However, even if these factors are all favourably aligned, there will be little progress without the capacity to implement desired programmes. Thus, international resource transfers and 'capacity-building' activities (i.e. the development of appropriate institutions and 'know-how') in weak or poor states may also be important.

The implementation of Agenda 21 is a dynamic process. Power relations, interests, values, knowledge and implementation capacities are not static. Rather, they are developing and mutually shape each other. Changes in one area will affect others. For example, developments in knowledge – either about the problem or about the consequences of adopted policies – could empower some groups relative to others, change actors' assessments of their interests, and increase their capacity to implement chosen policies. Decisions and actions by individual groups and organisations can affect the overall process, though often in unforeseen ways.

## 5.3   The international dimension of implementing Agenda 21

A discussion of the international politics of promoting sustainable development *per se* would be overwhelmingly broad. Agenda 21 touches on almost every aspect of human activity and, as outlined above, there is an important and complex international dimension to each of these aspects. Thus we focus on the operation and effectiveness of the main international processes and institutions specifically aimed at promoting the implementation and development of the Agenda 21 programme agreed at UNCED.

Even so, the field remains very broad. The forty chapters of Agenda 21 cover a vast range of issue areas. Each issue area has its own particular characteristics and associated international politics. The relative significance of power, interests, knowledge production and learning processes, resource transfers, and capacity-building varies. Also, not only does the identity of the main actors differ between areas, but so does the relative importance of different *types* of actors (states, international organisations, domestic actors [businesses, local authorities, communities, NGOs], transnational groups and so on). Moreover, there is a wide range of international organisations and mechanisms active in the area of environment and development; their roles, relationships and effectiveness vary according to the particular issue.

Thus, it is important to recognise the need for specific examinations of each issue. Nevertheless, as discussed in Section 2, UNCED assigned a number of institutions and processes with major roles in promoting the overall implementation of Agenda 21, and it is important to examine their significance and effectiveness. Pre-eminently, these were the Commission for Sustainable Development (CSD) and the Global Environment Facility (GEF), working in association with the UN General Assembly, UNEP, UNDP, and a range of other UN bodies and international institutions.

It is important to note from the outset that these international institutions and conferences could not conceivably directly implement Agenda 21, or oblige others to do so. No international body could have the power, knowledge or capacity to do this, and the CSD, for example, is a relatively weak international organisation. Thus, in focusing on the operation of the CSD and the other bodies listed above, we are really aiming to examine the international processes associated with them, and the ways in which the international institutions and conferences shape or focus these processes in a way that promotes sustainable development.

# 6   Developing institutions

## 6.1   The Agenda 21 institutions in historical context

UNCED marked the twentieth anniversary of the first global conference on the environment – the 1972 UN Conference on the Human Environment held in Stockholm. It is useful at this point briefly to compare their outcomes, and to note some developments in the intervening twenty years.

As outlined earlier in this chapter, the Stockholm Conference produced: a Declaration containing 26 principles concerning the environment and development; an Action Plan with 109 recommendations spanning six broad areas (human settlements; natural resource management; pollution; educational and social aspects of the environment; development and the environment; international organisations); and a Resolution on institutional and financial arrangements. These outcomes are similar in character to Agenda 21.

In the following years, dozens of international environmental agreements were achieved. However, apart from galvanising public concern and educating governments, the most enduring specific contributions of the Stockholm Conference are widely believed to be the following.

Firstly, Principle 21 acknowledged states' sovereignty over their natural resources but stipulated that states have 'the responsibility to ensure that activities within their jurisdiction or control do not cause damage to the environment of other states or of areas beyond the limits of national jurisdiction'. This provided a precedent for much subsequent environmental diplomacy.

Secondly, it led to the establishment of global and regional environmental monitoring networks, which have improved monitoring of environmental problems, such as marine pollution and ozone depletion, and indirectly stimulated action to tackle them.

Thirdly, the Stockholm Conference led to the creation of the UN Environment Programme (UNEP), which was charged with co-ordinating the environment-related activities of other UN agencies and promoting the integration of environmental considerations into their work. In practice, UNEP played a key role in:

- raising political awareness of environmental problems
- helping with the formation of scientific consensus on problems and responses to them
- facilitating negotiations (particularly for the protection of regional seas and the ozone layer)
- improving countries' environmental management capacities.

However, UNEP lacked the institutional weight seriously to co-ordinate other UN agencies, which typically protect their 'turf' vigorously. Moreover, and partly because of this, the development agendas included in the 1972 Action Plan and Declaration of Principles were never seriously followed up. Thus the 1987 Brundtland Commission, which put 'sustainable development' onto the international political agenda,

◄    *Mostafa K. Tolba,
Executive Director of UNEP,
addressing the Helsinki
Meeting in April 1989.*

recommended the creation of a powerful UN 'Board of Sustainable
Development' to provide necessary leadership and co-ordination. This was
to be chaired by the UN Secretary General, and was to be something akin to
the UN Security Council but charged with promoting sustainable
development.

However, this proposal gathered little support amongst states in the
run-up to UNCED. Indeed, the G7 major industrialised countries went to
the conference opposed to the creation of any new UN institution to
promote and co-ordinate follow-up to Agenda 21. This was partly due to a
general opposition to new UN bureaucracies and partly because they
thought that an existing organisation, pre-eminently UNEP, could play this
role. However, many countries, particularly from the developing
world, wanted a new dedicated institution to reduce the risk that the
'development' dimension of Agenda 21 would be lost or subordinated to
environmental protection as it had been before. Moreover, under the
charismatic but sometimes abrasive leadership of Mostafa Tolba, UNEP
had recently lost some developing country support in its energetic
promotion of certain environmental agreements.

Thus it was agreed to recommend to the UN General Assembly that the
CSD should be created, but as a lower-ranking and weaker body than that
envisaged by the Brundtland Commission. This new body would encroach
on the co-ordinating and catalytic role of UNEP, which was, however,
given some enhanced responsibilities in other areas, including an expansion
of its 'Earthwatch' monitoring programme. Moreover the CSD was to co-
ordinate the activities of other larger well-established organisations, such as
the Food and Agriculture Organisation (FAO), World Health Organisation
(WHO), the UN Development Programme (UNDP), UNESCO, the
International Labour Organisation (ILO), and the International Atomic
Energy Agency (IAEA). The promotion of sustainable development also
implied that the CSD would have to engage with the main economic
institutions in the UN system, including the IMF, World Bank and the
GATT/World Trade Organisation.

## 6.2   The role and significance of the Commission for Sustainable Development

### Constitution and main functions

The CSD consists of representatives of fifty-three states, elected for three-year terms in a way that ensures equitable geographical representation. It reports (via the UN Economic and Social Council) to the UN General Assembly and meets annually for three weeks, with each meeting ending in a two-day 'high-level political segment' when government ministers participate in final discussions and decisions are made. It is supported by a small Secretariat based in the UN Department of Policy Co-ordination and Sustainable Development, which was also created after UNCED and which is headed by a UN Under-Secretary General.

The CSD was given two main functions. The first was to monitor, review and promote the implementation of Agenda 21, and the integration of environmental and developmental goals, by all relevant UN agencies, organisations and programmes. It is assisted in this task by an Inter-Agency Committee on Sustainable Development (IACSD), which brings together representatives of the nine core agencies with particular responsibility for aspects of the implementation of Agenda 21 and is chaired by a representative of the UN Secretary General.*

The second main function given to CSD was to monitor and review countries' progress in implementing Agenda 21, mainly on the basis of national reports submitted to it by governments. In this role, it was hoped that the CSD would provide a forum for discussing and publicising states' performance. In this way, all involved in the process could learn from each others' successes and failures. Moreover, exposure of inadequate performance could help to generate domestic and international pressures on countries to take their commitments more seriously in the future.

The CSD has also developed a third important function. This is to carry forward the unfinished business of UNCED, and to promote the further development of international arrangements and commitments where opportunities arise to move beyond the general principles contained in Agenda 21. The unfinished business of UNCED included the development of international commitments to protect or manage both forests and stocks of highly migratory fish. It also included the effective development of regimes to combat desertification and protect biodiversity, and following up on the results of planned global conferences on issues including population, social development and urban development.

### The CSD process

The CSD met for the first time in June 1993, and has met once a year since then. In order to make each meeting a little more manageable and productive, a few themes were selected for review at each meeting. For example, the 1994 meeting focused on health, human settlements, freshwater, toxic chemicals and hazardous wastes. In 1995 the topics were land use, desertification, forests and biodiversity; and in 1996 the atmosphere, oceans and seas. In addition, a few 'cross-sectoral' issues were selected for consideration at each meeting including financial mechanisms, trade and consumption patterns, capacity-building and technology transfer. These meetings led up to an overall review of implementation of Agenda 21 in 1997.

*The nine core agencies on the IACSD are: FAO, UNESCO, WHO; WMO; WB; UNDP; UNEP; ILO; and IAEA. The chair is normally the head of DPCSD. The IACSD reports to the UN Administrative Committee on Co-ordination, which is chaired by the UN Secretary General.

Initially, according to one observer, the two weeks preceding the high-level Ministerial segment of CSD meetings resembled 'nothing so much as a party congress in which none of the deals between parties has been worked out beforehand, and where, under cover of old ideological compromises, major tensions between different interest groups are in the air'. Soon, the period between CSD meetings became filled with international conferences and meetings of working groups, ad hoc groups, preparatory groups and 'intersessionals' to consider issues that were on the agenda of the next CSD meeting. Indeed, the density of international meetings has sometimes been so great that clashes of timing could not be avoided. In addition, the major UN agencies were made 'Task Managers' responsible for receiving and assessing reports and for producing overviews, targets and objectives on particular themes due for consideration by the CSD. For example, WHO acted as Task Manager for health in 1994, and FAO dealt with forests and land use issues in 1995.

From the beginning, it was decided in principle that the operations of the CSD should be transparent to outsiders and run in a way that recognises and promotes the important role of non-state actors (NGOs, businesses, local authorities, communities and so on) and representatives of 'major groups' (such as women, youth, indigenous peoples and trades unions) in implementing Agenda 21. Thus, all such groups have access – as observers with discretionary speaking rights – to CSD meetings, to intersessional conferences and working groups, and also to associated documentation. Moreover, they can submit reports which are taken into account in the work of the CSD and its working groups.

This reflects an attempt to design the CSD to promote an international political process in which all groups interested in promoting sustainable development have access and influence, and which links UN institutions with domestic and transnational actors as well as with the governments of member-states.

Thus, before each CSD meeting there has typically been a number of preparatory workshops and reports, into which governments, non-state groups, UN agencies and other international organisations have all made an input. At the annual CSD meeting itself, the high-level segment is preceded by two weeks of intense negotiation between national officials in a forum wide open to presentations and lobbying by UN agencies and non-state groups. Significantly, in the first few years at least, national representation at the high-level segment was overwhelmingly at ministerial level (mainly Ministers of the Environment), lending the recommendations of the CSD political weight internationally and nationally. In a sense, each CSD meeting has been a mini Earth Summit.

## The role of non-state actors

In practice, NGOs have been the most active and organised of the non-state participants in the CSD process. Dozens of these participate in each meeting, and they have their own network of contact points, intersessional co-ordination centres, newsletters, Internet pages, and caucus meetings. Moreover, the capacity of NGOs should not be underestimated. On issues of particular concern to them, environmental NGOs in particular regularly field larger and more expert delegations than all but the largest developed states.

NGOs have a wide range of interests, values and priorities, and there are many tensions between them, particularly between groups primarily concerned with environment and development and between NGOs from the developed and developing countries. However, by the mid-1990s, they had learned how to manage these better and co-operated extensively. The main environment and development groups developed shared understanding of ways to integrate their concerns, and mechanisms evolved for relatively balanced and meaningful consultations between NGOs from North and South. Although northern NGOs have remained more numerous and better resourced in international activities than those from the South, the number and capacity of southern NGOs have developed greatly since the 1980s. Even the presence of a few NGO activists can have an effect. For example, a few activists from the UK, Caribbean and South Pacific with an interest in health reportedly played a major role in blocking a negotiated document that would have given health insurance and pharmaceutical industries pride of place in international programmes to promote environmental health.

In contrast, local authorities, community groups and most business organisations have been slower than NGOs to participate effectively in the CSD process. For example, in 1994 when the CSD was considering themes of core interest to local authorities (health; urban settlements; freshwater resources; management of sewage and solid, toxic and hazardous wastes), the only local authority representation was from the UK: one person from the UK's Local Government Management Board and another from Glasgow District Council. This was partly explained by a clash with another conference, in Aarlberg in Germany, on Local Agenda 21s. However, it also reflects the fact that local authorities have tended to be less aware of, or experienced in, international political processes. As a pressure group within CSD, they have also found it harder to forge transnational coalitions, tending instead to national perspectives, and thus have been less adept at agenda-setting.

It is also true, however, that for many local authorities and other non-state groups, the CSD has appeared distant and marginal to achieving their objectives. Thus it is worth asking how significant or effective the CSD process has turned out to be. We thus briefly examine progress in pursuing each of the three main functions of the CSD process.

## The three functions of the CSD

### 1   Assessing the CSD's co-ordination and integration role

In relation to the promotion and co-ordination of activities by the various UN agencies and international organisations, and the integration of their environmental and development programmes, progress has been mixed.

On a bureaucratic level, every UN institution has been aware of the advantages of being seen to be a leading agency in the implementation of Agenda 21 and in the CSD process, and of expanding its responsibilities where possible. However, the sustainable development agenda is often so broad and ill-defined that it is hard to determine the extent to which the progress reported by such agencies is simply the re-labelling of established programmes. UN agencies continue vigorously to try to protect and expand their status and areas of responsibility. Nevertheless, the IACSD gained real impetus in promoting inter-agency co-ordination in the mid-1990s. More substantively, the CSD process has provided a key focus for governments and NGOs to examine and influence the activities of well-established agencies dealing with environment and development, such as the FAO, UNDP and WHO. Numerous new programmes have been launched as a result.

The CSD's influence over the major economic institutions, such as the World Trade Organisation (WTO), has been much more limited and contentious. In this case, the main mechanisms for CSD influence is through governments. Government ministers who have examined issues at the CSD may subsequently change their governments' policies at home, leading to revised national negotiating positions within the governing councils of the WTO. Realistically, however, the Environment Ministers who attend CSD meetings often have little influence over their governments' ministries dealing with economics, finance, industry or trade. For this reason, there have been efforts to encourage economic ministers to attend CSD meetings, but with meagre results.

## 2   *Reviewing national implementation*

The second of the CSD's main functions – reviewing national implementation of Agenda 21 – is potentially the most important. However, the process began inauspiciously. Governments are generally reluctant to be held to account for their national performance, and many resisted obligations to provide regular reports on their progress in implementing Agenda 21. On the insistence of India and some other developing countries, the CSD watered down obligations to provide national reports on implementation so that reporting was voluntary: governments can decide whether to report and on the details and regularity of any reports they do choose to provide.

By the requested submission date for reports to the second meeting of the CSD (30 November 1993) only three national reports had been received (from the USA, Uganda and Cuba). Fortunately, fifty governments ultimately provided reports in time for the meeting. In subsequent years the number of countries providing reports increased, though they were of widely varying detail and quality.

*Q*   How could the stimulation of countries to produce national reports in itself promote implementation of Agenda 21?

*A*   The production of a national report in itself contributed significantly to the implementation of Agenda 21 in some countries, for a number of reasons. It requires the formulation of national policies relating to each of the main areas of concern; in many cases such policies did not previously exist. For example, China prepared a national programme for implementing Agenda 21 soon after UNCED, including establishing a regulatory framework for environmental protection and a number of priority projects.

Moreover the process of preparing the report promotes consultations between relevant ministries and agencies on ways in which their activities relate to Agenda 21 and on opportunities for promoting sustainable development. In some countries such preparations provided a focus for broader debates. The completed report helps to publicise government policies around the country. It provides a basis for interested domestic groups to scrutinise their government's plans, to hold it to account for its policies and performance, and to identify opportunities, gaps and problems. In the UK, for example, alongside the government's 'National Strategy for Sustainable Development' (published in 1994) a number of consultative and dissemination bodies were established with Department of the Environment support (including the Government Panel, a UK Round Table on Sustainable Development, and a Citizen's Environment Initiative). It was also used as a focus for activities by a wide range of NGOs and local authorities – many of which were directed at criticising and improving the government's declared strategy as much as at promoting its implementation.

It is important to recognise that many governments had not even formulated policies relating to key areas of Agenda 21 by 1992. Indeed, in some developing countries, the UNCED and CSD process was the stimulus for establishing a Ministry of the Environment. For example, Bolivia established a Ministry for Sustainable Development shortly after UNCED. This is perhaps even more true of local authorities, although, as Chapter 28 of Agenda 21 notes, '… because so many of the problems and solutions being addressed in Agenda 21 have their roots in local activities, the participation and co-operation of local authorities will be determining factors in fulfilling its objectives'. 'Local Agenda 21' initiatives were launched in several countries after UNCED, to promote the development of policies for sustainable development at local authority level. Although such initiatives have been slow to generate local authority engagement with the CSD at the international level, as noted above, they stimulated useful local networks and actions in some places.

A typical UK example is that of Leicester (see Box 5.11). Examples from elsewhere were provided by a 1993 conference on 'Partnerships in Practice', organised by the UK government as a follow-up to Rio. The central theme was that case studies of successful sustainable development should be presented. Table 5.1 gives three examples, in the (edited) words of their own presenters, which each has a distinctive style. It was recognised that many successful local initiatives already existed before Rio, and that these were developing into the Local Agenda 21 process.

---

### Box 5.11   Leicester

The only UK city represented at the Rio Conference, Leicester proclaimed itself 'environment city' in 1990, 'as a challenge rather than an accolade'. It is one of about two-thirds of UK local authorities claiming to be on course to implement Local Agenda 21 in 1996, although Michael Ashley, environment under-secretary at the Association of District Councillors, thinks many misunderstand it. It is not a set of policies for general environmental management – greening policies, planting trees and flowers – but rather a participatory planning process on issues such as jobs, homes and transport with environmental impact. Leicester claims to be leading the way with its 'Blueprint for Leicester', which drew on advice from an existing network of working groups from the city and county council, education, the voluntary sector and its charitable trust Environ supplemented by systematic consultation with local residents. The traditional apathy shown in a mere 1% response to questionnaires sent out with the local newspaper was overcome in a systematic survey of 748 representative households, activists from 88 community organisations and a number of businesses. The resulting findings cover everything from shopping and employment to crime prevention and traffic, with the council's role to 'influence and advise' in a partnership framework. As a big local employer, it has re-examined its own internal organisation – with environmentally friendly policies for staff transport including a bike mileage allowance, energy-efficient offices and 'green purchasing', and environmental implications considered in all policy discussions alongside financial, equal opportunities and other consequences. It does not call for higher spending, but different ways of spending existing money, with longer-term views of pay-back periods. It has launched a new pilot recycling scheme with special collection bins for plastics and cans, for which the present (1996) cost exceeds the returns, but this is expected to reverse with increased volume, long-term contracts with buyers of recycled material, and the projected increased cost of landfill from the new £7 per tonne landfill tax. Home energy grants meeting 80% of the costs of insulation and other energy efficiency measures saved fuel costs by £177 000 a year (the owner-occupiers' 20% of the costs was recovered in savings within seven months), reduced carbon dioxide emissions by 2295 tonnes a year, and helped the poor and old to keep warm. More long-term plans, such as those to attract manufacturers of environmentally friendly products to create 'green' jobs, may take longer to bear fruit – but 'this isn't a quick fix solution, it's about changing the decision-making process', according to Jane Morris of the local government management board.

Source: Based on an article in the *Financial Times*.

*Table 5.1   Three case studies from 1993. Responses are given for:*

1   Metro Manila Women's Garbage Recycling Programme from the Philippines

2   CAMPFIRE (Communal Area Management Programme for Indigenous Resources) from Zimbabwe

3   The Sustainable Seattle Project from the USA

**What are the aims of the initiative?**

1   To deal with the garbage problem in the town of San Juan in Metro Manila, and to make the local community aware of their responsibility to improve their environment. It aims to be 'sustainable' since the system should eventually reduce the need to exploit valuable natural resources in the manner that they have been.

2   To allow rural communities in marginal agricultural areas to utilise their natural resources and manage them sustainably. Local empowerment and participation through locally based institutions is encouraged.

3   It is made up of four initiatives: the Sustainable Seattle citizens' initiative, the Environmental Priorities Project, the Comprehensive Plan and the Waste Reduction and Recycling Programme.

The four initiatives share a similar, overriding objective: to promote more sustainable thinking and behaviour at all levels of society in Seattle: in homes, in neighbourhoods, in businesses, and within city government.

**Why do you think this initiative has wide applicability?**

1   It is universal by its very nature. The programme has been replicated in the 16 villages of Quezon City; six villages of the town of Pasig; one village in the town of Muntiniupa; two villages in the town of Paranaqua; one village in the town of Makati; one foreign bank in Pasig; the main headquarters of the Philippine National Police and the entire complex of the Department of the Environment and Natural Resources in Quezon City.

2   CAMPFIRE's emphasis on local institutions and participation has a wide appeal for rural development in developing countries, where community mobilisation is a vehicle for motivating local people. Governments in developing countries rarely provide resources for marginally settled communities.

3   They are addressing issues that are relevant, in varying degrees, to urban areas around the world: reducing urban sprawl, reducing over-dependence on the private car and reducing resource consumption. The key to all of these initiatives is to identify fundamental community values, and think strategically about what can and should be done – by government as well as non-government groups and individuals – to honour, preserve and promote those values.

**Who are the partners involved?**

1   Local village officials; citizens; local government; junk-shop owners. The local village officials help disseminate information about separating garbage while local government does very little, except in Quezon City, where the mayor recently donated T-shirts, paints and printed letters to citizens.

2   The Zimbabwe Trust, the Centre for Applied Social Sciences, World Wide Fund for Nature, the Ministry of Local Government, Rural and Urban Development, the Department of National Parks and Wildlife Management, CAMPFIRE Association and local authorities (District Councils).

3   All four efforts involve a partnership among various key sectors or constituencies within the city: the business community (e.g. the Chamber of Commerce (large), the Neighbourhood Business Council (small)), environmental advocates (e.g. the Washington Environment Council), neighbourhood activists and government officials (plus some 30 named organisations, from the University of Washington and Boeing to the Sierra Club, the Mukleshoot Tribe and Seattle Tilth, for the specific initiatives).

### When did the initiative start?

1   12 February 1983

2   Genesis of decentralisation was in 1975, but Appropriate Authority for the first two District Councils was achieved in 1989 (refers to enabling of statutory regulation to allow DCs to manage and utilise their own natural resources).

3   Sustainable Seattle began in November 1990 with one-day forum, more intensive volunteer effort from February 1991. Recycling and waste reduction began in earnest in 1988, with 60% recycling goal – since expanded and improved. Environmental Priorities in 1990; first Environmental Action Agenda adopted by City Council in 1992, implementation on-going. Comprehensive plan process began in 1990 (state Growth Management Act passed; community visioning conducted by Planning Department).

### If applicable, when did it finish?

1   The initiative is on-going.

2   The initiative is dynamic and on-going, with a total of 22 rural District Councils gazetted for CAMPFIRE by 1993.

3   All four initiatives are still on-going, with no set end-date. The Comprehensive Plan is scheduled to be adopted by Seattle City Council in June 1994, but the process of implementation will continue ad infinitum.

### Who was primarily responsible for getting the initiative going?

1   The Metro Manila Council of Women Balikatan Movement Inc., through its chairperson, Leonarda Camacho.

2   The Department of National Parks and Wildlife Management provided the enabling policy environment, but the Zimbabwe Trust and Centre for Applied Social Sciences were responsible for funding and research respectively.

3   Sustainable Seattle: a small group of dedicated volunteers (including representatives from Metrocenter YMCA, New Road Map Foundation, the Boeing Co., Seattle Planning Dept). Recycling and Waste Reduction: City's Solid Waste Utility (Director and then Mayor) with public encouragement (particularly environmental groups opposed to garbage incinerator). Environmental Priorities and Comprehensive Plan: Seattle Mayor Norman Rice, City Council and Planning Department.

### How were the other actors brought in?

1   The women's organisation manages the system, while local government assists in the dissemination of information. It is very easy to get the co-operation of the junk-shop owners as they are adequately motivated by profit, unlike the local officials who are indifferent. Users of secondary materials (i.e. the factories) are partners, as previous arrangements were made to supply them with used paper, plastics, bottles and cans; they then use less fuel and new materials. It is not a simple business transaction.

2   To assist with implementation, Zimbabwe Trust and Centre for Applied Social Sciences (knowledge and research experience with local communities) and WWF (wildlife knowledge) formed the Collaborative Group; Ministry of Local Government and CAMPFIRE Association were later entries, representing local community interests. These groups are government and non-governmental organisations who work in partnership with the CAMPFIRE Association partnership to facilitate CAMPFIRE development and undertake research at national and international levels.

3   Sustainable Seattle: informal and ad hoc; open advertised meetings, self-selected participation sustained by shared enthusiasm and belief.
Others: more formal, deliberate and resource-intensive. Mayor, Council and Planning Department set up advisory committees of representatives from major community sectors to ensure their input and participation; constant meetings with groups to sustain involvement.

### What have been the achievements of the initiative so far?

1   The system is operating successfully in the 21 villages of San Juan (population 250 000) separating 18 000 tons of paper, plastics, cans and bottles yearly, generating an income of $15–50 yearly for the families, some $1000 monthly net profit for 30 junk-shop owners, some $100 monthly for 200 eco-aides, unquantifiable savings for factories which buy recyclables and for government in imports of raw materials; most of all, relief to nature in trees, oil and minerals. The programme protects the environment and provides economic benefits.

2   Increased public awareness. Sustainable resource management through quotas for sustainable wildlife harvesting (e.g. if 100 male buffaloes, sustainable quota could be 4 for hunting purposes) with community involvement. Dept of NP&WM and WWF offer technical advice through aerial surveys; communities use own census methods and participate in quota-setting. Skills acquisition by locals in book-keeping, resource management and development of institutions. Monetary benefits to local communities, e.g. Kanyurira: $47 000 shared as dividends among 60 households in 1990; Muzarabani: $50 000 to buy maize for famine relief in 1992; Hurungwe: $400 000 shared among 11 wards in 1992; Mahenye: distributed $180 000 to 481 families in 1992.

3   Sustainable Seattle has sustained itself, given almost complete dependence on volunteerism; produced draft report on dozens of 'Indicators of Sustainable Community' to aid decision-making and reporting; organised discussions and expanded community dialogue on sustainability.

Comprehensive Plan has forged consensus around 'framework policies' to guide growth management with countless meetings and discussions; emerging strategy of twenty-year growth focused in 'urban villages' with improved public transport (light rail?), bicycle and pedestrian facilities.

Environmental Priorities Project has produced consensus, improved departmental co-ordination and community involvement and produced first Environmental Action Agenda.

Recycling and Waste Reduction is one of world's most successful, with 42% of solid waste recycled, and 90% household participation

### What have been the failures of the initiative?

1   The government, local and national, remains indifferent to the idea and does not want to adopt the method. It persists in making loans to build the so-called sanitary landfills and to buy unnecessary equipment. There are still some households who do not wish to separate their garbage and insist on throwing it in rivers and canals or giving it to the government trucks when available. 8000 out of 80 000 do not take part.

2   I would rather refer to constraints. Devolution of authority to the lowest level, the village, has still to take place. Diversication of CAMPFIRE to other natural resources such as trees and minerals is still slow. Staffing and financial constraints in expanding from 12 to 22 districts are being felt. Extension services and training all demand financial and human resources.

3   Premature to talk of outright failures, but challenges that could lead to failures if not properly managed: involving historically uninvolved sectors, e.g. minority groups; changing behaviour, about consumption, and individual freedom vs public good; bureaucracy: parochial thinking of individual City departments; building trust across sectors; affecting 'macro' changes in the economy to correspond to 'micro' changes in behaviour, e.g. 'secondary markets' for recycled products, only partly in City's control; integration in 'systems' approach of historically segregated planning, e.g. of economy and environment – separate government structures create barrier to more holistic thinking.

Sources: Department of the Environment (1994) *Partnerships in Practice*, London, HMSO, pp. 45–6, 31–2, 13–16.

Thus, the international requirement on countries to formulate and report on national programmes to implement Agenda 21 in itself stimulated important developments at national and local levels. However, as discussed above, the CSD was designed to review these national programmes at the international level in order to clarify their strengths and weaknesses and promote effective implementation. This review process proved difficult to get going. Even after many countries had submitted national reports, some governments were very suspicious about the prospect of having them reviewed at CSD meetings.

As a confidence-building process, five countries volunteered to have their national reports reviewed at the 1995 CSD meeting. The tone of these first reviews was supportive and constructive, identifying and learning from successes and failures and in some cases finding areas where international assistance might usefully be provided. It was generally treated by governments as an opportunity to learn and help rather than to expose and denounce. On the basis of this approach, the process of reviewing national reports has been accepted and has gradually developed.

However, the effectiveness of this process in promoting implementation depends on the availability of appropriate information about national performance and on critical and focused review. Governments are often keen to highlight successes and disguise failures, and other governments are often too diplomatic to provide undesired criticism. Moreover, a three-week CSD meeting with a full agenda is not the place for effective and detailed review of many national reports across a range of topics.

In practice, however, once the principle of reviewing national reports in the CSD was accepted, it could be followed up in the many more focused workshops associated with it and in small group meetings involving interested parties. Moreover, there are some incentives for countries to admit problems if there is the prospect of attracting international assistance.

The NGOs have also played an important role in sharpening up the reviews of national reports at the CSD. They have less inclination to be polite and uncritical, and indeed have sometimes relished the opportunity to reveal the inadequacies of their governments' performances in front of an international audience.

For example, a coalition of NGOs called CAPE 21 (Campaigns for Action to Protect the Earth) established a programme to monitor national implementation, called 'Earth Summit Watch'. For each CSD meeting, they focused on a few issues and published a detailed report on countries' progress in tackling them, based on national reports and any other information available to them. For example, in 1994 they focused on: reducing exposure to lead; cleaning up freshwater resources; ratifying and implementing the Basel Convention (limiting trade in hazardous wastes); and protecting natural rivers. This provided a focus for lobbying at the CSD meeting, and fed into all of the consultations and reports associated with the CSD process as well as domestic campaigns.

To facilitate review of progress in implementing Agenda 21, and to limit the amount of information requested from national reports, there were proposals to develop internationally agreed 'indicators of sustainable development'. A set of specific parameters indicating progress would clearly greatly streamline reporting requirements and, by providing targets, could provide a focus for policy development and implementation review.

The problem is to agree which indicators to use. People mean many different things by sustainable development, and they prioritise aspects of Agenda 21 differently – indeed it is because of this indeterminacy that so many countries could agree to the programme at UNCED. Moreover, developing countries want to avoid any process that allows industrialised governments to define international targets and to use them to assess whether poorer countries deserve GEF funding, for example. Efforts to develop indicators have continued within the CSD, and have perhaps encouraged participants to think about what they really want to achieve. In this context, they have also been useful at the local authority level, for example, in Lancashire, as a focus for policy development and for involving local communities in discussions about priorities. But the progress towards actually achieving international agreement upon a set of indicators has been slow.

### 3   *Following up UNCED's unfinished business*

The third main aim of the CSD process is to follow up the unfinished business of UNCED and to promote the development of more specific agreements on Agenda 21 items where possible. In this context, it is important to distinguish between issue areas for which:

(a)  specific regimes have already been established (such as climate change, ozone depletion, or oil pollution at sea);

(b)  conventions are under active discussion or negotiation (such as, in the mid-1990s, combating desertification);

(c)  the international community has yet to move beyond agreement on general principles (including most of Agenda 21);

(d)  even the principles are deeply contested (population).

What type of role might the CSD be able to play in following up each of these types of issue area?

For the first type of issue area listed above, the CSD can only play a secondary role, identifying links or gaps between international regimes and attempting to co-ordinate between them. This may be important. For example, an effective climate change convention would impact on virtually all areas of human activity, and it is important to have forums in which linkages with other regimes and programmes are considered. Similarly, while the Biodiversity Convention remains preoccupied with issues such as rights to genetic resources and biotechnology safety standards, it is important to have an international forum in which the core issue of maintaining biodiversity can continue to be addressed. In general, however, the international politics of responding to problems for which a regime has been established is focused around the institutions of the regime itself. This type of politics, and the development and implementation of specific environmental regimes, are discussed in Chapter 6, and will not be pursued further here.

For the second type of issue, the CSD could provide impetus to the negotiations and help to clarify where an international convention could be useful. The CSD provided a forum for considering issues associated with the development of the conventions on desertification and on 'straddling' and migratory fish stocks (i.e. fish that migrate across the borders between Exclusive Economic Zones and the high seas), which were agreed in 1994 and 1995 respectively. This was important to many African states, for example, who appreciated a major forum in which to raise their concerns about desertification at a time when they felt that it was hard to attract high-level attention to this issue in industrialised countries.

More centrally, the CSD process played a key role in restarting discussions around a possible international forestry convention. After discussions on this issue at intersessional working groups during the previous year, the 1995 CSD meeting was dominated by debates about how to take this process further. It was decided to establish an Intergovernmental Panel on Forests (IPF) under the auspices of the CSD itself. This Panel would have a similar task to the Intergovernmental Panel on Climate Change (IPCC), which played such an important role in the development of the Climate Change Convention (see Chapters 2 and 6).

For the third type of issue, the CSD comes into its own as a process through which guidelines for best practice and 'soft law' may be developed and disseminated, and international research and assistance programmes may be targeted. For example, an international campaign (involving NGOs, some states [such as USA, Sweden, Benin, Colombia, Egypt and India], WHO, scientists, and some multinational companies) to encourage the phasing out of lead additives in petrol was successfully launched through the CSD and has had significant international impact.

Similarly, for example, whereas it is not clear that it would be appropriate to try to promote the sustainable agricultural use of fertilisers through a negotiated treaty, it does make sense to develop international guidelines, information exchange, and education and technology transfer programmes on such an issue. Where potentially useful guidelines are being developed regionally (in this case, for example, in the Helsinki Convention for the Protection of the Baltic Sea Environment), it is useful to have a global forum where these can be considered for more general use.

The same applies to the development of international regulations that are non-binding – 'soft law'. Existing examples include the widely used system of prior, informed consent for trade in chemicals and pesticides, maintained by UNEP and the FAP, and the system for setting food safety standards in the *Codex Alimentarius* maintained by the WHO and FAO. The CSD has contributed to the development and dissemination of soft law relevant to Agenda 21 and to debates about whether and how to make it binding. Here again, its open access to non-governmental groups has been vital – industrial associations and NGOs have much expertise and influence in these issues.

Finally, the CSD is charged with following up on relevant global conferences and summits. There were several such conferences in the years after UNCED, for example on social development (1994), population and development (1994), women (1995) and urban development (1996). Such conferences galvanise much international activity and attract wide attention, including that of top political leaders and opinion-formers. They are events where principles and broad recommendations can be negotiated internationally, and around which international networks and coalitions can be formed and consolidated.

This was certainly the case in relation to the 1994 UN Conference on Population and Development (in Cairo) – an issue that had to be side-stepped at UNCED. The principles agreed at this conference provided a framework within which international follow-up programmes could be developed in the CSD and its associated international agencies in much the same way as for most other Agenda 21 issues. In practice, however, the CSD was not a particularly important focus for such follow-up in the mid-1990s.

In contrast to the population issue, where agreed principles were lacking, the 1996 Habitat Conference on Urban Development was dealing with an issue which was already firmly on the CSD's agenda, but where it was not clear how to prioritise and develop international programmes. In this sense, the 1996 Habitat Conference had something of the character of a massive top-level intersessional meeting, providing an agenda for further work in the CSD process. At the time of writing, however, the results of this were not clear.

## 6.3   *The Global Environment Facility*

As discussed in Section 4 of this chapter, the provisions for financial and resource transfers to the developing countries to help them implement Agenda 21 were a source of great North–South controversy at UNCED. It was finally agreed that the Global Environment Facility (GEF) would be the provisional financial mechanism for funding 'agreed incremental costs' of implementing the Climate Change and Biodiversity Conventions and for contributing to the implementation of Agenda 21.

In addition, there was agreement in principle that developed countries would aim to increase their overseas aid budget and facilitate technology transfer. In practice, however, there was little prospect of substantial overall increases in national aid budgets to fund implementation of Agenda 21, apart from ad hoc unilateral pledges. Moreover, most processes of technology transfer would continue to be driven by 'normal' international and national investment programmes. Agenda 21 endorses measures to promote free trade and access to markets as a way of facilitating such technology transfer, but no special measures to transfer technology for Agenda 21 purposes could be expected unless paid for by the countries themselves or through bilateral or international aid programmes. Thus GEF re-emerges as the central mechanism for international resource transfers specifically for implementing the UNCED agreements. The following paragraphs briefly examine the role and significance of the GEF in this context.

The GEF was established in March 1991, following a decision taken in the previous November, with a commitment of about \$1 billion over a three-year pilot phase. Three 'implementing agencies' managed the provision of project funding – the World Bank, UNEP and UNDP. The GEF was administered by the World Bank, which meant that its operation was essentially controlled by the World Bank's governing board, which was dominated by the major industrialised countries. Developing countries opposed such control, being generally suspicious of the Bank and the conditions it often attaches to its grants. Thus, they only agreed to accept the GEF as the main financial mechanism for the UNCED agreements on condition that it would be restructured and replenished with substantial additional funding.

Negotiations for restructuring the GEF were finally concluded in 1994. The new GEF was to be administered by a Council on which developing and developed countries would be equally represented. Decisions were to be taken by consensus or, failing that, by a majority of *each* of the two groups of countries. Thus, the groups of developed and developing countries each effectively had a veto. An additional \$3 billion was allocated to the fund for the period 1994–97.

Moreover, it was announced that the operations of the GEF and its implementing agencies were to become more transparent to NGOs and other interested groups, who would have increased access to meetings and documents and enhanced opportunities to participate in projects. In association with this, for example, the World Bank established a new Inspection Panel to increase its accountability to non-state groups, and UNEP announced new guidelines for NGO participation.

The GEF embodies a supposed 'bargain' between the industrial and developing countries, but in fact it is a focus of continual dispute. These disputes focus on both the overall levels of funding and the ways in which funds are disbursed. Developing countries argue that far more funds are needed, and that the criteria for funding projects are over-restrictive. Donor countries think that developing country demands are excessive, and generally regard the GEF as a mechanism to benefit the global environment, not to contribute to general development or poverty alleviation. Thus, they have insisted that funding for projects for implementing the Climate Change and Biodiversity Conventions should only be for the 'agreed incremental costs' of securing the relevant global environmental benefits. In practice, the calculations of such incremental costs are normally controversial. There is a continual war of attrition on these disputed issues, and this is reflected in the fact that financial issues have been high on the agenda of the CSD every year.

The philosophy of providing funds only for global environmental benefits clearly runs counter to the requirements of Agenda 21. Thus, it was agreed that projects addressing a broader range of issues within Agenda 21 would be eligible for funding and that the incremental costs criterion would be relaxed somewhat. Even so, the GEF funds made available to promote the implementation of Agenda 21 have been a tiny fraction of the total funds required. This would remain the case even if the GEF funds were increased by a factor of ten. As with the CSD, the effectiveness of the GEF in promoting sustainable development depends on the extent to which it can be used to shape wider and more powerful processes.

Numerous small GEF grants have been provided to developing countries and 'countries with economies in transition' (such as ex-Soviet bloc countries), mainly through UNEP or UNDP, to fund the preparation of national reports on the state of their environment and on strategies for implementing Agenda 21. As discussed above, the production of such policies and reports can itself contribute significantly to national implementation, as well as feeding in to the CSD review process. In many developing countries relatively small funds allocated for this purpose, supplemented by outside expert assistance, can make a big difference. In practice, lessons had to be learned. For example, where the preparation of the report was largely done by outside consultants rather than the national officials themselves, the benefits were found to be superficial and temporary.

Similarly, GEF grants have been provided for 'institutional capacity-building'. Many countries lack the institutions and the skilled personnel to monitor and implement their desired policies. Once again, modest funding and training programmes were found to make a substantial difference, when they were well targeted and when measures were taken to ensure that newly trained personnel were not rapidly moved to different responsibilities.

Furthermore, the possibility of securing GEF funding provided an incentive for countries to be open about their problems and needs at the CSD and elsewhere. Representatives for the GEF and its implementing agencies attended CSD meetings. This facilitated international reviews, and helped with the targeting of international measures or assistance. Even where funding was not forthcoming, this arguably contributed to transparency and learning processes.

Finally, substantial funding was provided for a number of large projects, particularly relating to land degradation, deforestation and desertification. These were normally managed by the World Bank, and often took a long time to come to fruition due to Bank procedures of establishing institutional frameworks and detailed co-operative arrangements with governments before disbursing funds.

However, valuable though they may have been, such contributions of the GEF must be recognised to be modest in comparison to the scale of the task of implementing Agenda 21 and also to the scale of the financial and resource transfers taking place in the wider international economy. Concerns about implementing Agenda 21 thus need to be fully integrated into the activities of the IMF, World Bank and WTO, as well as those of international society at large. This is clearly a long-term process.

*Activity 5*

Identify the main roles or functions of the CSD and GEF in promoting implementation of Agenda 21. On the basis of the information provided in this section, for each of these roles or functions examine whether these bodies have been effective and the extent to which this has been due to their effect on: power; influence; knowledge production and learning processes; values; resource transfers; or capacity-building.

# 7    Conclusion

The attempt to reconcile development needs and environmental protection in the concept of 'sustainable development' has evidently gathered some momentum in the last decade. New international institutions, national and local plans are proliferating. Perhaps it is the sum total of local initiatives, reflecting changed human perceptions, that will influence the future more than will any grandiose international schemes. But they are all operating within the existing social, political and economic framework, and are constrained by the very factors which have created the present situation – human population growth, sovereign states, international trade and finance, conflict over resources, thoughtless technology and so on. Whether a sufficiently radical reconstruction of these deep-seated practices can evolve from present policies must remain an open question: there is bound to be a wide diversity of views even among those who read this text, let alone the wider world population.

The final two chapters will take up this discussion at two levels: Chapter 6 will examine the various international regimes that have been discussed in this book from a more general point of view, seeking common features, and considering their interlocking influence on the way government operates; Chapter 7 will speculate on the possible future life-styles over the next century or so that might result from the changing perceptions of the present.

# References

BLUNDEN, J. (1996) 'The environmental impact of mining and mineral processing', Chapter 5 in Blunden, J. and Reddish, A. (eds).

BLUNDEN, J. and REDDISH, A. (eds) (1996) *Energy, Resources and Environment*, London, Hodder & Stoughton/The Open University (second edition) (Book Three in this series).

BRANDT, W. and SAMPSON, A. (eds) (1980) *North–South: a programme for survival* (The Brandt Report), Cambridge, MA, MIT Press.

BRAUDEL, F. (1979) *Civilisation and Capitalism: 15th–18th century* (English translation), London, Collins.

FINDLAY, A. (1996) 'Population and environment: reproduction and production', Chapter 1 in Sarre, P. and Blunden, J. (eds).

GOLDSMITH, E. *et al.* (1972) *A Blueprint for Survival*, London, Tom Stacey.

GRUBB, M. *et al.* (1993) *The Earth Summit Agreements: a guide and assessment*, Royal Institute of International Affairs, London, Earthscan.

HUTTON, W. (1995) *The State We're In*, London, Cape.

KROPOTKIN, P. (1899) *Fields, Factories and Workshops Tomorrow*; modern edition 1962, London, Allen and Unwin.

LOVELOCK, J. E. (1982) *Gaia: a new look at life on Earth*, Oxford, Oxford University Press.

LOVELOCK, J. E. (1989) *The Ages of Gaia*, Oxford, Oxford University Press.

MADDOX, J. (1972) *The Doomsday Syndrome*, London, Macmillan.

MEADOWS, D. H. *et al.* (1972) *The Limits to Growth* (A Report for the Club of Rome's Project on the Predicament of Mankind), London, Earth Island.

MEADOWS, D. H. *et al.*(1992) *Beyond the Limits*, London, Earthscan.

OLIVIER, D., ELLIOTT, D. and REDDISH, A. (1996) 'Sustainable energy futures', Chapter 2 in Blunden, J. and Reddish, A. (eds).

O'RIORDAN, T. (1981) *Environmentalism*, London, Pion.

REDCLIFT, M. (1987) *Sustainable Development: exploring the contradictions*, London, Methuen.

REDDISH, A. (1996) 'Environment and development', Ch. 4 in Sarre, P. and Reddish, A. (eds).

REDDISH, A. and RAND, M. (1996) 'The environmental effects of present energy policies', Ch. 2 in Blunden, J. and Reddish, A. (eds).

SARRE, P. and BLUNDEN, J. (eds) (1996) *Environment, Population and Development*, London, Hodder & Stoughton/The Open University (second edition) (Book Two in this series).

SARRE, P. and BROWN, S. (1996) 'Changing attitudes to Nature', Ch. 3 in Sarre, P. and Reddish, A. (eds).

SARRE, P. and REDDISH, A. (eds) (1996) *Environment and Society*, London, Hodder & Stoughton/The Open University (second edition) (Book One in this series).

SCHUMACHER, E. F. (1973) *Small is Beautiful: economics as if people mattered*, London, Abacus.

SILVERTOWN, J. (1996) 'Inhabitants of the biosphere', Ch. 6 in Sarre, P. and Reddish, A. (eds).

TAYLOR, A. (1996) 'World trade and its environmental effects', Ch. 6 in Sarre, P. and Blunden, J. (eds).

UN COMMISSION (1983) *Common Crisis: North–South cooperation for world recovery* (The Brandt Commission), Cambridge, MA, MIT Press.

UN WORLD COMMISSION ON ENVIRONMENT AND DEVELOPMENT (1987) *Our Common Future* (The Brundtland Report), Oxford, Oxford University Press.

## Further reading

BERGESEN, H. and PARMANN, G. (eds) (annual) *The Green Globe Yearbook of International Cooperation on Environment and Development*, Oxford, Oxford University Press.

CAMPIGLIO, L. *et al.* (eds) (1994) *The Environment after Rio: international law and economics*, London, Graham Trotman/Martinus Nijhoff.

GRUBB, M. *et al.* (1993) *The Earth Summit Agreements: a guide and assessment*, (Energy and Environmental Programme, RIIA), London, Earthscan

KEOHANE, R. and LEVY, M. (eds) (1996) *Institutions for Environmental Aids: pitfall and promise*, London, MIT Press.

POOLE, J. and GUTHRIE, R. (eds) (annual) *Verification 199X: arms control, environment, and peacekeeping*, Oxford, Westview Press.

THOMAS, C. (1992) *The Environment in International Relations*, London, RIIA.

THOMAS, C. (ed.) *Rio: unravelling the consequences*, London, Frank Cass.

VOGLER, J. and IMBER, M. (eds) *The Environment and International Relations*, London, Routledge.

A useful and accessible monthly journal is *Environment* published by Heldref Publications (Washington, DC, USA).

## Answers to Activities

### Activity 1

The last 10 000 years is the period since the end of the last glaciation, with comparatively stable global temperatures now threatened by predictions of global warming. Herding and settled agriculture, with domestication of particular plants and animals, have gradually replaced hunter–gathering. Developing craft skills led to widening trade, settlements and competing empires, later to industry and science. Population has increased by some 1500 times. Large-scale destruction of forests, cultivation, over-grazing, soil modification, water management, and waste disposal have significantly modified much of the natural landscape.

The last 1000 years have seen separate world cultures increasingly brought together into a linked global economy by expanding trade, improved transport, colonialisation and industrialisation. Population has increased by about 20 times. Urbanisation and industrial pollution have increased local environmental change.

The last 100 years have seen massive technological changes in transport, communication, materials and destructive weapons; population has increased by about four times. Resource depletion, deforestation, desertification and pollution have accelerated and widened in scale, with acid rain and other transboundary effects, damage to the ozone layer and some evidence of climate change.

### Activity 2

Stewardship implies care and conservation of the natural world, violated then and now by unsustainable resource use, and waste disposal or pollution beyond the rates of natural regeneration.

Imperialism regards natural resources, including other humans, as freely available for exploitation by the sufficiently powerful and determined, whether in industry, colonialism or the present disparity in wealth and resource use within and between nations.

Utilitarianism seeks to maximise human happiness, and provides a basis for the liberal economic order as a means of optimising distribution, whether in nineteenth-century industry or modern free trade agreements.

Romanticism values Nature for its own sake, seeking 'unspoiled' landscape or wilderness as a reaction against industrial towns in the last century, or in environmentalist or New Age movements now.

### Activity 3

Choosing an Earth Council would be a controversial and formidable task – who is best qualified to represent the diversity of human interest and aspiration, as well as that of non-human life and the inanimate? Opinions will differ about the best representation of men and women, old and young, rich and poor, urban and rural, North and South, expert and commonsense. If the choice is to be of *expert* advisors, with all the questions that raises about the legitimacy of their status and influence, a personal view about (some of) the disciplines to be represented would be:

- philosophy, about the nature and limitations of human knowledge and ethics
- mathematics, about logical implication and the 'chaos' limits on computer models
- science, concerned with understanding the world – perhaps in five aspects: the physics of planetary structures and atmospheres, the biogeochemistry of material diversity and change, the evolution of life and ecosystems (all over some $10^{10}$ years), the anthropology and sociology of human institutions (over $10^6$ years), the history of the present interglacial period (some $10^4$ years)
- technology, concerned with action to provide for human needs now, again in at least five aspects: energy, food, construction (craft and industry), communication (information, transport), survival (medicine and security)
- current value systems: economics (markets, planning, environmental and inter-generational effects), government (law, international regimes)
- and perhaps at least one artist (musician, dramatist, painter, film-maker . . .) to provide a human aesthetic counter to the narrower specialists.

At least their debates should be interesting.

### Activity 4

Unsustainable consumption in the North, and population policy in the South, were only able to obtain muted expression.

Environmental accounting for transnational corporations was not accepted. The implications of the Conventions on the atmosphere for changes in life-style and energy policy were contested, issues of land ownership and distribution were avoided, as was formal regulation of international business. Most controversial of all was the issue of funding and technology transfer to the developing countries.

All of these contribute to the maintenance of the status quo in the world economic order, with high levels of consumption in the North, international business and finance unregulated and little acceptance of a substantially changed view of 'development', particularly if it imposes significant costs on the richer countries.

# Chapter 6  The development and implementation of international environmental regimes

## 1  Introduction

Tackling environmental problems is difficult enough at a national level, where there are typically established government and state structures available for formulating policy options, deciding between them, and promoting or enforcing their implementation. At the international level, the challenges multiply. There is no world government. International society is 'anarchic', in the sense that decision-making authority is decentralised – at least amongst states which cannot legally be forced to obey an international law or decision to which they do not subscribe. Yet co-ordinated and

While you are reading the chapter, look out for answers to the following key questions:

• What approaches have been developed for responding to international environmental problems?

• How have the Climate Change, Biodiversity and Desertification Conventions developed since Rio?

• What general features can be recognised in agenda setting, negotiating commitment and implementation of such conventions?

"The picture's pretty bleak, gentlemen. ... The world's climates are changing, the mammals are taking over, and we all have a brain about the size of a walnut."

sustained international action is useful or essential for tackling many global or regional environmental problems.

Fortunately, experience shows that it is possible to develop and implement co-ordinated international responses to such problems, as has been illustrated in the previous chapters of this book. This chapter aims to examine the political processes and institutions involved in such international efforts, and the factors that have contributed to success and failure. It seeks to build upon the discussions in earlier chapters, and to discuss some developments since UNCED in 1992. In this respect, it picks up from where the last section of Chapter 5 ended. That section examined the international politics of implementing the 'Agenda 21' programme of action for sustainable development. This chapter examines the development and implementation of international environmental regimes established to address specific environmental issue areas.

The chapter examines the development of the three main Conventions associated with UNCED 1992: those on climate change, on biodiversity and on desertification. It then explores some key issues in the development and implementation of international responses to particular environmental problems, and the ways in which these relate to the broader international context. First, however, it is necessary in Section 2 to provide an introduction to some of the main challenges, issues and actors involved in responding to international environmental problems, and to the character and significance of international agreements or 'regimes' that may be developed to promote co-ordinated international actions. The different *types* of international actors, and the major factors affecting the outcomes of international political processes, were briefly identified in Section 5.2 of Chapter 5, which you may wish to review at this point.

# 2   Responding to international environmental problems

Decisions about ways to tackle environmental risks depend partly on understandings of the nature of the 'problem' and of the human activities that are causing it. These vary greatly according to the particular issue and context. However, it is convenient here to refer to one particularly influential model which aims to explain why human societies may over-exploit shared environmental resources even where people know that they are doing so: the **'tragedy of the commons'** (see *Reddish*, 1996). This not only illuminates a way in which environmental problems may be generated, and some potential responses. It also helps to introduce some of the particular problems of tackling *international* environmental problems.

*Activity 1*

If you are familiar with the term, describe and illustrate what is meant by the 'tragedy of the commons'. If not, turn to the answer at the end of the chapter.

The notion shows how it is possible that 'rational' individual actions can lead to 'irrational' collective practices resulting in catastrophic over-exploitation of common resources. The 'tragedy of the commons' is that this depletion of common resources can continue remorselessly to their destructive conclusion, even if each organisation involved is well-intentioned, well-informed and exercising only its traditional and legal rights. Unilateral acts of public-spirited restraint are insufficient to tackle the problem. If the rest of the community continues in its old ways, the public-spirited suffer along with the selfish without even having benefited from the 'good times' in the interim.

In principle, a range of types of responses to such over-exploitation is available. One traditional response is to 'exploit and move on'. This has been the approach taken by 'slash and burn' agricultural communities in the tropical forests, cattle herdsmen in regions of Africa, and many international timber companies. Increasingly, however, this is no longer an option. The environment cannot recover (or is given insufficient time and space in which to do so), and there are fewer places to move on to.

Another type of response is 'privatisation'. In relation to the overgrazing of common land, for example, if ownership of the common grazing land were divided amongst the herdkeepers, each of these would have a direct interest in maintaining the value of their own land by grazing it at sustainable levels. Each would bear the full costs of any unsustainable practices, and each would have the ability to control how the land is managed. In principle, this approach could play a significant role in improving resource management of the global commons. For example, the new Law of the Sea, discussed in Chapter 1, transferred effective ownership of much of the world's ocean resources to coastal states, with a broad obligation on these states to manage their Exclusive Economic Zones (EEZ) sustainably.

However, in order for this approach to be effective, the private owners would have to have a clear interest in the long-term conservation and management of the resources under their control, and have the capacity and knowledge necessary for effectively carrying out their management role. In practice, such conditions would often not be met (as shown in *Sarre and Blunden*, eds, 1996). For example, without state regulation, it is not clear that private forest-owners can be relied on to manage their forests sustainably rather than sell the timber and invest the proceeds in other businesses.

Moreover, there are many resources or environmental problems, such as the atmosphere, river or sea pollution, that by their nature cross boundaries. Similarly, fish do not respect any artificial division of a marine area. Thus, whereas some species of sea fish may remain within a particular EEZ throughout their life-cycle, many do not. The new Law of the Sea was therefore not sufficient to provide a framework for managing fish that straddle or migrate across EEZ boundaries. This resulted in conflicts (for example, between Canada and Spain in 1995), and led to the negotiation of the UN Convention on Migratory and Straddling Fish Stocks, through which co-operative and joint international fishing rules and quotas relating to such fish can be established.

This brings us to the third type of approach to environmental conservation and management: the establishment of new regulations or policies to prevent unsustainable or damaging practices. This approach is probably the most generally applicable. But it is clear that any system of regulations and taxes to tackle environmental problems is bound to be controversial. Experience with attempts to prevent overfishing, for example, has shown that some fishers can be expected to deny that there is an overfishing problem. Others might dispute the maximum sustainable yield. Moreoever, the ways in which fishing quotas or the burdens of implementing taxes or regulations are distributed amongst the community are also sure to be controversial. The benefits or costs of any environmental policy or regulation are bound to be distributed unevenly, leading to disputes about which regulations or policies to adopt and also to possible compliance problems in the future.

Such disputes and problems are characteristic of all attempts to tackle environmental problems or problems of unsustainable practices. Establishing and implementing appropriate regulations or policies at the local or national level is thus hard enough. But at least there is a state structure for taking decisions and enforcing them on dissenting groups. This is not true at the international or global level. In the absence of world government, authority for legislation and enforcement is dispersed amongst over 180 sovereign states. Moreover, the atmosphere and much of the Earth's oceans do not come under the jurisdiction of any one state.

In this context, efforts to tackle global or regional environmental problems must mainly focus on developing and implementing international co-operative arrangements to which all participating states must agree, and in the absence of international regulatory or enforcement systems that can oblige a state to participate.

Such arrangements involve developing appropriate international institutions – in the form of sets of principles, rules, shared understandings, organisations, consultation processes and the like – governing or shaping uses of environmental resources. They can take a variety of forms, ranging from legal treaties through political commitments to tacit understandings, and from the establishment of stringent rules to the promotion of open-ended general principles and processes. At one time, many doubted that this is possible in the absence of state-like government and enforcement systems. But studies of local communities have shown that such structures are not always essential. Experience shows that groups of local communities sharing a lake or forest can sometimes develop amongst themselves agreed rules for sustainable management, and can successfully implement and maintain them. Moreover, we now have many examples of international agreements which have had some success.

The international institutions and processes established to promote implementation of Agenda 21 examined in Section 6 of Chapter 5 were an example of a relatively open-ended process governed by general or relatively weak commitments relating to a wide range of issues. However, there are many examples of more specific or binding international agreements.

### Activity 2

Review this book, and list the international agreements of this kind that have been discussed.

The international politics of responding to particular environmental problems can often most usefully be discussed in terms of the formation and operation of international environmental 'regimes'. The exact definition of an international 'regime' is still contested amongst specialists. Nevertheless, a broad understanding of an **international regime** as 'an international social institution with (more or less) agreed-upon principles, norms, rules, procedures, and programmes that govern the activities and shape the expectations of **actors** in a specific issue area' probably provides a good enough basis for proceeding to examine the international politics of responding to particular environmental problems.

Note that the notion of an institution is broadly conceived in this definition: it refers to established and structured social practices rather than simply to an 'organisation'. This means that a regime is understood to be broader than a specific organisation or treaty, because it includes all formal or informal international institutions, guidelines, rules, programmes and practices directly associated with its operation. On the other hand it is narrower than a broad international convention (such as the liberal international economic order): it is constituted by relatively specific norms, procedures or institutions developed to address a particular issue area. The reference to 'shaping the expectations' of actors in the definition reflects the fact that a regime provides a framework for the interactions of actors involved in the area, and that they take the regime into account in their activities and plans.

Although the definition of a regime may seem a little obscure in the abstract, identifying one in practice is often fairly straightforward. For example, the ozone layer protection regime consists of the principles, rules, procedures, institutions, guidelines and programmes established through the 1985 Vienna Convention, the 1987 Montreal Protocol (and its subsequent amendments), and the decisions of the meetings of the parties to the Protocol (see Chapter 2). It also includes the international, transnational and domestic groups, coalitions, institutions and programmes (involving scientists, NGOs, industrial groups and international organisations, as well as governments) established in direct connection to the development or implementation of the Protocol.

Effective international environmental regimes, such as the Montreal Protocol and the International Whaling Commission (IWC), typically involve a range of different types of influential international and domestic actors, and are key ways in which international co-operation can be organised and maintained. Such situations can provide an important framework for the interaction between international and domestic actors and processes, and between 'knowledge', power and interests relevant to a particular issue area. Moreover, a regime provides a focus for the formulation and implementation of policies to tackle a particular transnational environmental problem, including the organisation of relevant resource transfers and capacity-building activities.

---

*Activity 3*

Review Chapter 2, and list the different types of actors and institutions involved in the development and implementation of the Montreal Protocol, and the ways in which the Protocol and its associated institutions affected interactions between international and domestic actors, and between knowledge and interests.

---

In general, people's expectations of the role and significance of international regimes vary according to their understanding of the fundamental character of international politics. Some are more sceptical than others about the extent to which international institutions in themselves can substantially affect the behaviour of powerful actors or shape the course of major political or economic processes. There are now at least 120 multilateral environmental agreements, and hundreds more bilateral agreements. Some of these must be regarded as 'dead letters', in that international rules and procedures have been agreed but with little or no follow-up or implementation. Nevertheless, case-study research has established numerous examples of 'effective' regimes – that is, regimes that have shaped the behaviour of a range of international, transnational and domestic actors, affected outcomes, and at least to some extent helped to tackle the problem for which they were established. Thus the development and effectiveness of international regimes form an important area of study.

The following sections briefly examine the development since 1992 of each of the three main UNCED conventions – the Framework Convention on Climate Change (FCCC), the Biodiversity Convention, and the Convention on Combating Desertification – to complete the information necessary for the subsequent more general discussion of the development and implementation of international environmental regimes.

# 3   The development of the Climate Change regime

The agreement to establish the Framework Convention on Climate Change (FCCC) was perhaps the most important achievement of the 1992 Earth Summit. As outlined in Chapter 3, 150 states signed the Convention at the summit. In less than 18 months it had been ratified by over fifty states, bringing the Convention into force on 21 March 1994. The first Conference of the Parties (CoP) took place in Berlin one year later, on 28 March–7 April 1995. Since negotiations on the Convention only began in earnest in early 1991, this represented remarkably rapid progress by normal international standards. Figure 6.1 shows the progress on the FCCC.

*Activity 4*

Review Chapter 3, and identify the objectives and main commitments of the FCCC.

## 3.1   From Rio to Berlin

In order to achieve consensus in time for UNCED, contentious or complex issues had been sidestepped in the final pre-Rio negotiations. The FCCC imposed only vague obligations on parties to limit their greenhouse gas emissions. In practice, the Convention allowed each party to adopt whatever national programmes and commitments to limit greenhouse gas emissions it deemed appropriate. Developed countries were each obliged regularly to submit national communications detailing: inventories of their greenhouse gas emissions; projections of such emissions in 2000; and

| | Climate Change | Biodiversity | Desertification |
|---|---|---|---|
| | 1979 First World Climate Conference | 1971 Ramsar sites | 1977 Nairobi Conference |
| | 1988 Toronto | 1972 Heritage Sites | UNEP & African states pressed for convention: |
| (pre-Rio) | IPCC | ICDPs     MAB | |
| | 1990 Second World Climate Conference | 1973 CITES | |
| 1992 | **Framework Convention on Climate Change** (150 states signed) | **UNCED 'Earth Summit', Rio** **Biodiversity Convention** | agreement to set up: |
| 1993 | ratified by >50 states in force March | in force December | negotiations for |
| 1994 | first *developed* countries' reports | **1st CoP** (Bahamas) Montreal Secretariat and SBSTTA set up | **Convention to Combat Desertification** (Paris) |
| 1995 | **1st CoP** 'Berlin Mandate' * | **2nd CoP** (Indonesia) | signed by >100 states |
| 1996 | **2nd CoP** guidelines: *developing* countries' reports | ratified by 133 states (conflicts over genetic resources and biotechnology) | ratified by >50 states in force December? |
| 1997 | **3rd CoP** (Japan) new protocol? | first national reports due | **1st CoP**? |
| 1998 | | binding Protocol on biosafety and LMOs? | |
| 1999 | | | |
| 2000 | first emissions deadline (developed, restore 1990 level) | | |
| 2005 | potential further emission reduction deadlines | | |
| 2010 | | | |
| 2015 | (*developed* countries) | | |
| 2020 | (*developing* countries) | | |

▲   Figure 6.1   *The chronology of the three international regimes.*

Note: *See Figure 6.2 for the 'alliance' that led to the agreement in Berlin.

specific estimates of the effects of their declared policies and measures on their net emissions. These national communications were to be regularly reviewed by the Conference of the Parties. However, most of the details relating to rules, procedures and institutions established in the Convention remained vague and unresolved.

The committee which had negotiated the Convention was therefore charged with developing these details in time for them to be approved at the first CoP, a contentious and complex task. Progress was made on a number of issues, including the financial arrangements, guidelines for compiling national inventories of greenhouse gas emissions, as well as procedures for reviewing the first national communications of the developed country parties. But on other issues major differences remained about the rules of procedure for the CoP (including voting rules) and 'Joint

Implementation' schemes (in which an emission reduction project based in one country is carried out with the assistance of another state).

Furthermore, the first CoP would have to decide whether or not the existing commitments were adequate to achieve the declared aims of the Convention. If not, a new protocol containing strengthened obligations would be needed. Throughout the original negotiations for the FCCC, governments had disagreed profoundly in their assessments of whether the risks of anthropogenic climate change were sufficient to merit stringent binding commitments to limit their country's greenhouse gas emissions, bearing in mind the potential political and economic costs of implementing any such commitments. Disagreements continued between 1992 and 1995.

As would be expected, governments adopted positions mainly according to their perceptions of their national interests and of how the burdens of responding to climate change should be distributed around the world. The likely costs of meeting a given limit on emissions vary between countries, as do the capacity and likely costs of adapting to projected changes in climate.

National assessments of these factors resulted in a variety of different negotiating positions. For example, the US Administration's position was informed by beliefs that: the costs to the USA of adapting to climate change would be less than those of stabilising or reducing US greenhouse gas emissions in the near future: any attempts by the Administration to implement measures that would substantially limit $CO_2$ emissions from fossil-fuel burning would be politically unpopular and would probably be blocked by Congress; and it would be unfair to expect the USA to bear a major share of the burden of limiting global greenhouse gas emissions without most other countries (including some developing countries such as China) also accepting some of burden. The German government appeared to believe that: the overall risks of climate change are great; the costs of substantially reducing its national greenhouse gas emissions are manageable (thus its unilateral pledge to cut its emissions by 25% by 2005); and that it was fair at this stage for developed countries to bear the burden of limiting global emissions.

The Saudi government and other Organisation of Petroleum Exporting Countries (OPEC) governments appeared to believe that the risks posed to their countries by climate change were less than the risk that international commitments to cut $CO_2$ emissions would result in substantial reductions in their income from oil exports. The Chinese and Indian governments believed on balance that climate change posed substantial risks and that international actions to limit greenhouse gas emissions were necessary, but that, for the foreseeable future, developing countries such as themselves should not be asked to bear any of the burden of achieving these limits. Small Island States (represented by the Alliance of Small Island States, or AOSIS) regarded climate change and associated sea-level rises as a threat to their very existence. However, lacking the capacity to make any difference to greenhouse gas emissions through their own actions, they have been determined to do all that they can to promote stringent international commitments.

In view of these differences, it is not surprising that it proved impossible to agree on whether existing commitments were adequate and how to strengthen the Convention. As would be expected in a global environmental negotiation, there was often a strong North–South dimension to the debates. However, the divisions within the developing country group and within the industrialised country group were often as great as the divisions between them.

Thus, in practice, the Organisation of Economic Co-operation and Development (OECD) tended to split into at least two opposing camps in climate negotiations. On the one hand, there was a block led by the European Union (complete with its own internal divisions) which generally supported measures to strengthen the Convention, including legally binding limits on developed countries' emissions of $CO_2$ and other greenhouse gases. On the other hand, there was a group led by the USA, Canada and Australia, with which Japan and New Zealand are normally associated. These were known as the JUSCANZ countries (from the initial letters of the countries' names), and they generally took the position that the risks of climate change do not yet justify strengthened obligations to limit or reduce emissions and that promotion of voluntary measures is in any case preferable.

The group of 'countries in economic transition', from Eastern Europe and the former Soviet Union, also share certain interests and concerns. At least some of this group were major exporters of fossil-fuels, and thus inclined towards the JUSCANZ camp. However, their substantial industrial decline in the early 1990s and the associated economic restructuring processes meant that their greenhouse gas emissions had fallen well below 1990 levels and were likely to remain so for some time to come. Thus, they did not expect to be greatly constrained by any limits that were likely to be negotiated in the near future, and tended to be neutral (if sceptical) in most debates on commitments.

The main negotiating block of the developing countries, the Group of 77 (G77) and China, was similarly split. At one extreme was AOSIS, who supported measures to strengthen the Convention and introduced a draft protocol advocating a 20% cut in carbon dioxide emissions by 2005 for Annex 1 (developed country) parties only. They were often supported in this by other low-lying countries such as Bangladesh and Egypt. On the other extreme were the OPEC countries, usually led by Saudi Arabia and Kuwait, who opposed strengthening developed countries' commitments that might reduce demand for fossil fuels. The rest of the G77 countries mostly took the position that emission reductions by developed countries would be desirable, so long as no additional obligations were imposed on them at this stage. In spite of their differences, the G77 and China made great efforts to present a common front in climate negotiations, and more or less maintained this latter position up to the first CoP.

## 3.2   The First Conference of the Parties

The first CoP thus convened in Berlin on 28 March with a widespread sense of foreboding. By that stage, some 127 countries and the European Community had ratified the Convention, and almost all of these were represented at the meeting, together with fifty other states and 200 other organisations attending as observers. However, the negotiating committee had failed to resolve key issues, and nothing had happened in the meantime to indicate that compromise was more likely to be achieved this time. There was a real possibility that the CoP would be widely perceived to be a failure. In line with this, progress during the first week of the Conference seemed dismal. After repeated attempts to achieve agreement on the rules of procedure, it was finally agreed to side-step the issue: the CoP would proceed without agreed voting rules, and by default operate according to UN consensus rules. By the time the government ministers arrived in the middle of the second week, little had been achieved.

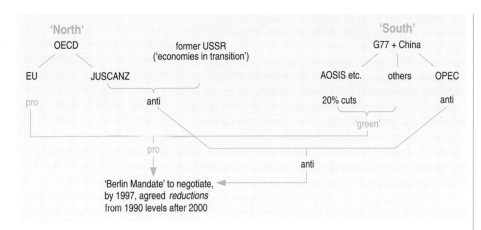

◄ *Figure 6.2*
*Negotiating positions (before and during 1995 Berlin CoP) about new emission limits after 2000.*

Then the negotiations began to progress rapidly. Within the G77 and China group, a 'green group' emerged (essentially the G77 and China minus OPEC) which forced movement in developing country positions beyond the lowest common denominator. (See Figure 6.2.) The negotiations on Joint Implementation began to be resolved on the basis of a 'green paper' submitted by India on behalf of this coalition, which soon led to a deal being struck. The CoP also endorsed the earlier agreements on: the preparation and review of national communications; interim financial arrangements (entrusted for four more years to the recently reformed Global Environment Facility).

The discussions on adequacy of commitments proved much harder to complete in the CoP, and continued day and night in closed meetings. The EU and the 'green group' of developing countries formed a coalition pressing for a mandate for negotiations for a protocol, while the USA, Canada, Australia and others from OPEC and the group of countries from eastern Europe and the former Soviet Union were still resistant.

However, the negotiating group finally achieved agreement that the existing commitments in the Convention were inadequate and that a 'process' should begin to enable the CoP to take 'appropriate action for the period beyond 2000, including the strengthening of the commitments of the Parties included in Annex 1 to the Convention … through the adoption of a protocol or another legal instrument'. As a concession to the countries that were reluctant to move to a new protocol, this was described as a 'process', not a 'negotiation'. The aim of the process was, by 1997, to:

> … elaborate policies and measures, as well as to set quantified
> limitation and reduction objectives within specified time-frames, such
> as 2005, 2010 and 2020, for their anthropogenic emissions by sources
> and removals by sinks of greenhouse gases not controlled by the
> Montreal Protocol. [The process will] not introduce any new
> commitments for parties not included in Annex 1 [i.e. developed
> countries], but reaffirm existing commitments and continue to
> advance the implementation of these commitments in order to achieve
> sustainable development.

This agreement on the process and guidelines for developing and strengthening commitments was named the 'Berlin Mandate'. The open-ended ad hoc group of parties charged with pursuing this process met repeatedly from the summer of 1995, with a view to developing further commitments in time for agreement at the third CoP due to take place in Japan in 1997.

In view of the preceding debates and uncertainties, the establishment of the Berlin Mandate was a remarkably successful conclusion for the first CoP. The emergence of the coalition between the EU and the 'green group' of developing countries was critical to achieving this outcome. Much credit was also due to the diplomatic skills and determination of Angela Merkell (the President of the Conference, and Environment Minister of Germany – the hosts to the Conference), Raoul Estrada Oyuela (Chair of the INC and of the Conference of the Whole, and the Argentinean Ambassador to China), and representatives of the countries that took a leading role in brokering a deal, such as India.

Those who had wished for a declaration on new targets and timetables for greenhouse gas emissions at the first CoP were disappointed. Certainly a strong case could be made that much more urgent and stringent actions to reduce greenhouse gas emission are necessary to reduce the risk of dangerous anthropogenic interference with the climate system. This view was reinforced by a new Intergovernmental Panel on Climate Change (IPCC) assessment finalised on 15 December 1995. This concluded that there was now substantial empirical evidence that measurable, anthropogenically induced climate change has already occurred, which reinforces the conclusions from scientific modelling. Moreover, the impacts on human society and ecosystems from climate change that could result from a doubling of atmospheric $CO_2$ concentrations above pre-industrial levels were judged to be sufficiently damaging to warrant urgent consideration of preventive actions.

However, the problems of developing an effective international response to an emerging environmental problem such as climate change are very great, particularly when decisions have to be accepted by consensus. In view of the variety of known interests and concerns of the countries involved, the first CoP could not realistically have been expected to do more than begin a process of negotiations for a protocol containing more stringent commitments. This it did, along with establishing the other institutions, mechanisms and processes required for the further implementation and development of the Convention. Thus, the first CoP was widely judged to have been a success.

## 3.3   *Implementing and developing the Framework Convention*

Pending the negotiation of protocols with more stringent commitments, the effectiveness of the Convention depended on the extent to which it stimulated individual countries or groups of countries to take measures to limit their greenhouse gas emissions. The main mechanism by which this was to be achieved was through a combination of the obligation on states to formulate and submit national communications, and international reviews of national performance based largely on the data provided in these communications.

How could this process of compilation, submission and international review of national communications contribute to achieving the aims of the Convention? (See Chapter 5, Section 6.2 for a discussion of a similar issue in relation to Agenda 21.)

This mechanism can contribute in a number of ways. The obligation to compile official national reports may in itself stimulate governments to compile information on their greenhouse gas emission inventories and their emission projections, providing key information for the formulation of national targets and plans as well as for international scientific assessments. It also requires governments to formulate national policies and measures, which they might otherwise not do, and provides a focus for inter-agency consultations and domestic debates about the content and adequacy of the proposed measures.

The international expert review of the national emission inventories and projections should help to promote the reliability, transparency and consistency of the submitted inventories, providing a basis for assessments of national performance in limiting emissions. The prospect of international review, and wide discussion, of proposed national policies and measures may increase the incentive to adopt relatively ambitious policies. Subsequently, international scrutiny of countries' progress in implementing their declared policies and meeting their emissions targets can increase the international and domestic political pressures on governments to ensure effective implementation. By highlighting successes and problems, international review can facilitate learning about how best to implement emission reductions and stimulate timely and effective international responses to emerging implementation problems.

Developed countries were obliged to provide their first national communications by the end of 1994. It was agreed that these should include detailed inventories of their national anthropogenic emissions in 1990 of the three main greenhouse gases (carbon dioxide, methane and nitrous oxide), compiled and presented according to the internationally agreed guidelines and methods prepared by the IPCC in 1994 to promote reliability, transparency and comparability. According to these guidelines, the emissions data were to be disaggregated into various sub-sectors according to their source (for example, seven subcategories of sources from fuel combustion [such as transportation or power production] or agriculture [such as animal wastes or enteric fermentation from ruminants such as cattle and sheep]). 1990 was adopted as the 'baseline' year with which future national emissions would be compared.

In fact, the reporting and reviewing process got off to a good start. By 1995 most developed country parties had submitted quite comprehensive and detailed reports on their emissions in 1990, emission projections, and policies and measures for emission abatement. Moreover, procedures for international review of the communications were established at the first CoP, including in-country visits by international review teams. These guidelines provided the potential for developing an effective and detailed in-depth review system. The key was to achieve this potential, and to establish good precedents. In practice, several developed countries immediately offered themselves for in-depth review, including detailed in-country interviews. By the end of 1995, 15 in-depth reviews of developed countries had been undertaken, and several reports by the review teams had been published. The key questions on what responses are made to any problems identified in these reviews will only become clear after a number of years and particularly as the target dates for implementing emission limitation commitments approach.

As outlined in Chapter 3, by the end of 1994 all OECD states except Turkey and Mexico had made unilateral commitments to limit greenhouse gas emissions: mostly to stabilise emissions at 1990 levels by the year 2000, but in some cases (such as Germany and the Netherlands) to achieve significant reductions below the baseline. In their national communications, they described (in varying degrees of detail) how they aimed to achieve these targets. However, in many cases there was wide scepticism that the declared policies would be sufficient to achieve the targets that countries had set for themselves.

By 1996 it was becoming clear that several countries would have difficulties even achieving stabilisation by 2000. The EU countries had collectively committed themselves to the stabilisation target, but without agreeing how to share the burden of achieving it. Countries such as Spain, Greece, Ireland and Italy made it fairly clear that they expected their own emissions to increase, and were relying on north European EU countries to achieve the reductions necessary to balance such increases. Germany, the Netherlands, the UK and some Nordic EU states were on course for modest emissions reductions: in the German case this was helped by the unification of Germany and the collapse of industrial production in East Germany in the early 1990s; and in the UK it was helped by the major restructuring of its energy industry away from coal-burning. However, it was becoming clear that new and additional measures would be needed in the EU to achieve overall stabilisation.

Similarly, the USA, Canada and Australia appeared to be on a trajectory where emissions in 2000 would exceed 1990 levels. In eastern Europe and the countries of the former Soviet Union, the great reductions in industrial production and economic transition after 1990 meant that their emissions in 2000 were unlikely to exceed 1990 levels. However, few measures were being put into place to ensure that greenhouse gas emissions would not continue to increase as economic production recovered.

Thus, international responses to the risk of climate change in the industrialised countries had had some impact by the second half of the 1990s. The climate change regime had:

(a) helped to stimulate declarations of emissions targets and the formulation and implementation of modest but useful mitigation measures;

(b) generated and reviewed baseline data on national emissions inventories for 1990; and

(c) established review procedures that could help to identify problems and promote pressures to improve national performance.

However, much more needed to be done if they were to achieve the stabilisation of atmospheric concentrations of these gases at a level that prevents dangerous interference in the climate system.

As far as developing countries are concerned, it was accepted that their emissions could be expected to increase for the foreseeable future in order to meet their development needs. Thus the burden of achieving overall stabilisation would mostly fall on developed countries. Helped by funds and expert resources provided by the Global Environment Facility (GEF) and national programmes such as the US Country Studies programme, several developing countries had made progress towards compiling national greenhouse gas inventories. In 1996 the guidelines for submitting national communications from developing countries were agreed. These were compatible with those applying to developed countries, though requiring less detail. Moreover, many developing countries were determined to ensure that there was nothing in the guidelines to imply that they had any obligations to limit their greenhouse gas emissions at this stage. The first national communications from developing countries were due by 1997.

## 3.4   *The development of further commitments*

The debates about whether and how to develop a new protocol quickly resumed the patterns established before the first CoP, with the same divisions and interest groups. The key elements of any new protocol were outlined at an early stage by the EU, to provide a framework around which discussions for a protocol could focus, given that many states, including the JUSCANZ group, regarded the AOSIS proposal as far too radical. The EU proposal included no specific limitation or reduction objectives. However, several EU states made it clear that they were thinking in terms of quite short time-frames (2005 or 2010) with reductions in the range of 5% to 10% compared with the 1990 baseline. At the time of writing (early 1996), it is not clear whether a protocol for the FCCC will be ready for signing in 1997, nor what it will contain. It seems likely that progress will depend on the re-emergence of the coalition between the EU and the 'green group' in the G77, which was so important to the success of the first CoP.

However, it is clear that there will be a need for further and more stringent commitments even after a new protocol is agreed. At some stage, developing countries will also have to accept commitments to limit the increase in their greenhouse gas emissions: their emissions are projected to swamp those of the Annex 1 states after 2025. The task of negotiating substantial limits on emissions from developing as well as developed countries is clearly awesome. Success in stabilising atmospheric concentrations of greenhouse gases at less than double pre-industrial levels (measured in terms of $CO_2$-equivalents) depends on being able to achieve the adoption of sustainable ('climate-friendly') patterns of development at an early stage – for example, while the countries of Asia, Latin America, eastern Europe and the former Soviet Union are in the process of rapid economic transition and are investing heavily in new plant and equipment. To achieve this, the whole range of the actors, institutions and processes associated with the climate change regime will need to contribute, by: increasing knowledge and concern (about the problems and about implementing emissions reduction measures); changing patterns of power, influence, interests and values; promoting appropriate resource transfers; and building appropriate institutional capacities. These processes can be developed even before agreements to stringent and legally binding commitments are achieved.

# 4   *The development of the Convention on Biodiversity*

The signing of the UN Convention on Biological Diversity by 155 states (and the EU) was another major achievement of the Earth Summit. It came into force on 29 December 1993. The historical process leading to this Biodiversity Convention, and its main aims and commitments are described in Chapter 4 (particularly Section 6), and you should review this discussion before proceeding further; also look back at Figure 6.1 for an overview of the process.

By some indicators, the early development of the Biodiversity Convention might be regarded to be very promising. By the end of 1995 it had been ratified by 133 countries, and two meetings of the Conference of the Parties (CoP) had been held (in the Bahamas on 28 November– 9 December 1994, and in Indonesia in November 1995) – remarkable by the normal standards of global conventions and comparing favourably with the FCCC.

A Subsidiary Body on Scientific, Technical and Technological Advice (SBSTTA) has been operating with a defined work programme since the first CoP in 1994, as has a clearing-house mechanism to promote and facilitate technical and scientific co-operation, focusing particularly on the needs of developing countries. GEF had been accepted as the interim financial mechanism for the Convention, making biodiversity one of the GEF's main funding areas (the others relating to climate change, ozone depletion, protection of international waters, and land degradation – primarily desertification and deforestation). A Secretariat had been established in Montreal, under the auspices of UNEP. Guidelines prepared by SBSTTA had been agreed for the preparation of national reports relating to the protection of biodiversity, and the first reports were due to be submitted to the Secretariat by June 1997. Moreover, a comprehensive 1016-page *Global Diversity Assessment* was released by UNEP at the second CoP, serving similar functions for the Biodiversity Convention as the IPCC Assessments have done for the FCCC. Numerous countries, including the UK, had already published reports detailing their national biodiversity strategy, and biodiversity country studies were under way in several developing countries, helped by GEF funding.

These were indeed promising developments. Nevertheless, the Biodiversity Convention was widely perceived to be in trouble by the end of 1995. There had been major disagreements about the core principles, objectives and means of achieving them during the original negotiations (indeed, it was sometimes suggested that one of the reasons for the wide diplomatic support for the Convention amongst developing states was the initial refusal of the USA to sign). After 1992 profound disagreements about priorities and means of achieving them continued. In the process the development of international measures to conserve biodiversity *in situ* appeared to be increasingly sidelined, to the consternation of those for whom preservation of natural habitats was a primary objective of the Convention.

Q   What are the main objectives of the Biodiversity Convention, and what are the main ways in which it seeks to achieve them? Identify the main disagreements about its core principles, objectives and means of achieving them.

A   For the industrialised countries, the main objective of the Convention was to promote conservation of biodiversity, largely through the preservation of natural habitats, of which many of the most important are located in developing countries. Developing countries had a wider agenda including: securing international financial and technological transfers to assist in both *in situ* and *ex situ* conservation; and gaining access to a share of the economic benefits of biodiversity in general and biotechnology in particular. They aim to achieve the latter objective by securing ownership of, and access to, biological resources from their territory (including intellectual property rights over the genetic resources and products made from them). This has been an anathema to the USA and several other developed countries, who want to preserve open access to natural genetic resources while protecting the international property rights of their chemical and biotechnology companies over their products.

North–South disputes about the scale and operation of international funding mechanisms have become familiar in global environmental regimes, but this makes them no less divisive and difficult to resolve. In general, industrialised states are reluctant donors, and they have only been prepared to fund substantial financial transfers to developing countries when they are to be used to finance the incremental costs of implementing agreed environmental commitments to achieve results from which they expect to share the global benefits. Developing countries want to secure as much as they can in financial and technology transfer, and have been bargaining hard to get it. They want direct payment for conserving 'their biodiversity resources', which the developed countries appear to value, and without it are unwilling to agree to any restrictions on their rights to dispose of these resources as they want.

In this context, the agreed compromise principle on which the Convention is based – that biodiversity is a 'common concern of humankind' – left core issues unresolved. In principle, negotiators could aim to achieve a 'package deal', in which trade-offs could be obtained across a range of disparate issue areas. However, the experience with the Law of the Sea Convention indicates that negotiations to stitch up such a package can be extremely difficult and long-lasting, and the resulting inflexible deal may in any case need to be unstitched in order to achieve implementation (see Chapter 1).

Such a comprehensive package-deal approach goes against the general philosophy of the framework convention approach to regime-building, which emphasises flexibility and the negotiation and amendment of protocols as opportunities arise to make progress. In line with this, in 1995 the second CoP agreed to begin negotiations to develop by 1998 a legally binding protocol on biosafety, and especially the transboundary movement of 'living modified organisms' (LMOs). The basis for this objective is contained in a controversial element of the Biodiversity Convention directing the parties to 'establish or maintain means to regulate, manage, or control the risks associated with the use and release of living modified organisms resulting from biotechnology'.

The linkage between biodiversity and biosafety was made late in the pre-UNCED negotiations, through a proposal from Malaysia which immediately received wide support from developing countries and some NGOs. It was not clear scientifically that the transboundary movement of LMOs, however important it may be in its own right, was important to the preservation of biodiversity – the *Global Biodiversity Assessment* scarcely referred to the issue. However, the linkage certainly provided developing states and other interested groups with political leverage over an issue on which developed states with a strong bio-technology industry (such as the USA) cared greatly. Such developed states want to avoid anything they believe would lead to heavy-handed international regulation of the biotechnology industry, and prefer to develop non-binding international guidelines on the transport and use of genetically engineered products in a separate forum. Developing states rightly perceived that pursuing this issue within the Biodiversity Convention could not only give them more influence on the biosafety regulations but could provide a way of pursuing their arguments for access to biotechnology or a share in its economic benefits. At least, it might provide opportunities for trade-offs with other areas.

The prospects for successfully negotiating a protocol on biosafety within the Biodiversity Convention by 1998 did not appear very good. However, there was a danger that this issue would remain at the top of the agenda in the Convention during this period, sidelining issues of more direct and immediate importance to the conservation of biodiversity. As discussed in Chapter 5 (Section 6), in 1995 an Intergovernmental Panel on Forests was established by the Commission on Sustainable Development, providing a possible alternative forum in which the conservation of forest habitats might be addressed. The Biodiversity Convention itself may make its most important contribution to the preservation of biodiversity *in situ* through its role in promoting monitoring programmes, information exchange, and the development and implementation of national biodiversity strategies, facilitated by GEF, the World Wide Fund for Nature and national funding programmes.

# 5   The development of the Desertification Convention

Although it was not signed at UNCED, the UN Convention to Combat Desertification (CCD) should be regarded as the third international environmental convention arising directly from the 1992 Earth Summit. As a result of strong lobbying from African countries and UNEP, it was agreed at UNCED that the UN General Assembly should be requested to create an intergovernmental negotiating committee to develop a Desertification Convention so that it would be ready for signature by June 1994. This was achieved. The CCD was agreed in Paris in June 1994, and by 1995 it had been signed by over 100 states. It was due to come into force after being ratified by fifty states, a process that was expected to take until the end of 1996. (Look back at Figure 6.1 for an overview of the process.)

The CCD aims to promote co-ordinated actions to address problems of 'land degradation in arid, semi-arid, and dry sub-humid areas resulting from various factors, including climatic variations and human activities' – the definition of desertification used in the Convention. It consists of a main text of forty articles, supplemented by four regional annexes for Africa, Asia, Latin America and the Northern Mediterranean region. The Convention provides a 'code of good practices' for governments of affected countries and for donors in relation to the management of marginal drylands. It provides a framework for co-operation between local land-users, governments, NGOs, international organisations and funding agencies, and northern donor countries.

The governments of affected countries are obliged to develop National Action Programmes to Combat Desertification, building on existing environmental frameworks where they exist. Donor countries that join the Convention are committed to provide flexible and long-term support to affected country governments, and to co-ordinate amongst themselves. Multilateral funding mechanisms have yet to be formally decided: this will

▲   *Desertification in Burkina Faso: this area had been heavily wooded ten years before the photograph was taken.*

be a matter for the first CoP expected to take place in 1997. However, it seemed likely that the GEF would be established, at least on an interim basis, as the primary global mechanism, alongside the many bilateral and regional aid programmes that already address desertification issues in some way.

The CCD needs to be seen in historical context. Concerns about 'desertification' were placed on the international agenda after the widespread drought in Africa in 1972/73. A UN Conference on Desertification (UNCOD) was held in Nairobi in 1977 to address the problem. As a result, the newly created UNEP was given responsibility for co-ordinating follow-up activities, including monitoring desertification processes and developing a global plan of action to combat desertification. It set up a group to raise funds, but donors preferred to continue to channel funds through their established bilateral aid programmes or through the World Bank or UNDP, rather than adopt the UNEP agenda of which they had become sceptical. They tended to regard desertification as a widespread but local problem of land degradation in arid regions due to human activities, rather than a problem of advancing deserts due to regional or global climate changes.

Debates about definitions of desertification have been prominent in all subsequent international discussions about how to respond to dry land degradation, because they had a bearing on which countries had access to multilateral aid programmes and whether the problem should be addressed as a global or regional environmental issue or as a local development problem. In the lead-up to UNCED, UNEP and African states were determined to push the issue high on the agenda, arguing for a convention devoted to the issue in addition to a chapter in Agenda 21. Western developed states supported them in this goal in return for African support for an international Forests Convention. In the event, the efforts to establish a Forests Convention failed, but the CCD was ultimately negotiated.

The particular definition of desertification adopted in the CCD (see above) is noteworthy. 'Dry sub-humid areas' are included alongside arid and semi-arid lands, allowing a wider range of countries potential access to international assistance than would previously have been the case. Moreover, it includes problems arising from climate variations as well as human activities, opening the possibility for claims linking land degradation with anthropogenic climate change, and thus to claims on the GEF which was established to facilitate responses to global environmental problems. Northern donor countries argued against defining desertification in this way, but later compromised in the course of global negotiations to restructure the GEF (see Chapter 5, Section 6.3).

Nevertheless, in the mid-1990s the future development of the CCD remained in doubt. Some major northern donor countries remained sceptical that an international convention was a necessary or appropriate response to problems of land degradation in dry areas. They continued to perceive the problem mainly as a problem best addressed through local sustainable development projects, within the framework of Agenda 21 or traditional aid programmes. Moreover, unlike the problems of ozone depletion, climate change or biodiversity loss, most developed countries do not perceive themselves to have a direct environmental interest in tackling dry land degradation in sub-Saharan Africa. It therefore remained unclear whether the CCD would attract any additional funds. However, it might provide an institutional framework through which international concerns about the problem can be maintained and increased, and to stimulate the development and implementation of regional and local action programmes.

*Activity 5*

Extend the chronology in Figure 6.1 as new developments unfold.

# 6   The development and implementation of environmental regimes: post-UNCED reflections

We are now in a position to explore some key issues in the development and implementation of international environmental regimes, drawing where appropriate on the experience since UNCED.

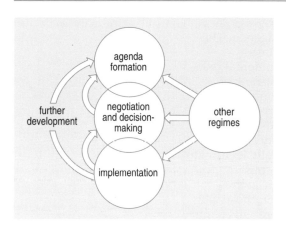

◀   *Figure 6.3*
*Regime development.*

In principle, the processes of regime development can be divided into several phases: see Figure 6.3. The three main stages can roughly be characterised as:

(a)  agenda formation;

(b)  negotiation and decision-making; and

(c)  implementation.

In practice these often overlap and the process is not so linear. Effective regimes need to be dynamic: agenda-setting, negotiation and further development processes typically continue alongside implementation. Moreover, regimes do not develop in isolation from one another: since UNCED at least, international responses to environmental problems take place in the context of multiple and overlapping regimes and institutions. In this section, we discuss each of these issues and phases (including further development) in turn.

## 6.1   Agenda formation

The **agenda formation** stage includes the processes by which the problem becomes recognised, emerges onto the political stage, is framed for consideration and debate by the relevant policy communities, and rises high enough on the international political agenda to initiate negotiations and decision-making processes.

Each of these processes is complex in itself, and experience shows that many issues do not even make it through the agenda-formation stage. For environmental issues, it can often be difficult even to secure recognition that there is a problem. Problems such as water and air pollution, depletion of fish stocks, decline in biodiversity or habitats, extinctions of species, ozone depletion and climate change, may emerge slowly and not become obvious until it is too late to prevent major impacts or even disaster.

This is a major reason for the particular importance of science and 'knowledge production' processes in environmental politics. Careful scientific monitoring or modelling often plays an important role in identifying and assessing emerging environmental problems, and such scientific findings are used in attempts to place the issue on the political agenda and to frame the debates about possible responses.

However, scientific findings of emerging environmental risks do not automatically mean that the issue emerges onto the political agenda. The science enters a political arena, and is in any case often uncertain. Alongside groups that want to highlight the apparent environmental risks, there may be groups who wish to deny the existence or seriousness of a problem. Often this is because they suspect that addressing the problem could be economically or politically costly to them, or that possible responses could be against their interests or run counter to existing priorities. However, it is important also to recognise that many environmental problems emerge from long-standing and wholly legitimate practices which did not seem to have caused a problem in the past. Suggestions that these practices have now become damaging can thus appear to be counter-intuitive and scare-mongering. Moreover, those who have been intimately engaged in such activities for years can resent claims to superior knowledge by an unfamiliar group of environmentally concerned scientists and activists. For all of these reasons, the 'agenda-setting' processes by which an environmental issue emerges onto the political agenda, and then moves sufficiently high in this agenda to stimulate serious negotiations and decision-making, typically involve a complex mixture of politics and science.

*Activity 6*

Referring to earlier chapters in this book, identify some key elements in the agenda-formation process in the development of one or more of the following regimes:

- the new Law of the Sea
- the North Sea environmental protection regime
- the ozone layer protection regime
- the Biodiversity Convention
- the Climate Change regime
- the Convention to Combat Desertification.

One of the interesting lessons from experience with the early formation of environmental regimes is that it is so various. Whereas all of the factors outlined above usually play a role, the ways in which they interact and the resulting agenda-formation process is highly contingent.

For example, Molina and Rowland's (1974) scientific paper, showing that CFCs *might* deplete the ozone layer, rapidly ignited heated scientific and political controversy, leading to the 'spray-can war' and domestic legislation in the USA. In contrast, in most of western Europe, the issue did not 'take off' until the detection of an 'ozone hole' over Antarctica in 1985. The issue was forced onto the international political agenda in the late 1970s through the combined efforts of North American and Scandinavian governments, some individual scientists and environmental NGOs.

In relation to climate change, the scientific hypothesis that anthropogenic carbon dioxide emissions could lead to 'global warming' was first proposed in the nineteenth century, certainly by Arrhenius in 1896. With the development of global climate monitoring systems and increasingly credible climate models since the 1960s, the international scientific community became increasingly concerned about the risk, but it was not until 1979 that the First World Climate Conference was held, where scientists and government representatives discussed the issue. International agenda-formation activities continued through a series of international conferences through the 1980s, involving scientific bodies (including the Scientific Committee on Problems of the Environment [SCOPE] of the International Council of Scientific Unions [ICSU]), 'expert policy entrepreneurs', NGOs and governmental representatives, culminating in the 1988 Toronto Conference involving scientists and government representatives from forty-eight states.

In the absence of a dramatic equivalent to the 'ozone hole' (or the disastrous *Torrey Canyon* oil spill in 1967 that provoked the development of the MARPOL Convention for the prevention of pollution from ships), it had proved difficult for the risk of climate change to capture wide public and political attention up to that point. However, by the end of the 1980s, wide awareness of the global environmental problems had developed, sensitised by the problems of acid rain and ozone depletion. In 1988 the fact that seven of the ten years of the 1980s were the warmest since records began, the drought in the USA, and severe storms around the world, combined to place the climate change issue high on the agenda in most developed countries. After the Toronto Conference, the IPCC, convened under the auspices of WMO and UNEP, were given responsibility for co-ordinating preparatory work for an international response to the risk of climate change. As discussed in Chapter 3 Section 3.3, the IPCC report to the Second World Climate Conference in 1990 provided an essential basis for the decision to initiate intergovernmental negotiations for a Climate Change Convention.

These experiences illustrate at least two important further points. One is the significance of vivid events or discoveries for mobilising concern and placing emerging environmental issues high on the political agenda. Some issues are intrinsically more amenable to the mobilisation of public concern than others, but the potential is to a large extent culturally determined and affected by the way the problem is 'framed'. The image of being exposed to UV radiation from space was already one that resonated with the general public in the late twentieth century, but one can nevertheless wonder whether timely international action to prevent the depletion of the ozone layer would have taken place if there had been no ozone holes. Public concern in the UK about North Sea pollution was only mobilised sufficiently to persuade the government to support more stringent international action when it was linked to an epidemic amongst the (photogenic) seal population. Campaigns to stimulate international action to protect biodiversity tend to focus on the plight of whales, tigers and elephants rather than the bugs that are at least equally important from an ecological or biodiversity perspective. In the 1970s UNEP felt obliged to present the problem of desertification in terms of images of desert sand dunes advancing on farming areas in order to stimulate international action.

A second point relates to the mechanisms through which issues can be formally placed on the international agenda, and to whom can access or participate in them. Even with relatively high levels of concern, it can be hard to initiate international discussions on response strategies unless there are readily available mechanisms and institutions through which this can be done. On the other hand, once formal discussions on an issue have begun in international institutions, they tend to carry on even if wide public and political concern dies away. For this reason, there is a tendency where possible to address new issues through established international frameworks rather than try to create new ones. For example, problems of oil pollution and concerns about extinctions of whale species were tackled within the International Maritime Organisation and International Whaling Commission (IWC) respectively, even though they started out as ship-owners' and whalers' clubs. Moreover, the open rules of access to the IWC meant that several non-whaling countries could join in the 1960s, changing the character of the regime.

In this context, the UN often plays an important facilitative role, providing an established framework of procedures and rules within which negotiations to tackle a new problem of concern can proceed. The institutions created at the 1972 Stockholm conference (particularly UNEP) and at UNCED are similarly important. Before 1972 there were few recognised international institutions through which environmental issues of emerging international concern could be framed for possible international negotiation. After UNCED there are several, including the CSD. Broad framework conventions such as the Biodiversity Convention can also in practice sometimes be used for this purpose. UNEP is an international agency with a mandate to facilitate such international agenda-formation processes.

Access and participation in many such environmental bodies has become quite wide. In addition to states and relevant international organisations, NGOs and local authorities at least have access as observers and can lobby and make statements. In the 1970s even such access would have been controversial. Principle 10 of the 1992 Rio Declaration calls for wide public participation and consultation. The role of NGOs in the agenda-formation process has often been important, playing a key role in generating public concern and political attention. However, their role has sometimes gone beyond this. In many cases, implicit coalitions have developed between NGOs, concerned scientific bodies and sympathetic governments, not only in raising general awareness of the issues, but also in framing them for negotiation and constraining obstructive activities by reluctant governments who might be inclined to block progress if there were no risk of adverse publicity at home. Such coalitions were seen, for example, during the agenda-formation processes relating to ozone depletion, climate change and dumping waste at sea. For wildlife conservation and biodiversity issues, the role of NGOs such as the World Wide Fund for Nature and the IUCN has often been prominent and formally recognised (as discussed in Chapter 4); they have sometimes been allocated substantial responsibility in the preparation of assessments or for co-ordinating the process.

## 6.2   *Negotiation and agreeing commitments*

The stage of **negotiating and agreeing commitments** takes an issue from the point where it becomes a priority item on the agenda of relevant policy-making or negotiating bodies to the point where international decisions are made about which policies and rules will be adopted to address the issue. It is at this stage that choices are made about commitments, policies and measures. In principle, there are normally a number of possible ways to respond to a given environmental problem. The ways in which the main policy response options are actually framed, considered and assessed are a key part of environmental politics, and constitute another important dimension to the relationship between policy, science and 'knowledge' (inverted commas are used here to indicate that understandings of the problem and possible solutions may not actually be 'true' or complete, and are often contested).

Typically, a wide variety of different types of policy approaches are potentially available, including: bans; regulations; market instruments;

technology and resource transfers; changing property rights; voluntary agreements; developing new codes of 'best environmental practice'; research and education. Alternative policies (or combinations of policies) can be compared according to their relative overall costs and benefits, and according to the extent to which they are appropriate to the problem and are likely to be effective. Such assessments already involve economic, political, institutional and social judgements as well as judgements from natural or environmental science. However, policies also differ in the ways they distribute the costs and benefits amongst different social groups and actors, and this also has a profound effect on the policy-making process and on final decisions. For example, even if the overall benefits of a set of commitments to policy greatly outweigh their overall costs, it may not be adopted if the benefits would be spread thinly amongst a wide range of weak or unmobilised groups while the costs would be borne by a small but highly mobilised number of politically powerful groups. This is true at the domestic as well as the international level. However, to persuade sovereign states to join an agreement, they each have to believe that it would be to their overall benefit, or at least that the costs were acceptable of 'doing the right thing' morally.

Negotiations to establish an international environmental agreement regime may be regarded as successful if the key participants agree upon commitments that, if implemented, would help to achieve the desired objectives. Not surprisingly, agreement is most easily achieved on commitments that all parties already intended to implement anyway, or that are sufficiently vague or ambiguous that points of contention are not resolved and compliance would be hard to assess. Some governments may find such symbolic agreements useful for international or domestic political purposes, but they are not effective: such agreements do not alter the behaviour of environmentally relevant actors or contribute to tackling the problem they address.

Effective agreements can normally be expected to be harder to achieve, because it means persuading countries to agree to change existing practices, create new programmes or accept new restrictions. At the least, participants' understandings of the problems and the most appropriate responses to them need to be brought into alignment. Often concessions need to be made and 'package deals' need to be put together. The problems of achieving agreement often multiply as the number of participants and the variety of their interests increase. Thus questions inevitably arise about which interests it is important to try to accommodate, and which interested parties it is necessary or desirable to include. Moreover, participants need to consider carefully what they realistically want and need to achieve through the negotiation process, referring to long- and short-term goals.

## Activity 7

Review the main issues, objectives for the negotiations to establish at least two of the following agreements:

- UNCLOS
- Montreal Protocol
- FCCC
- Biodiversity Convention
- Convention to Combat Desertification.

Identify which were the participants amongst which agreement was most important to achieve, and also the incentives for going beyond these to achieve global participation as far as possible.

Developing a regime, and successfully negotiating an effective agreement, typically requires leadership. In any set of negotiations, it is normally possible to identify *leader states* that want an agreement and work hard to get one through a combination of active diplomacy, promoting the production and dissemination of relevant knowledge, or (informal or formal) sanctions or 'side payments'. In this way *laggards* – states which are reluctant to achieve agreement or agree to effective commitments – may be persuaded to sign. Further co-ordination and persuasion can be achieved of the (often large number of) states that are willing in principle to join an agreement provided it is not too costly, but are uninterested in devoting substantial efforts towards achieving one.

When powerful states or groups of states, such as the United States or the EU, adopt a leading role, the prospects for achieving an agreement improve greatly. US leadership in achieving a whaling moratorium in the IWC was critical, as was the EU (and German) role in achieving agreement at the first CoP of the FCCC. However, with skilful diplomacy and 'policy entrepreneurship', small states, international organisations or individuals can also play effective leading roles. For example, UNEP (and its Executive Director Mostafa Tolba) played an important role brokering agreement on the Montreal Protocol and the CCD, as did Canadian Maurice Strong and Argentinean Paulo Estrada at critical phases of FCCC negotiations. Particularly in some biodiversity and wildlife conservation agreements, environmental NGOs such as the World Wide Fund for Nature and IUCN have played an important leading role as participants in the negotiating process, even if they are not formal parties to the final agreement.

Naturally, leader states aim to shape the commitments in line with their interests. However, for most environmental issues, they do not have the capacity to force agreement entirely on their own terms. Lasting multilateral agreements cannot be achieved without all or most participants perceiving them to be achieving a common interest. Moreover, in any given issue area, there are likely to be *veto* states, without whose agreement and participation an effective regime cannot be established. Malaysia, Brazil and other developing countries in whose territories the main tropical and sub-tropical forests are located are veto states in negotiations for global forestry and biodiversity conventions, and they successfully exercised their veto in negotiations in the early 1990s.

*Activity 8*

Identify lead, veto and laggard states in negotiations for an effective Climate Change Convention.

The management and definition of technical and scientific issues, as distinct from political issues, have proved particularly important in international environmental negotiations. In this context, few technical or scientific questions are without potentially important policy implications. However, where they want to make progress, diplomats have become adept at agreeing to define certain issues as 'technical', in order to gain as much benefit as possible from scientific and technical expertise and to simplify the negotiating terrain. In the Montreal Protocol, for example, the Technology and Economic Assessment Panels and the Science Assessment Panels have played a key role in helping parties to agree on objectives and to frame rules. The focusing of objectives and the framing of rules is a particularly critical activity: it helps to focus institutional interventions in extremely complex social and economic processes, and to define the areas where international regulations should apply and where choices of implementation measures can be left to the discretion of governments and relevant non-governmental actors. In this context, transnational 'knowledge-based' communities of experts with shared understanding of the problem and preferred policy responses (known as **epistemic communities**) have proved particularly influential.

## 6.3    *Implementation*

The **implementation** phase includes all of the activities involved in implementing the decisions and policies adopted in response to the problem. This can include: the incorporation of international commitments into domestic law; the development and operationalisation of agreed programmes; and all other measures aimed at appropriately changing government, social and economic practices.

This stage is typically no less complex than the other two. On the contrary, experience shows that it is one thing to agree to international obligations, and quite another to bring them into operation and to achieve the desired effects on the behaviour of relevant actors. Those charged with implementing the decisions may lack necessary commitment or resources, and will interpret the decisions in their own ways. In practice, some countries tend to take legal obligations very seriously, whereas others tend to regard them as symbols of general intentions and a stage in an on-going negotiation process not to be interpreted too literally. Actors whose interests are substantially affected by the changes in policy can be expected to continue to try to influence the policy and the ways in which it is implemented. Compliance may leave much to be desired, and in any case the actual effects of decisions can be very different to the expected ones.

Whether or not international agreements are implemented can depend greatly on the nature of the commitments themselves. Governments may not try hard to implement if they believe themselves to have been coerced into an unfair agreement, which is a reason why it is important that agreements should be regarded as legitimate and, on balance, fair or in each participant's overall interest. The will to implement may be weak if parties suspect that others may not be complying, and are attempting to 'free-ride'.

Moreover, implementation is more straightforward for governments to carry out, monitor and enforce for some commitments than others. For example, national consumption of CFCs was carefully defined in the Montreal Protocol as 'production plus imports minus exports minus approved destruction', because these were most amenable to government regulation and monitoring (an example of skilled framing of commitments, discussed above). Similarly, compliance with MARPOL rules to limit oil pollution from tankers became much better when they focused on technological regulations relating to tanker construction and port facilities instead of operational rules relating to the conduct of ship operators at sea, who have stronger interests in non-compliance and whose behaviour is harder to monitor. These examples illustrate the importance of factors such as transparency and 'monitorability' of implementation of commitments.

The structure and role of the international secretariats or institutions associated with the regime can also be important. For example, a professional and relatively autonomous secretariat can play a significant role in promoting effective implementation, by increasing awareness of obligations, identifying emerging problems, and stimulating timely discussions amongst parties and other relevant groups on effective responses. Processes by which information on parties' progress towards implementation is regularly collected and internationally reviewed can play a similar role. This is the reason that so much attention has been devoted to the development of review procedures in the Climate Change Convention, for example, as discussed in Section 3. Similarly, the design of consultative and dispute resolution mechanisms can make a big difference to whether problems or concerns tend to be neglected or to escalate damagingly, or whether they are routinely tackled in a timely and constructive way. For example, an Implementation Committee was established for the Montreal Protocol, to address compliance problems in a non-judicial and relatively non-confrontational way. Without it, the problems of non-compliance with CFC phase-out by Russia and other countries in the former Soviet Union after 1995 might not have been tackled constructively, and the ozone protection regime would have been greatly undermined.

Mechanisms for resource transfers amongst participants can critically affect patterns of interest and capacity amongst participants – as is clear, for example, from the experience with the ozone protection regime. The development of the Multilateral Fund was critical to the development of the Montreal Protocol. Similarly, the GEF has been a key component of the developing UNCED framework conventions.

## 6.4   *Further development*

The experience of trying to implement an agreement typically shows up problems which need to be addressed by changes and developments in the institutions and resources of the regime, and in the commitments themselves. In a dynamic and effective regime, therefore, the further development of the regime is intimately connected with experiences from implementation.

However, even if implementation goes smoothly, knowledge about the problem was normally incomplete during the negotiations. This is even more true of understanding amongst national and international policy-makers about the most effective ways of responding to the problem. Moreover, the commitments that were made initially were likely to have been the result of a somewhat unsatisfactory compromise between different interest groups. Thus initial commitments and adopted policies may come to be regarded as inadequate, stimulating negotiations for changing or widening commitments, or making them more stringent.

Further, to be effective, the institutions, principles, rules and procedures of a regime must directly or indirectly alter relevant patterns of power, interest, knowledge, values, resource transfers and institutional capacity in the outside world. The extent to which it does so depends on the ways its 'internal' characteristics and design (rules, procedures, obligations, resource transfer mechanisms, dispute resolution systems etc.) engage with this 'external' world. Since the external situation is constantly changing, an institutional design that is effective at one time will need to adapt and develop continually in order to remain useful. A regime should therefore be able to adapt to external developments – such as changes in science or technology, or new opportunities to promote implementation or improve its rules for environmental protection. The regime may itself have helped to bring about some of these external changes or opportunities.

Thus a key characteristic of an effective regime is flexibility and the capacity to adapt its institutions, rules and procedures in a timely way in the light of experience with implementation and as patterns of power, interest, knowledge, capacity and concern develop. Since the 1972 Stockholm conference there has been increased awareness of the importance of the ways in which rules and institutions are designed for their capacity to adapt and thus maintain or improve their effectiveness. Awareness also increased about the varieties of institutional approaches that are potentially available.

Some of the international environmental conventions agreed before the Stockholm conference seem ill-equipped to achieve this. They have no provisions for regular meetings of the parties, and the procedures for review or for revising procedures and commitments are often unclear. For example, the 1940 Washington Convention on wildlife preservation in the Western Hemisphere effectively has no mechanisms to review progress and no provisions for meetings of the parties. Its Annex of endangered species is incomplete and almost permanently out of date: no species have been added since 1967.

In contrast, the parties to the IWC meet annually to review national reports of whale catches and the reports of expert groups set up to monitor whale populations or other aspects of whaling activities. These expert groups can be set up or terminated by majority vote at any meeting of the parties. Commitments can easily be revised in the same way by amending the Schedule to the Convention. Similarly, CITES has several expert standing committees reviewing and advising the Conference of the Parties, which meets every two years. There are also provisions for additional meetings, which are occasionally held, and commitments are frequently and easily revised by altering the Appendices listing endangered species.

CITES and the IWC both share some of the essential flexible characteristics of **framework conventions**. These are conventions which establish the basic principles, norms and procedures of a regime, including provisions for regular reviews of implementation and the adequacy of commitments, and for the subsequent negotiation of protocols. The routine reviews facilitate learning and institutional development, and provide a forum for agenda-setting. Using these protocols, parties can develop or revise procedures and commitments as they see fit. The UNEP regional seas agreements (such as the 1976 Barcelona Convention for the Protection of the Mediterranean Sea Against Pollution) and the 1985 Vienna Convention for the Protection of the Ozone Layer are widely regarded as paradigms of this institutional approach. In contrast with the IWC's somewhat alarming potential for dramatic and sudden changes in rules, once the Montreal Protocol came into force, parties could only revise its main rules or commitments by separately negotiating and ratifying amendments or adjustments (as occurred in this case with the amendments agreed in London (1990), Copenhagen (1992) and Vienna (1995)).

By the early 1990s framework conventions were widely regarded as the most appropriate model for the design of new conventions to tackle a wide range of global or regional environmental problems. The main conventions emerging from the 1992 Earth Summit – the Framework Convention on Climate Change, the Biodiversity Convention, and (subsequently) the Convention to Combat Desertification – were all framework conventions. However, as has been shown, the framework approach includes many variants and is certainly no panacea. Framework conventions can sometimes emerge, initially at least, with little more institutional focus and strength than Agenda 21 and the CSD. Moreover, particularly in relation to biodiversity and climate change, the scope and complexity of the issues they are trying to address make the challenges of effective development qualitatively greater than for focused regimes such as the Montreal Protocol and the IWC.

## 6.5    *Environmental regimes in their international context*

Much of the preceding discussion has focused on the development and implementation of individual regimes. But it is important to recognise the importance of the interrelationships between different environmentally relevant regimes, institutions and political processes, of which there are now hundreds. The scope of different regimes can and does often overlap significantly. The Climate Change and Biodiversity Conventions, for example, overlap with a wide range of global and regional agreements. Moreover, different regimes are often serviced and supported by the same institutions and international bodies. UNEP serves as the Secretariat for several conventions, including the ozone protection regime and some regional seas agreements. GEF serves the CSD, CCD, FCCC, Biodiversity Convention, and also the Montreal Protocol (for assistance to countries of the former Soviet Union and eastern Europe).

Regulatory frameworks increasingly overlap. The experiences and knowledge from one regime and or institution are frequently carried over into others. This is particularly true of strategies for designing and developing a regime's commitments, procedures and institutions. Successive flora and fauna and regional seas agreements have been modelled on each other, and the Montreal Protocol is frequently held up as a possible model for the UNCED conventions. Moreover, the boundaries between domestic and international regulations are becoming increasingly blurred and complex. EU regulations have the force of law in EU member-states, in many areas even in states that did not endorse them.

This provides many opportunities for mutually supportive activities, but also for incoherence, confusion and destructive interference. This becomes even more true when one recognises the relationship between policies and regimes relating to the environment, economic development and trade. The actors involved in each of the regimes and institutions bear a major responsibility for ensuring compatibility and coherence between them. Frequently, the same states or NGOs are involved in clusters of related regimes. However, this task is becoming increasingly daunting, and less ad hoc approaches may be needed.

Through the 1960s awareness and concern about international environmental problems increased substantially, particularly in western Europe and North America. There was increasing interest in underpinning fragmentary responses to particular environmental problems with a more integrated approach based on generally agreed principles. The institutional products of the 1972 Stockholm Agreement and UNCED, including UNEP and the CSD, were in part explicitly established to facilitate co-ordination between various agencies and environmental and development goals. Moreover, since the mid-1990s discussions have been developing within the World Trade Organization on the principles for resolving potential conflicts between free-trade regimes and the subsidies and trade restrictions involved in several environmental regimes.

Since UNCED, therefore, the interrelationships between international regimes and institutions have become as much of an issue as the development of responses to particular environmental problems.

The increasing density and complexity of interrelated institutions and policy processes since the 1960s may have major implications for the overall character of international politics and practice, particularly in relation to the environment. There is no sign of the development of the structures of centralised world government. Decentralisation and 'anarchy' (as defined in the introduction to this chapter as well as in the normal sense) remain facts of life. Nevertheless, new decentralised patterns of international governance are emerging.

An international regime is itself a decentralised governance system relating to a particular issue area, almost by definition (see Section 2). As we have seen, as an individual international institution, a single regime may be more or less effective in co-ordinating co-operation and shaping or constraining the behaviour of relevant international or domestic actors. Clusters of linked regimes can reinforce each other. After UNCED, linkages between expanding complexes of international institutions relating to the environment and development, and to trade, finance, industrial standards, working conditions and human rights, have been growing rapidly. By the mid-1990s there were few areas of domestic or international activity relating to these broad issue areas that were not closely bound up with sets of international principles, norms, rules, procedures, mechanisms, consultations and organisations, or shaped by transnational communities of scientists, experts and NGOs with an interest in promoting effective international co-operation. This was particularly true of OECD countries, but increasingly so also in the former Soviet Union and developing countries.

In this context, international political and policy processes may be profoundly changing. For the foreseeable future, power and influence and wealth appear likely to remain unevenly and inequitably distributed, and politics will continue to involve a complex and dynamic interaction of power, interests, knowledge and values. States will continue to remain legally sovereign. But international society – involving a mix of state and non-state actors – will operate in a context of dense and complex institutions for international governance that, though fragmented and decentralised, shape and constrain the behaviour of even the most powerful groups.

## References

MOLINA, M. J. and ROWLAND, F.S. (1974) 'Stratospheric sink for chlorofluoromethanes: chlorine atom-catalysed destruction of ozone', *Nature*, Vol. 249, pp. 810–12.

REDDISH, A. (1996) 'Environment and development', Ch. 4 in Sarre, P. and Reddish, P. (eds) *Environment and Society*, London, Hodder and Stoughton/The Open University (second edition) (Book One in this series).

SARRE, P. and BLUNDEN, J. (eds) (1996) *Environment, Population and Development*, London, Hodder and Stoughton/The Open University (second edition) (Book Two in this series).

## Further reading

Suggestions for further reading on the regimes discussed in previous chapters in this book are provided in the relevant chapters.

For regular articles and chapters on progress in the development and implementation of the UNCED conventions, and a range of other environmental regimes, see the monthly journal *Environment*, and the annual Yearbooks, *The Green Globe Yearbook* and *Verification: arms control, environment, peacekeeping*.

Useful books include:

HAAS, P., KEOHANE, R. and LEVY, M. (eds) (1993) *Institutions for the Earth: sources of effective international environmental protection*, London, MIT Press.

THOMAS, C. (1992) *The Environment in International Relations*, London, RIIA.

## Answers to Activities

*Activity 1*

The notion of the 'tragedy of the commons' can be illustrated using a hypothetical example of the use of common fish resources. Consider a sea or large lake on which many local fishing communities depend as a source of food and income. Each fisher has an immediate interest in making as large a catch of fish as s/he can sell or eat, in order to improve her or his standard of living. For centuries, this arrangement has worked satisfactorily. Human populations were sufficiently low, and fishing technologies were sufficiently primitive, that there was no over-fishing. Gradually, however, living conditions improved and human populations grew, increasing the number of people fishing and also the demand for fish. At the same time, fishing technologies improved. In recent years, the sea or lake has been fished at unsustainable levels, and the total fish stock is falling.

In spite of this, each individual fisher continues to have an interest in maintaining or improving their catch. Each fisher gains the full extra benefit of catching additional fish, but bears only a small part of the extra cost of fishing a depleted fish stock because this cost is shared throughout the whole community. Even concerned and environmentally aware fishers may be sorely tempted to continue to make large catches: they know that even if they desist, others are likely to continue to maximise their own catches while they can. The 'tragedy of the commons' is that this depletion of common resources can continue remorselessly to their destructive conclusion, even if each organisation involved is well-intentioned, well-informed and exercising only its traditional and legal rights.

Many environmental problems of industrial society appear to have a similar structure. In an unregulated society the owners of a factory have an interest in continuing to produce goods in the cheapest way, even if that involves dispersing untreated pollutants into the rivers or atmosphere. They gain most of the benefits of cheap production, while the pollution costs are uncertain and in any case shared by the whole 'downstream' community and other species of life. In this way, governments used to be relatively tolerant of sulphur emissions from power-stations in their territory, since the resulting acid rain was dispersed over a number of downwind states. Moreover, the damage caused by acid rain to buildings and forests typically does not appear in power-generation budgets, whereas the costs of cleaning the emissions would do so. Likewise, an operator of a modern fishing fleet has an interest in maximising fish catches, even beyond sustainable levels, in order to cover immediate maintenance costs and repayments of interest on borrowed capital.

*Activity 7*

In relation to UNCLOS, these questions are quite directly addressed in the discussion in Chapter 1. In the original negotiations for the Montreal Protocol, the main objectives were to achieve agreement amongst the main producers and consumers of CFCs to reduce their production and consumption. In 1987 these were mainly developed industrialised states, and so these were the key participants at this stage. Production and consumption of CFCs were sufficiently spread amongst parties that unilateral action by the USA or a sub-group of developed states would not tackle the problem, particularly since this might give other non-participating developed states a competitive advantage (these would avoid the costs of using potentially expensive CFC substitutes, and also companies producing or using CFCs might migrate to the territories of non-parties). In the negotiations for the London Amendment 1990, when the main objective was to negotiate a phase-out of consumption of CFCs amongst industrialised countries, it became all the more necessary to include all major CFC producers and consumers. At this stage, it also became important to industrialised states that developing states also joined the agreement and agreed to phase out CFCs at some stage, because their CFC production was set to grow in relative importance (due to developed country phase-out) and to prevent the re-location of CFC production plant in developing countries. To achieve the participation of developing countries, industrialised countries were prepared to pay the additional costs they would incur in switching to ozone-friendly substitutes and to accept delays in developing countries' phase-out dates.

In relation to the Convention to Combat Desertification, the countries that were most affected by dry land degradation had the greatest interest in achieving an agreement, to provide a framework within which their problems might receive greater attention and to provide a mechanism for generating additional international assistance and facilitating research and the exchange of information and 'know-how'. It was critical for these countries that donor states also join the agreement, and to achieve this they were ultimately prepared to water down commitments. Wider participation was also desirable to strengthen the authority and legitimacy of the regime and perhaps increase the political and moral pressures on donor countries to provide more assistance.

The main objectives of the negotiations to establish the FCCC were to establish a set of principles, rules, procedures and institutions on the basis of which international responses to climate change could be developed as deemed necessary over the medium and long term, and also, if possible, to negotiate near-term limits on the greenhouse gas emissions of the major historical and current emitters – the industrialised countries. In relation to the latter goal, the key participants were the developed countries themselves, but developing countries that would be potentially badly affected by climate change (such as the Small Island States) or by emissions reduction measures (such as the OPEC countries) wanted to participate to ensure their interests were taken into account (and to obtain international assistance if available). In relation to the former objective, global participation was clearly important in a regime to tackle such a global and long-term problem. Moreover, global participation would bolster the authority and legitimacy of any agreed principles, rules and institutions.

# Chapter 7   Environmental futures

## 1   Introduction

The aim of this final chapter of this book (and the series as a whole) is an immodest one: it is to explore what society and environment might be like a century into the future. The reason for doing this is in no way to claim that I can predict over such a long span of human history, but to take seriously the environmentalist commitment to think long-term, and to use this as a way of overcoming the fatalism that dominates one-year or even ten-year predictions. In the short term it is likely that the future will be much like the present, but a century is long enough to open up the possibility of substantial change. In view of the range of serious environmental and social problems identified throughout this and earlier books in the series, and the slow and confused progress towards sustainable development, I hope that the longer view will rekindle your enthusiasm to work for a better future.

Any discussion of the future involves two aspects. First, there needs to be a *prediction* about which features of the present will survive, which trends will continue and what new developments will occur. Such predictions may claim to be objective, but usually turn out to be wrong as trends encounter resistance and unexpected events change the whole system. The second aspect involves explicit or implicit *evaluation*: optimists imagine that the future will be something they would like, pessimists that it will exceed their worst fears. The best-known depictions of the future appear to be intended more to influence change than to predict it: from Sir Thomas More's *Utopia* onward, authors have tried to persuade readers to seek or to avoid a future that they see as plausible. It may say something about the late twentieth century that **dystopias** like George Orwell's *Nineteen Eighty-Four* and Aldous Huxley's *Brave New World* have had a much wider readership than **utopias**, whether based on humanistic psychology (Huxley's *Island*), behaviourism (B. F. Skinner's *Walden 2*) or environmentalism (Callenbach's *Ecotopia* – see Section 3). The genre of fiction most concerned with futures – science fiction – tends to be much more preoccupied with drama and conflict involving individuals and small groups than with the social, technological or environmental detail that provides the context. Perhaps because of the requirements of drama, the futures depicted seem to be characterised by social inequality, political repression and environmental degradation, whether on Earth (for instance, *Blade Runner*) or in space (the *Alien* film trilogy), though the *Star Trek* series offers a continuing example of individual virtue and legitimate authority. These examples suggest that recent years have seen more emphasis on the future as entertainment than on the implications for policy.

As you read this chapter, look out for answers to the following key questions:

- What factors explain long-term social and environmental change?
- What sorts of environmental futures have been proposed?
- What are their strengths and weaknesses in relation to society (developed and less developed) and nature?
- How might they happen?

◄ *Woodcut of More's* Utopia *from the Basle edition of 1518. Sir Thomas More established the genre of writing about ideal societies – though later scholars have pointed out that his description of the island of Utopia cannot be mapped. In that sense it is 'no-place' as much as 'the good place'.*

*Activity 1*

Think about your own experience: have you read books or seen films which predict or depict life on Earth a century from now in a way which you find compelling? If so, perhaps you should reflect on what it is that attracts or repels you, and consider what the implications might be for environmental policy.

My intention is to explore alternative scenarios in order to encourage value judgements about what we would like or dislike rather than to claim to predict objectively. My belief, perhaps I should say my hope, is that a wide range of futures are possible and that the course of the future is at least partly open to influence by ordinary citizens. I find that the accounts of possible futures help clarify what I personally would like to pursue or to avoid, and I hope they will do the same for you.

This then raises further questions about how feasible they are, and how the course of change can be influenced. These kinds of questions depend on our understanding of long-term change, as indeed does the selection of possible futures, so my first step is to look back at what the series has had to say about long-term change and ask what has influenced it.

◀  *Still from the film,* Blade Runner. *One of the features which has made Ridley Scott's film,* Blade Runner, *into a classic was the nightmare vision of a future Los Angeles, contrasting the luxury pyramids of the corporate headquarters with the teeming streets and derelict tenements inhabited by a polyglot population living on their wits.*

## 2   *Looking backward*

One reason for looking back is to remind you (if you are familiar with them) of some earlier parts of the series which set today's environmental problems into a long-term context; in particular, *Reddish* (1996) identified ten time-scales for looking at events on Earth, *Simmons* (1996) gave an account of human impacts on the environment over 10 000 years and *Sarre and Brown* (1996) outlined attitudes to Nature over the same period.

Q   If you have read this chapter, can you remember what factor *Simmons* (1996) put at the centre of his account of human impacts on environments?

A   Technology, especially as it gave access to energy for agriculture, industry and transport.

However, Simmons stresses that,

> . . . the explanatory point that needs to be made is that technology does not work independently. Technology has been advancing in societies in which many other political and economic developments were occurring. Industrial technology has been applied in societies where economic motives were increasingly central and perhaps only outweighed by military considerations [...] The lesson of the past is that application of technology has been bound up with economic and political factors and influenced by cultural values. (1996, p. 86)

*Reddish* (1996) suggests that different factors are apparent over different time-scales and that in the last one hundred years there has been rapid technological change, fast population growth and substantial change in political structures and economic activity, especially as a result of the two World Wars. Over this period there was conflict between capitalist and communist blocs and a growing recognition of environmental problems. The thousand-year time-scale shows that this modern world has evolved from a very different world of separate agricultural civilisations.

*Sarre and Brown* (1996) briefly allude to the different kinds of attitudes to Nature which existed in hunter–gatherer, agricultural and industrial societies. They also identify a key change three centuries ago: the establishment of a **modernist** vision of society in which science, economic growth and legitimate authority would replace superstition, subsistence and autocracy. This seems to be the obvious period on which to focus an investigation of the factors which influence social and environmental change – except that they also suggest that the modernist period may have ended in the last decade or two, with a new era being ushered in. If this is so, it makes sense to look both at the longer time-scale and at the more specific characteristics of the modernist period.

Michael Mann, a leading British sociologist, summed up the long-term perspective in a major book which analysed the development of human society from the first city-states of Mesopotamia to the dawn of industrialisation (Mann, 1986). This showed that history was more complex and less predictable than is often suggested by conventional views of an almost inevitable progress based on reduced superstition, increased knowledge, improved technology, greater affluence and wider political participation.

Mann's account of the past has a number of implications for the future: technological and economic development have been important, but are far

---

### Box 7.1    Influences on social change

Mann's analysis recognises the enduring influence of *ecology* (for example, in providing a basis for the ability of alluvial agriculture to yield a sufficient surplus to feed the first cities) and of *technology* (notably in the superiority of water transport over land transport through most of human history), but argued that events reflected four other sets of influences. *Politics* and *economics* are factors which will surprise no one who has studied social and environmental change, but Mann also identifies major influences from the *military* sphere (for example, Greek infantry techniques as the basis of democracy, or medieval cavalry as the centrepiece of feudal Europe) and from *ideology* (most notably religion, but also increasingly from science).

An example of ideology playing a powerful role is the spread of Christianity through literate but relatively powerless groups in the Roman Empire, and its persistence over a millennium or more after the fall of the Empire as the basis for social cohesion and economic cooperation in a politically fragmented Europe. Environmentalists often argue that this had negative consequences for the environment because of Christianity's injunction to 'multiply and have dominion'. The rise of Protestantism has long been seen as contributing to the emergence of capitalism.

More recently Mann (1988) has argued that the late twentieth-century coexistence of capitalism and democratic politics did not come about because the free market automatically encourages political liberty, but because the democracies proved more effective than fascism at mass mobilisation in the Second World War, emerged victorious and both imposed democracy on the losers and embraced it at home more fervently than before or since. In turn, the new mood of democracy spread to the colonies and led to the break-up of the European empires and the political independence of the less developed world. Mann's analysis makes the recent emergence of newly industrialised countries with authoritarian regimes presiding over a booming capitalist economy much less surprising than it is to those who assume that capitalism promotes democracy.

from inevitable. Political structures and military events have often diverted or even reversed the direction of change. Peripheral areas have overtaken the centre on many occasions and in different ways: Akkad over Sumer; Greece over Persia; Germanic tribes over Rome; USA over UK. Throughout these millennia, ecology has exerted constraints and most people have had to live within those constraints most of the time (often developing appropriate ideologies for the time and place, as described by Sarre, 1995). There have always been periodic disasters and famines but these had relatively local effects and could be avoided by migration or compensated by the development of new areas. This 'localisation of nature' has broken down in the modern era, both in practical terms and in ideas about Nature.

During the process of globalisation, European ideas and economic power came to dominate much of the world, as argued by Roberts (1985). In turn, most of the non-European environmental ideologies, whether those of single tribes or of sub-continents like India or China, have been marginalised by western practice, which has been predominantly imperialist, though subject to the critiques outlined by *Sarre and Brown* (1996, Section 4). With the possible exceptions of Marxism and Islam, which are no more favourable to wild nature than are Christianity or capitalism, indigenous philosophies and religions have been pushed aside by western values of economic growth and consumerism, though some, notably Buddhism, persist as an inspiration to environmentalists in the developed as well as the less developed world. Hence it is western values which are likely to dominate the next one hundred years, unless either Japan becomes world economic leader and rediscovers some of the environmental values of Zen Buddhism, or the values of modernism lose their dominance in the new era. (See Box 7.2.)

Toulmin (1990) has recently reinterpreted the origins of modernism in a way which throws new light on the relations between interpretations of Nature and events in society. He challenges the simplistic view that rediscovery of classical texts in the Renaissance sparked a greater interest in science which led in turn to the Reformation, a more secular society, new scientific discoveries, modernism and the beginnings of industrialisation. His account of the origins of modernism puts great stress on conflicts between Catholicism and Protestantism, and especially the Thirty Years War, in which nearly a third of the population of Germany was killed. He shows that it was this context which spurred Descartes to set out to construct a philosophy based only on rationality, to establish a way of thinking on which everyone could agree. Although the Cartesian method had questionable assumptions, it seemed to be vindicated when Newton used it as a basis for his laws, which successfully predicted planetary motion and the behaviour of falling bodies.

The questionable assumptions include the separation of humans from Nature, of mind from body, and of rationality from feeling. Moreover, Nature was depicted as a mechanical domain where events took an inevitable course as defined by God at the creation. Such a Nature could easily be treated as a mere resource base, to be used to satisfy human needs, with consequences described by *Simmons* (1996) and this series.

Toulmin goes on to show that science had in effect demolished all Descartes' assumptions by the end of the nineteenth century. James Hutton had shown that the Earth is much too old to be just a created mechanism (a mechanism would have broken down or worn out, whereas the Earth renews itself); Darwin that Nature evolves and that humans are closely related to the apes; Freud that mind and body, rationality and feeling are intimately connected; physicists were beginning to discover that matter can

## Box 7.2   Buddhism

*Laughing Buddha (Budai), Chinese, Ming Dynasty. This image of the Buddha eloquently expresses the life-affirming aspirations which have attracted environmentalist towards Buddhism rather than Christianity or the other monotheistic religions.*

Buddhism originated in India as a reaction against the suffering evident under Hinduism. It shares some of the key assumptions of Hinduism, notably the unity of humanity and nature and the importance of non-violence. However, it avoids the Hindu preoccupation with deities and ritual in favour of a focus on 'the noble eight-fold path' – living in an optimum relationship with other people and the environment.

Buddhism recognises that humans are often motivated by greed, hatred and self-delusion and encourages individuals to eliminate these motives from their own behaviour and to achieve a state of Nirvana, or enlightenment. Such individuals would then live non-violently and without taking anything which has not been given. However, Buddhism does not encourage extremes of self-denial as it recognises that starvation and squalor are self-destructive and no more conducive to right livelihood than extremes of luxury and power.

Instead, people are urged to seek 'the middle way' – a lifestyle that is materially comfortable enough to live well without having to harm other people or the environment. Strictly speaking, Buddhism does not prescribe a future, since it focuses on living each moment in the best possible way. However, it seems reasonable to deduce that a world inhabited only by enlightened people would be a gentle world in which people were concerned mainly with relationships, co-operation to improve living environments, art, science, recreation and self-improvement. One view of such a society is the description of Pala in Aldous Huxley's novel, *Island*.

Buddhism has exerted an influence on western thought, and especially that of environmentalists for well over a century: Thoreau was familiar with Buddhist as well as Greek ideas from his time at Harvard, and his self-imposed exile to live in the woods and grow beans has a distinctly Buddhist flavour. More explicitly, environmentalist poet Gary Snyder lived in Japan, studied and became a Buddhist. Directly or indirectly, Buddhist ideas continue to present an enlightening alternative. The tragedy is that, even in Asian countries with a Buddhist majority, capitalism and consumerism have become dominant.

**Riprap**
Lay down these words
Before your mind like rocks.
　　　　　placed solid, by hands
In choice of place, set
Before the body of the mind
　　　　　in space and time:
Solidity of bark, leaf, or wall
　　　　　riprap of things:
Cobble of milky way,
　　　　　straying planets,
These poems, people,
　　　　　lost ponies with
Dragging saddles –
　　　　　and rocky sure-foot trails.
The worlds like an endless
　　　　　four-dimensional
Game of *Go*.
　　　　　ants and pebbles
In the thin loam, each rock a word
　　　　　a creek-washed stone
Granite; ingrained
　　　　　with torment of fire and weight
Crystal and sediment linked hot
　　　　　all change, in thoughts,
As well as things.

*Gary Snyder*

behave in very unmechanical ways. Yet an essentially modernist world-view was to survive for nearly another century. Once more, the explanation Toulmin gives is conflict and war: the First World War not only caused vast destruction to people and the environment in Flanders, it stimulated a group of philosophers in Vienna (capital of the war's major losers, the former Austro-Hungarian empire) to reinvent rationalism in the form of **logical positivism.** In the face of the threats posed by the Second World War and the Cold War, rationalism maintained its dominance. For example, ecology, which had begun with organic and even Romantic interpretations of Nature became by the 1950s the scientific study of energy flows in ecosystems and matters which could not be quantified became uninteresting (Sachs, 1993).

Just as ecology later abandoned this restrictive view to define a new paradigm which stresses the complex and unpredictable patterns of change which can result from natural hazards, human pressures and change in species competition, so a whole range of changes in science and in society has contributed to a widespread belief that the era of modernism has ended. Whether characterised as post-modernity, late- or advanced modernity, this new era is seen as less mechanical and less purely rational than modernity, and therefore seems to be even less predictable. Some accounts welcome a more holistic world-view, because it seems to take environmental issues more seriously; others regard post-modernity as a form of mystification which merely serves to conceal continued exploitation of the environment and of disadvantaged groups. Toulmin's account makes it possible that there could be a qualitative change in society's view of and treatment of Nature and the environment, but does little to suggest what the changes will be.

### Activity 2

On the basis of this brief summary of past change, look back at the last few centuries in the context of the last few millennia. What factors would you expect to determine the course of the next century?

At one level, this is a totally impossible question, analogous to asking someone in 1699 to predict the eighteenth century. Yet, although the possibility of a new era tends to emphasise possible discontinuities, the same six broad issues tend to recur in all the accounts discussed, so at least we know where to look.

We can be fairly confident in expecting continued *technological* development, and that much of this will involve more refined methods of carrying out existing activities. The precise direction to be taken is likely to depend on *economic* and *political* factors. At present, the trajectory seems to be towards the reduction of state influence and globalisation of the economy, but this could change if the US were to respond to its loss of economic pre-eminence by reverting to economic protectionism and political isolation. This seems improbable at present because the US retains military pre-eminence, but that is unlikely to be sustained through a century of relative economic decline. *Military* events could well disrupt present expectations, either in established conflict zones like the Middle East, or possibly more significantly in the rapidly growing Pacific Rim, where a variety of territorial claims could cause conflict between countries of growing world significance. As suggested above, the emergence of the

Pacific Rim could generate new *ideological* developments, though the
conflict of interest between developed and less developed worlds may
continue to stress conflict rather than co-operation.

Most of the analysis in the series suggests that *environmental factors* will
play an increased role in the next century, as population expansion and
economic growth place increased stress on land space, resources and
ecosystems via clearance or pollution. These factors could change
ideologies and politics towards pursuit of sustainability, but might also
provoke conflict and military confrontation. Much depends on the degree
to which political and technological change can prevent overt crises of
resource availability and pollution: this is a question where optimists and
pessimists disagree sharply. The earlier chapters of this book add a further
factor which has not been relevant to earlier periods: anthropogenic climate
change seems certain to become apparent over the next century, adding a
new element of uncertainty about the incidence and extent of
environmental problems. This is a factor which has not been anticipated by
any of the environmental futures to which we now turn.

# 3   *Alternative futures*

It is in some ways anachronistic to write about futures, especially in the
spirit of prediction, because it presupposes that the world is mechanical
and predictable rather than creative. Studies of futurology – the self-
proclaimed science of the future – were most common in the 1970s but have
since been taken less seriously, though Frankel (1987) argues that a utopian
strand runs through critical social theory. This is certainly true of
specifically environmental futures, where 'blueprints' for the author's
desired future continue to appear, and for science fiction books and films.
I will consider several of these alternatives, using a combination of my
summary and evaluation and quotations from the original authors. If these
futures are to 'speak' to you, they are more likely to do so in the author's
words than my own.

I will structure my survey in the light of one of the more immodest
pieces of futures writing. Wagar (1992) claims to identify 'the next three
futures' – the 'techno-liberal', the 'radical' and the 'countercultural'. He
treats them as sequential, a prediction I regard with some scepticism, but I
am prepared to accept that they are in order of probability, since they
require progressively more fundamental forms of change. The techno-
liberal future is a projection of present orthodoxy, the radical alternatives
relate to current and past critiques and programmes, while the
countercultural are attempts to transcend past and current politics in favour
of innovative lifestyles.

A constant difficulty when writing about varieties of environmental
belief is in relating them to conventional political positions like 'left' and
'right'. Environmentalist utopias may combine elements usually regarded
as mainstream with others that are seen as 'far left' and 'far right'. Petra
Kelly, of die Grünen, claimed that green politics are 'neither left nor right
but ahead', but in my view any environmentalism which aims to become

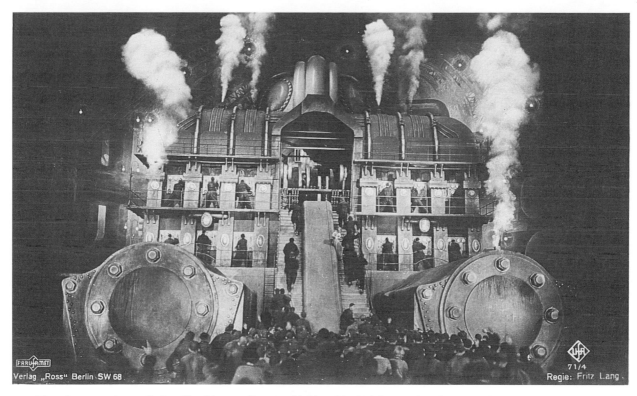

▲   *The pioneer science fiction film,* Metropolis, *was highly critical of the tendencies in society towards huge industrial cities, social divisions and dehumanised work.*

part of mainstream politics has also to deal with social issues and hence take up a recognisable position on them, which can be categorised as left or right. One indication of this is that the most ecocentric environmental thinkers tend to concentrate on changing individual values and lifestyle, and offer no political programme to reach their desired future. Less ecocentric thinkers find it easier to relate their goals to familiar political positions, but as a result are more exposed to critics of those positions. In this section, I will both present a range of possible futures in their own terms and identify how they are seen by supporters and critics. I have chosen these particular examples simply to illustrate the possible range and with no claim that they either exhaust all possibilities or that they are the best examples of particular approaches. Anyone seeking a more systematic and detailed survey of environmentalist positions should consult Pepper (1996).

## 3.1   *Techno-liberal*

This is the most familiar and apparently likely prospect, because it is largely a business-as-usual future based on an American faith in free enterprise and minimal government. There is no doubt that many powerful individuals and institutions are working to this end, but the major question is whether it is sustainable, as it is precisely the projection of the 'growth is progress' view which has been blamed for increased rates of environmental

damage. One of the most influential contributors to this school of though
has been Herman Kahn, working from the Hudson Institute in the USA.
His more recent work with Julian Simon (Simon and Kahn, 1984) has been
more explicitly environmental, but the most direct specification of this kind
of future remains his book, *The Next 200 Years* (1976).

This was presented as a scientific enquiry into the future, using the best
possible projections of current trends, the best analyses of resources and
prudent argument and assumptions. Kahn recognised and documented a
range of views from 'neo-Malthusian pessimist' to 'technology and growth
enthusiast'. He characterised himself as a guarded optimist, but in fact his
predictions read very optimistically: continued economic growth and
technological progress are seen as overcoming all foreseeable obstacles and
so providing wealthy lifestyles to both the poorer members of developed
societies and, within fifty years or so, to less developed societies worldwide.
Kahn envisages that technological progress and economic growth will
produce an extremely affluent world in which most people pursue
'quaternary' activities – recreation, ritual, tourism, art and public works – as
opposed to the primary (agriculture/mining), secondary (manufacturing) or
tertiary (managerial) activities which have predominated in past societies.
In case such a world is unattractive to high achievers, he points to space
exploration and settlement as 'the next frontier':

> While our scenario for America and the world is generally optimistic
> for the long term, we do recognize the real possibilities of serious
> anomalies, dislocations and crises in the short term, any one of which
> could greatly complicate the process of getting from here to there.
> Among these potential difficulties are regional overpopulation,
> retarded economic growth, energy shortfalls, raw materials shortages,
> local famines, short-run but intense pollution, environmental
> surprises and (most fearful of all) large-scale thermonuclear war.
> While we offer no solutions that will guarantee the avoidance of these
> problems, we do believe that acceptance of our position presents the
> best hope of both reducing the possibility of their occurrence and
> mitigating the consequences if any do occur. (Kahn, 1976, p. 211)

The book has chapters which analyse these 'known problems of the short
term' and propose solutions which the authors claim to require little
political change to implement, although the analyses in *Sarre and Blunden*
(eds, 1996) and *Blunden and Reddish* (eds, 1996) suggest that solution of
agricultural, energy and minerals provision problems in fact require
changes to the fundamental organisation of current world society. It is only
at the very end of Kahn's book that questions of politics and goals are
squarely identified:

> The postindustrial world we foresee will be one of increased
> abundance, and thus hopefully of reduced competition; it will be one
> of greater travel and contact, and thus possibly one of diminished
> differences among its peoples. But it will also be one of enormous
> power to direct and manipulate both man [*sic*] and nature; and thus
> its great issues will still be the very questions that confront us now,
> though enlarged in range and magnitude: Who will direct and
> manipulate, and to what ends? (1976, p. 226)

Kahn does not explain why politics and goals will only come to the fore
after rational and technical solutions have been found to the problems he

expects. However, Greene's discussion of international environmental politics in the previous chapter does little to support the view that recognition of a problem leads quickly to adoption of policies to solve it, let alone quick and effective implementation of those policies. Most other schools of thought have seen the need for more palpable change, either of political institutions or of ways of life, or of both.

## 3.2   Radical: authoritarian

Like many social theorists, many environmentalists have advocated radical institutional change to bring about a more desirable environmental future. Most of these proposals have been for more equitable and democratic societies, but one notable contribution has argued that a more authoritarian society is needed.

William Ophuls accepted many of the environmentalist arguments considered in this series, and indeed took what Kahn would identify as a neo-Malthusian line in concluding that ignorance and vested interests

◀   Frontispiece from Leviathan, *1657. Thomas Hobbes argued that only the surrender of power to a single sovereign, which he called Leviathan, could save society from the consequences of human passions.*

would prevent society from reacting adequately. Ophuls returned to Thomas Hobbes, one of the key philosophers of modernism, to argue that to overcome the problems of equity and environment individual citizens would have to give sovereign power to the state (known to Hobbes as 'Leviathan') and accept that it was in their best interests that the state should require everyone to conserve more and pollute less:

> Examining the nature of agreements between men [*sic*], Hobbes concludes that human passions make good intentions and voluntary compliance totally unreliable. Only a common power standing above the parties to a contract can make men fear the consequences of breaking their word enough to meet their obligations unfailingly. Although the laws of nature are always binding in conscience, 'covenants without the sword are but words', and only enforcement by a civil government can put them into practical effect. Hobbes says that the only way to erect a common power sufficient to guarantee peace among men is for them to come together and by majority decision 'confer all their power and strength upon one man, or upon one assembly of men that may reduce all their wills, by plurality of voices, unto one will.' In this way, they become a 'multitude united in one person' or a 'commonwealth' ruled over by a 'sovereign' holding all the power originally possessed by the multitude of 'subjects'. (Ophuls, 1973, p. 218)

Ophuls saw Hobbes as providing philosophical backing for the conclusions that Garrett Hardin (1968) had drawn from his analysis of the tragedy of the commons (see *Reddish*, 1996): that new, more authoritarian, social arrangements are needed to make us take the environment more seriously.

> Again, the similarities to Hardin's argument are striking. Hardin says we are faced with the necessity to 'legislate temperance' – i.e., limit certain natural freedoms we now have. However, we need to ensure that all individuals and groups in society are equally temperate lest the irresponsible continue to breed and pollute. Because conscience is likely to be self-eliminating where irresponsibility is unrestrained, we have to have definite social arrangements that produce responsibility, and these are necessarily coercive. In sum, we must have 'mutual coercion, mutually agreed upon by the majority of the people affected'. Hardin maintains that coercion need not imply 'arbitrary decisions of distant and irresponsible bureaucrats'. However, even if this danger exists, 'an alternative to the commons need not be perfectly just to be preferable'. Indeed, 'injustice is preferable to total ruin'. (Ophuls, 1973, p. 219)

### Leviathan Or Oblivion?

If scarcity is not dead, if it is in fact with us in a seemingly much more intense form than ever before in human history, how can we avoid reaching the conclusion that Leviathan is inevitable? Given current levels of population and technology, I do not believe that we can. Hobbes shows why a spaceship earth must have a captain. Otherwise, the collective selfishness and irresponsibility produced by the tragedy of the commons will destroy the spaceship, and any sacrifice of freedom by the crew members is clearly the lesser of evils. (Ophuls, 1973, p. 224)

In his later work, Ophuls (1977) argued that many of the necessary changes could be made through an ethic of voluntary simplicity, but maintained that domination by Leviathan is preferable to allowing democracy to fail to deal effectively with environmental crisis.

## 3.3   Radical: libertarian

At the opposite end of the political spectrum from Hobbes, Hardin and Ophuls stands anarchism, a movement which seeks to eliminate the state as a form of domination and establish a libertarian society. This strand of radical thought has a long-standing affinity with environmentalist thought; indeed Kropotkin had already outlined a series of policy proposals and William Morris had sketched an essentially anarchist utopia well over a century ago. Anarchism is often criticised as undesirable or impracticable, but rests on values of self-determination which are widely shared. It is sometimes presented as an extreme left political position, possibly because in practical politics anarchists have often vied with Marxists in attempts to overthrow the existing state, but in its demand for elimination of government it now seems to anticipate the position of neo-liberals like Hayek, who provided inspiration for Thatcherism and Reaganism.

◄ Frontispiece of the Kelmscott Press edition of News from Nowhere, 1892. Although William Morris is known as a pioneer socialist, his utopian novel depicts a future England which has adopted anarchism: the Houses of Parliament are used for storing dung and people live idyllically in rural communities engaging in craft production and organic cultivation.

The most sustained advocate of anarchism allied to environmentalism has been Murray Bookchin. He diverged from America's old left over the centralisation and authoritarianism of Marxism, seeing the Soviet system as the logical outcome of Marxist thought rather than as a perversion of it. He argues that anarchism and ecology share a common hostility to domination of any form:

> Ecology, in my view, has always meant social ecology: the conviction that the very concept of dominating nature stems from the domination of human by human, indeed, of women by men, of the young by their elders, of one ethnic group by another, of society by the state, of the individual by bureaucracy, as well as of one economic class by another or a colonialised people by a colonial power. (Bookchin, 1980, reprinted in Nash, 1990, p. 295)

> It cannot be emphasized too strongly that the anarchist concepts of a balanced community, a face-to-face democracy, a humanistic technology and a decentralized society – these rich libertarian concepts – are not only desirable, they are also necessary. They belong not only to the great visions of man's [*sic*] future, they now constitute the preconditions for human survival. The process of social development has carried them out of the ethical, subjective dimension into a practical, objective dimension. What was once regarded as impractical and visionary has become eminently practical. And what was once regarded as practical and objective has become eminently impractical and irrelevant in terms of man's development towards a fuller, unfettered existence. (Bookchin, 1971, pp. 69–70)

> If the ecological community is ever achieved in practice, social life will yield a sensitive development of human and natural diversity, falling together into a well balanced, harmonious whole. Ranging from community through region to entire continents, we will see a colourful differentiation of human groups and ecosystems, each developing its unique potentials and exposing members of the community to a wide spectrum of economic, cultural and behavioural stimuli. Falling within our purview will be an exciting, often dramatic, variety of communal forms – here marked by architectural and industrial adaptations to semi-arid ecosystems, there to grasslands, elsewhere by adaptation to forested areas. We will witness a creative interplay between individual and group, community and environment, humanity and nature. The cast of mind that today organizes differences among humans and other life forms along hierarchical lines, defining the external in terms of its 'superiority' or 'inferiority', will give way to an outlook that deals with diversity in an ecological manner. Differences among people will be respected, indeed fostered, as elements that enrich the unity of experience and phenomena. (Bookchin, 1971, pp. 81–2)

Bookchin's vision of a world without domination is based upon the success of the direct democracy of the Greek *polis* (*Hague et al.*, 1996) and his analysis of the effectiveness of direct democracy during periods of revolution in Russia, France and Spain. It is noteworthy that direct democracy was also practised in the 'town meetings' of New England, his home region. Bookchin's ideas have been subjected to a range of criticisms. These include: assertions that he is in practice still anthropocentric and hardly gives environment a significant place in his thought; questioning whether the principle of local autonomy is consistent with environmentalism, since a locality might opt for nuclear power or some other ecologically harmful activity; asking how small-scale communities can be a solution to the problems of a highly integrated world; observation that small-scale communitarianism has historically been associated with conservatism and social control rather than with liberation and creativity. Bookchin and his followers have vigorously refuted these criticisms and argued, for example, that community autonomy can be built into systems of 'municipal federalism' which combine the benefits of voluntary co-operation with those of local autonomy.

## 3.4   Eco-socialist

Ever since they became politically active, environmentalists have had very mixed feelings about socialists. They share hostility to capitalism and both look to the state to regulate and tax in favour of their cause. Yet they often find themselves in conflict because allegedly socialist economies have tended to be just as concerned to increase production as have capitalist – and in many countries have been less able to control pollution. One of the major works of early environmentalism was concerned to analyse the complex but ultimately irreconcilable relationships between 'the red and the green'. It was written by Rudolf Bahro (1984), who moved from a base in East German socialism to argue the fundamentalist cause in the West German Green Party and ended by arguing that socialism was antithetical to environmentalism.

The most elaborate attempt to explore what it would feel like to live in an environmentally sound society is James Callenbach's (1978) novel, *Ecotopia*. I have classified *Ecotopia* as an eco-socialist future because this fictional scenario is brought about by state action: the governments of Northern California, Washington and Oregon supposedly join together to secede from the USA and establish an ecologically sound society.

> What was at stake, informed Ecotopians insist, was nothing less than the revision of the Protestant work ethic upon which America had been built. The consequences were plainly severe. In economic terms, Ecotopia was forced to isolate its economy from the competition of harder-working peoples. Serious dislocations plagued their industries for years. There was a drop in Gross National Product by more than a third. But the profoundest implications of the decreased work week were philosophical and ecological: mankind, the Ecotopians assumed, was not meant for production, as the 19th and early 20th centuries had believed. Instead, humans were meant to take their modest place in a seamless, stable-state web of living organisms, disturbing that web as little as possible. This would mean sacrifice of present consumption, but it would ensure future survival – which became an almost religious objective, perhaps akin to earlier doctrines of 'salvation'. People were to be happy not to the extent they dominated their fellow creatures on the earth, but to the extent they lived in balance with them. This philosophical change may have seemed innocent on the surface. Its grave implications were soon spelled out, however.

[...]

Tens of thousands of employees were put out of work as a
consequence and the new government made two responses to this.
One was to absorb the unemployed in construction of the train
network and of the sewage and other recycling facilities necessary to
establish stable-state life systems. Some were also put to work
dismantling allegedly hazardous or unpleasant relics of the old order,
like gas-stations. The other move was to adopt 20 hours as the basic
work week – which, in effect, doubled the number of jobs but virtually
halved individual income. (There were, for several years, rigid price
controls on all basic foods and other absolute necessities.) (Callenbach,
1978, pp. 43, 45)

Callenbach's description of Ecotopia is of a fertile country with organic
farms set in the context of forests and mountains. Substantial cities remain,
but private cars do not exist and efficient public transport systems leave
space for an intensive street-life of shops, stalls and entertainers. Natural
materials are widely used, both because Ecotopians enjoy working with
them and because the society insists on recycling all waste. Although
working-hours are short, work is efficiently organised and includes
factories, offices, farms and forestry camps.

Ecotopians are in touch with their own feelings and discuss them
openly with work or living groups. They practise free love, and women are
fully equal to men in all ways. All this does not make them weak, however;
they not only play ritual war-games where participants are wounded or
even killed, but turn out to have repulsed a secret military attack by the
USA.

Perhaps the most remarkable aspect of Ecotopia is that, although
brought into being by government action and facing a continuing external
threat, there is substantial personal and local autonomy. As a result, the
society seems to have some of the more positive features of eco-anarchist
utopias and even countercultural aspirations. In this context, enlightened
democratic socialism seems to have prevented any negative features of
localised societies without imposing the colourless uniformity that critics
always expect in socialist regimes. In this respect this novel anticipates by
nearly two decades David Pepper's (1993) powerfully argued case that only
an alliance with socialism gives environmentalists any chance of
overcoming capitalism as the major determinant of the future.

## 3.5   Countercultural environmentalist

This is the most difficult form of environmental future to pin down in a
short account, since the main thrust of countercultural movements is not
just to oppose current practices but to reject current conflicts and concepts
in favour of fundamentally different values and lifestyles. There is a variety
of approaches, ranging from mystical 'New Age' practices through to deep
ecology, where the philosophy is carefully articulated and the practice less
so. For most counterculturalists, the main priority is to live their beliefs
rather than analyse them or publish them. Environmentalist elements may
be more or less emphasised against other goals like gender and racial
equality. Typically counterculturalists are more concerned to break away
from current forms of domination, like violence, commodity relations and
state power than to spell out what life would be like if domination were

eliminated. However, the section will sketch some of the implications of three approaches: deep ecology, ecofeminism and bioregionalism.

*Deep ecology* is a movement which explicitly claims to have developed a more profoundly ecocentric position than the shallow forms of environmentalism which are widespread. It seems to have developed in countries where wilderness is a key issue and to have its organisational centre in California, though its major thinker is Norwegian philosopher, Arne Naess. Perhaps the most effective introduction is to reproduce the basic principles of deep ecology as formulated in 1984 by Arne Naess and by George Sessions, one of the central figures in the development of deep ecology in California:

1   The well-being and flourishing of human and nonhuman life on Earth have value in themselves (synonyms: intrinsic value, inherent value). These values are independent of the usefulness of the nonhuman world for human purposes.

2   Richness and diversity of life forms contribute to the realization of these values and are also values in themselves.

3   Humans have no right to reduce this richness and diversity except to satisfy *vital* needs.

4   The flourishing of human life and cultures is compatible with a substantial decrease of the human population. The flourishing of nonhuman life requires such a decrease.

5   Present human interference with the nonhuman world is excessive, and the situation is rapidly worsening.

6   Policies must therefore be changed. These policies affect basic economic, technological, and ideological structures. The resulting state of affairs will be deeply different from the present.

7   The ideological change is mainly that of appreciating *life quality* (dwelling in situations of inherent value) rather than adhering to an increasingly higher standard of living. There will be a profound awareness of the difference between big and great.

8   Those who subscribe to the foregoing points have an obligation directly or indirectly to try to implement the necessary changes.

(reproduced in Fox, 1995)

These principles, especially 3 and 4, require dramatic changes from present practices and future expectations. Satisfaction of just vital needs suggests very low-impact lifestyles with only food, warmth and shelter as justifiable demands on natural systems. The requirement to reduce human populations is in stark opposition to the expected doubling or tripling of world population over the next century. It implies that deep ecological lifestyles are only available to a small percentage of world population until substantial reduction in world population has occurred, which would take two or three centuries unless some social or natural catastrophe occurs. The final principle gives deep ecologists an obligation to act to achieve their goals, but this seems to be interpreted as a lifestyle of voluntary simplicity, participation in non-violent protests against damaging developments and rational argument rather than in any form of political organisation. Indeed, some of the leading thinkers of deep ecology, notably Warwick Fox (1995), regard self-realisation as its major imperative, requiring a move towards a transpersonal psychology – an extension of self to incorporate parts of the environment. Critics of deep ecology see this as reverting to anthropocentrism while claiming to pursue ecocentrism, though Fox would claim to have overcome the dualism and gone beyond it.

A similar two-stage argument has taken place in *ecofeminism*. At a commonsense level, many feminists have claimed that women are closer to nature than men because of their caring roles and reproductive functions. As a result,

> The story of a land where women live at peace with themselves and with the natural world is a recurrent theme of feminist utopias. This is a land where there is no hierarchy, among humans or between humans and animals, where people care for one another and for nature, where the earth and the forest retain their mystery, power and wholeness, where the power of technology and of military and economic force does not rule the earth, or at least that part of it controlled by women. For usually this state is seen as a beleaguered one, surviving against the hostile intent of men, who control a world of power and inequality, of military and technological might and screaming poverty, where power is the game and power means domination of both nature and people. Feminist vision often draws the contrasts starkly – it is life versus death, Gaia versus Mars, mysterious forest versus technological desert, women versus men. (Plumwood, 1993, p. 7)

Plumwood is very critical of this portrayal of female virtue: she argues that it accepts precisely the derogatory characteristics attached to women in masculinist thought – that women represent body rather than mind, feeling rather than rationality, object rather than subject. She argues that it is necessary to challenge these prejudicial dualisms, which she traces back to Plato and an assumption that the purpose of life is beyond the Earth. In her view, the real contribution of feminism is that it has taken the lead in theorising the implications of the fact that all human beings have bodies as well as minds and that we are therefore a part of nature as well as culture. Such challenges to philosophical assumptions like the separation of reason from nature open up new possibilities for the future, but possibilities which cannot be foreseen:

> The reason/nature story has been the master story of western culture. It is a story which has spoken mainly of conquest and control, of capture and use, of destruction and incorporation. This story is now a disabling story. Unless we can change it, some of those now young may know what it is to live amid the ruins of a civilization on a ruined planet. The power to direct, cast and script this ruling drama has been in the hands of only a tiny minority of the human race and of human cultures. Much inspiration for new, less destructive guiding stories can be drawn from sources other than the master, from subordinated and ignored parts of western culture, such as women's stories of care. Those of us from the master culture who lack imagination can gain new ideas from a study, undertaken in humility and sympathy, of the sustaining stories of the cultures we have cast as outside reason. If we are to survive into a liveable future, we must take into our own hands the power to create, restore and explore different stories, with new main characters, better plots, and at least the possibility of some happy endings. (Plumwood, 1993, p. 196)

This argument suggests that the most innovative visions of the future should come from non-white feminists in the less developed world. Such perspectives do exist, but they tend to focus on critiques of the mainstream (Shiva, 1991) rather than on happy endings. Since the purpose of this section is to indicate what some possible endings might look like so that you can judge whether they are happy or not, I will move on to explore a third countercultural form of environmentalism.

*Bioregionalism* has not only been advocated in principle (Sale, 1985) but is also being lived out by dozens of groups in the USA. The heart of bioregionalist doctrine seems to be the notion that living in small-scale units according with natural variations in landscape will tend to correct the exploitation of nature. In those small-scale contexts, people are asked to follow four principles:

- 'liberating the self' – from consumerism and bureaucracy
- 'developing the region' – towards self-reliance
- 'knowing the land' – through direct contact
- 'learning the lore' – finding out about history, folklore and 'traditional wisdom'.

At the basis of bioregionalist thought is the notion that living in small-scale units will make people aware of the effects of their actions, both on other people and on the environment. As a result of this stress on practice, it is hard to be sure how such societies would develop. Sympathetic critics like Pepper (1993) point out that there are strong resemblances between bioregionalist ideas and those coming from eco-anarchist sources. But, like small-scale anarchist communities, bioregions would be at risk of mirroring traditional conservative values like those advocated by Goldsmith (1978) or even German fascists in the 1930s – where tradition and stress on the 'organic' links between people and land were used to provide justification for highly regulated lives and social structures.

## 3.6   Environmental futures: general issues

Even this short and selective account makes it clear that there is a wide range of possible futures when thinking in terms of a century or so. Since each future contains a variety of ideas, a number of more general themes emerge from these particular cases. A sharper awareness of these may make it possible to envisage new combinations.

A central issue is the degree to which social and environmental goals are compatible. Increasingly efficient technology seems to hold out the possibility of supplying a good quality of life to more people, but not of doing so for an exponentially growing population or for one in which material consumption and profit outweigh environmental considerations. The size of population and level of consumption which can be sustained not only relate to each other, but to the extent to which reservation of wild areas is seen as part of the quality of life. One can envisage a totally humanised world which is sustainable, but most environmentalists seem to crave for wild areas where they can commune with nature, study science or hunt wild animals. The extent of such wild areas, and the activities permitted in them, are a key issue for the world, though often not addressed when considering how human society should be organised.

As with any programme for change, these futures involve a pervasive debate about the proper roles of the state, the market and the locality. Each of these has problems: historically, many states have oppressed their populations and exploited the environment; indeed liberalism argued that the market could reduce oppression and increase efficiency. Today, most environmentalists see the market – at least as it has developed under capitalism – as a prime cause of environmental degradation and social inequality. Anarchists, bioregionalists and conservatives concur on the need for a stronger role for localities to resist both state and market, but it is unclear how local autonomy can always be expected to yield positive social and environmental outcomes when they have historically been associated with stable, hierarchical societies in which the majority lived lives of toil and want.

A key issue connected to this is interpretation of the role of technology. Some see it as the primary means of liberating people from both toil and want, with the potential also of overcoming environmental problems. Others see technology as the principal cause of social oppression and environmental exploitation, with nuclear weapons as the extreme case of both. Others argue that there are appropriate technologies and inappropriate technologies … and so on.

Underlying attitudes to technology are different views of growth: again, some see indefinite growth as the ultimate good – and one which makes it unnecessary to worry about inequality; others argue for pursuit of a steady-state economy, but here there are differences between those who see this as requiring the reduction of inequality and others who regard inequality as tolerable – and in the case of Ophuls as essential to impose correct behaviour on unwilling citizens.

Two underlying issues in all this are the kind of assumptions that are made about human nature and about the possibility of change coming through changing attitudes or through changed social organisation. Conservatives tend to assume that human nature is inherently wicked and that it must be held in check by authority. Liberals and socialists believe in the improvability of human nature, or at least human behaviour, but disagree over the means. The former believe in the merits of markets as a form of co-ordination, the latter believe in state intervention. Each is deeply suspicious of the other. Finally, counterculturalists believe that changed attitudes can subvert established organisation, that humans will show new and better characteristics in new circumstances, but prefer to invite us to experience these improvements than to specify what they will be like.

Because they vary in these and other respects, and because they have appeared in a variety of genres, these futures vary in terms of how they could be brought about. The next section addresses how this might happen.

*Activity 3*

Look again at Chapter 5 and outline the principal features of the future as prescribed by Brundtland and implemented by the Rio conference. Can you categorise it as techno-liberal, radical (and, if so, authoritarian, libertarian or socialist) or countercultural? Or does it combine elements of more than one position?

# 4   From here to there

This section attempts to 'map' possible futures, drawing on both the set of futures described in Section 3 and some of the more familiar ones in fiction. The account works from the present and indicates a ramifying set of possibilities, as indicated in Figure 7.1.

The trajectory of the last twenty years, accelerated by the outcome of the Uruguay Round of GATT (*Taylor*, 1996), seems to be towards Kahn's techno-liberal society, moving towards super industrialisation in the developed world with some parts of the less developed world beginning to follow suit but others stagnating. Over the next decade this suggests an increasingly market-dominated world, which is likely to experience even greater inequality than the present. In Europe there remains support for a higher level of social welfare, with some redistribution to achieve at least minimum standards for all, but at present this seems to be in retreat. Since an extremely unequal world opens up the prospect of mass migration, terrorism and wars over resources, it may be that some degree of welfare provision will be seen even by market enthusiasts as the price of stability.

As Kahn recognised, the techno-liberal world will face short-term problems including resource shortages and pollution problems, and these may act as a significant brake on economic development, in some areas at least. Given the extent of such problems over the last 50 to 100 years, it may take more than one decade before the effects are regarded as significant by the developed world and the multinationals.

A decade ago it was believed that Marxism provided the basis for an alternative future to rival the techno-liberal mainstream or the Malthusian nightmare (see, for example, Sandbach, 1980). After the events of 1989, this seemed quite impossible, but with former Communist parties beginning to win elections in Eastern Europe, it may have regained some possibility of influencing future events, though it seems more likely that these regimes will preside over mixed economies than the state-planned economies of the past.

*Fifty years* from now, a wider range of futures seems possible. At the centre of the spectrum is some variant of techno-liberalism, possibly Kahn's vision of an affluent post-industrialism, with a range of possibilities from a

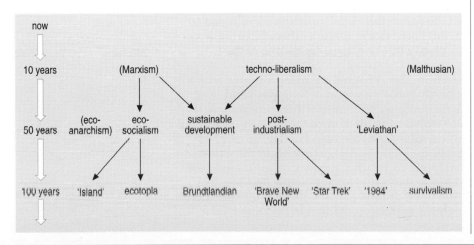

◀ *Figure 7.1 Ramifications*

free global market with vestigial state activity to a much more welfarist accommodation of international agencies, states and multinational corporations. Brundtland's sustainable future, especially the rather weak version negotiated at Rio, would be one of the options here, provided that the United Nations was able to move the international system in that direction and that the economy could continue to grow while conserving resources and redistributing socially.

Plausible alternatives could by then have come into being both to the 'left' and to the 'right' of techno-liberalism. A greater commitment to equity and redistribution, more like that of the Brundtland Report itself, might involve an internationalised socialism, or at least social democracy. This could be explicitly environmentalist, at least in relation to resource conservation and 'human welfare ecology' (*Sarre and Brown*, 1996), but seems likely to be predominantly anthropocentric, given the magnitude of the problem of international inequality. On the other hand, if Malthusian problems grow in severity, Ophuls' argument that the necessary solution to growing environmental problems lies in Leviathan seems persuasive in principle. How such a 'solution' could be achieved globally is more difficult to envisage: perhaps an international environmental protection agency with powers of surveillance and enforcement could metamorphose into such a global totalitarian regime, but it would need extremely serious problems and/or an extremely compelling belief system to control a world of 8–10 billion people. The effects of global warming are likely to be apparent in fifty years' time, but they may not be sufficiently clear to prompt major political change.

The dominance of a truly ecocentric regime seems improbable within a fifty-year span. Eco-anarchism, like other forms of anarchism, relies on local organisation and collaboration with other localities. As such, it is vulnerable to larger-scale power structures. The most likely route to such a situation seems to be an initial move towards a globalised economy with liberalism encouraging the withering away of the state system. Such a world would leave some room for local organisation, perhaps initially in areas of little interest to global capitalism as a result of lack of resources or purchasing power. History contains many examples of the emergence of lower levels of organisation to prominence when centralised authorities fail, but whether a decentralised global system like capitalism can fail is more speculative.

Looking at the *100-year time-horizon*, most things become speculative: think for a moment about how much of the twentieth century could have been anticipated at the time of the Boer War. Yet many of the seeds of future events already exist and the future depends on which are allowed to flourish and which are rooted out. Fortunately, many authors have contributed visions which may clarify how we feel about some of the possibilities.

If we project forward the techno-liberal future, we reach a future which in Kahn's description sounds rather banal – a world of long weekends and frequent holidays in which recreation is more common than production. It does not take much cynicism to wonder whether, once the majority are not needed for production, this future might not resemble Huxley's *Brave New World* – a world where subtle manipulation makes people want the games, easy access to drugs and sexual freedom which the system encourages. Nor is it hard to envisage the strict hierarchical division of people and lifestyles. Perhaps it is of some consolation to environmentalists that it is the outsider, the savage, who can see through the surface enjoyment to the underlying hollowness.

◄   *The* Star Trek *series celebrates an optimistic vision of a future in which human decency negotiates fair outcomes far from central authority. Viewed more cynically, it represents a projection of the expansion of American frontier, with its exploitation of the environment and subjections of native peoples, and is achieved more by industrial technology than by individual enterprise.*

Oddly enough, the one aspect of Kahn's 'next 200 years' which promises an element of striving is the one which looks most improbable in the short-term view – the exploration of space. Many science-fiction writers have argued that the growth of the world population and economy will provide both the means and the motivation to explore and then colonise space. However, although a future as an officer of the Starship Enterprise may look agreeable, and although space travel had the initial effect of making us aware of the Earth as a beautiful but lonely sphere in the vastness of space (see Plate 1), this emphasis on the next frontier seems likely to perpetuate past attitudes of exploiting the here and now as there will always be somewhere else to go. In fact it seems improbable that more than a tiny proportion of the Earth's people, matter and energy could be transported across space. For the majority, the future will be terrestrial, though in a cosmic context.

One of the most famous predictions of the future, Orwell's *Nineteen Eighty-Four*, resonates both with Kahn's worst nightmare, thermonuclear war, and Ophuls' advocacy of Leviathan. It also interestingly resembles the contemporary world division into three major economic blocs – Europe, North America and the Asian Pacific Rim. Orwell's three Leviathans – Eurasia, East Asia and Oceania – need external enemies to justify internal repression and so are at war incessantly and, since they change alliances if one side looks too powerful, with no prospect of an end. In this nightmare world, surveillance cameras and communication screens are everywhere, standards of living are low – except for the Party elite – and the environment is laid waste. It seems improbable on the surface – until one recalls that the world is more heavily armed than ever, that wars have been continuous since 1939 and that a recent best-seller in the USA dealt with the 'coming war with Japan'.

A related strand of American thinking points towards one of the most paradoxical futures of all – socially extremely pessimistic but

environmentally tempting. Projection of a Malthusian kind, as done famously by the Club of Rome (Meadows *et al.*, 1974), leads to exponential growth and then, around a century hence, to collapse of energy, production and population. Such a global collapse is exactly what misanthropic US survivalist cults are preparing for. Indeed Manes (1990) actually advocates the destruction of civilisation as the first step to establishing a new radical environmentalist world from the ashes of the old. Few people are prepared to advocate such a catastrophe, but many environmentalists seem to feel a yearning for a less crowded and complicated world. The question posed by this future is whether the end justifies the means and, if not, whether there is a more socially acceptable way of reaching a similar end.

Ophuls (1977), in his more considered work, in fact suggests that it may be possible to avoid the need for Leviathan if people generally would adopt a lifestyle of voluntary simplicity. This substitution of ideology for authority seems to move him far across the spectrum to a sort of Buddhist future and is consistent with the widespread environmentalist sympathy for Buddhism. It might well resemble Aldous Huxley's optimistic utopia, the Island of Pala, which has an environmentally conscious and psychologically healthy way of life based on Buddhism tempered by a commonsense Scottish doctor. Unfortunately, this regime depends on isolation from the rest of the world and Pala has large oil reserves and is coveted by both international oil companies and the neighbouring dictator. Slightly more persuasive is the case of Ecotopia, because its scale is larger and it is willing to arm and fight for self-determination. Perhaps this is an indication that eco-socialism, whatever its intrinsic compatibility with

▲  *A scene from the film of* Nineteen Eighty-Four. *George Orwell's novel described a near future based on the worst features evident in 1948 – citizens subject to surveillance, propaganda and ruthless coercion, justified by the demands of constant conflict between three major powers.*

ecocentrism, is a possible medium-term route towards an environmentally sustainable world. The key, however, is that such regimes must be democratic so that they cannot transform into Leviathan, the autocratic state. The problem for forecasters is that nation-states not infrequently change from democratic to autocratic and vice versa: politics may well be harder to predict than is economic or technological change.

Ecotopia also seems to be a possible route towards an eco-anarchist future where the Deep Green lifestyles of Arne Naess or Warwick Fox (1995) could flourish. This would resemble the 'voluntary simplicity' of Ophuls, but appears more life-affirming. It seems to rest on an optimistic view of human nature as content to live simply, socially and spiritually, whereas Ophuls tends to imply that humans will only adopt this kind of lifestyle for fear of something worse.

---

### Activity 4

Review this discussion of possible changes over the next century in the light of the factors identified by Mann in Section 2 as the main influences on long-term social change. Have all the factors been considered explicitly and adequately?

---

*The longer term* becomes dramatically more unpredictable than a decade or a century, not least because global warming will by then be readily apparent. However, one further point needs to be added: earth scientists have recently recognised that the Earth is from time to time subject to major natural events with dramatic short-term effects. Volcanic eruptions big enough to prevent harvest for two or three years are now known to occur, indeed there have been twenty-five major blasts within the last 10 000 years. Dramatic asteroid impacts also occur from time to time, most notably the one believed to have caused the extinction of the dinosaurs. Such events, or even a major earthquake in Tokyo, could subject global society to catastrophic change and, as time-scales grow longer, so the probability of these events increases. The Earth has demonstrated its ability to survive these shocks and evolve new species and ecosystems. Human society needs to be able to do the same if it is not to be a very short-lived event in the Earth's history. The existence of these mega-hazards is symbolic as well as practical: they are a reminder that the environment is not just a constraint on human development, but an awesomely powerful set of natural processes within which we have to live and which should be respected as well as used.

## References

BAHRO, R. (1984) *From the Red to the Green: interviews with New Left Review*, London, Verso.

BLUNDEN, J. and REDDISH, A. (eds) (1996) *Energy, Resources and Environment*, London, Hodder and Stoughton/The Open University (second edition) (Book Three in this series).

BOOKCHIN, M. (1971) *Post Scarcity Anarchism*, Berkeley, CA, Ramparts Press.

BOOKCHIN, M. (1980) *Toward an Ecological Society*, Montreal, Black Rose; reprinted as pp. 294–8 in Nash, R. (ed.) *American Environmentalism*, New York, McGraw-Hill.

CALLENBACH, J. (1978) *Ecotopia*, London, Pluto.

FOX, W. (1995) *Towards a Transpersonal Ecology*, Dartington, Green Books.

FRANKEL, B. (1987) *The Post-industrial Utopians*, Cambridge, Polity.

GOLDSMITH, E. (1978) 'The religion of a stable society', *Man–Environment Systems*, Vol. 8, pp. 13–24.

HAGUE, C., RAEMAEKERS, J. and PRIOR, A. (1996) 'Urban environments: past, present and future', Chapter 7 in Sarre, P. and Blunden, J. (eds).

HARDIN, G. (1968) 'The tragedy of the commons', *Science*, Vol. 162, pp. 1243–8.

HUXLEY, A. (1932) *Brave New World*, London, Chatto and Windus.

HUXLEY, A. (1972) *Island*, London, Chatto and Windus.

KAHN, H. (1976) *The Next 200 years*, London, Associated Business Programmes.

MANES, C. (1990) *Green Rage: radical environmentalism and the unmaking of civilisation*, Boston, MA, Little Brown.

MANN, M. (1986) *The Sources of Social Power, Vol. 1: A History of Power from the Beginning to AD 1760*, Cambridge, Cambridge University Press.

MANN, M. (1988) *States, War and Capitalism: studies of political sociology*, Oxford, Blackwell.

MEADOWS, D. H. *et al.* (1974) *The Limits to Growth*, London, Earth Island.

MORE, T. (1516) *Utopia* (London, Dent, Everyman's Libary, new edn, 1974).

MORRIS, W. (1891) *News from Nowhere*, London, Reeves and Turner.

NASH, R. F. (1990) *American Environmentalism: readings in conservation history*, New York, McGraw-Hill.

OPHULS, W. (1973) 'Leviathan or oblivion?', pp. 215–30 in Daly, H. E. (ed.) *Toward a Steady State Economy*, San Francisco, Freeman.

OPHULS, W. (1977) *Ecology and the Politics of Scarcity*, San Francisco, Freeman.

ORWELL, G. (1949) *Nineteen Eighty-Four*, London, Secker and Warburg.

PEPPER, D. (1993) *Ecosocialism: from deep ecology to social justice*, London, Routledge.

PEPPER, D. (1996) *Modern Environmentalism: an introduction*, London, Routledge.

PLUMWOOD, V. (1993) *Feminism and the Mastery of Nature*, London, Routledge.

REDDISH, A. (1996) 'Environment and development', Ch. 4 in Sarre, P. and Reddish, A. (eds).

ROBERTS, J. M. (1985) *The Triumph of the West*, London, BBC Publications.

SACHS, W. (1993) 'Environment', pp. 26–37 in Sachs, W. (ed.) *The Development Dictionary: a guide to knowledge as power*, London, Zed Books.

SALE, K. (1985) *Dwellers in the Land: the bioregional vision*, San Francisco, Sierra Club.

SANDBACH, F. (1980) *Environment, Ideology and Policy*, Oxford, Blackwell.

SARRE, P. (1995) 'Paradise lost, or the conquest of the wilderness', in Sarre, P. and Blunden, J. (eds) *An Overcrowded World? Population, Resources and the Environment*, Oxford, Oxford University Press/The Open University.

SARRE, P. and BLUNDEN, J. (eds) (1996) *Environment, Population and Development*, London, Hodder and Stoughton/The Open University (second edition) (Book Two in this series).

SARRE, P. and BROWN, S. (1996) 'Changing attitudes to nature', Ch. 3 in Sarre, P. and Reddish, A. (eds).

SARRE, P. and REDDISH, A. (eds) (1996) *Environment and Society*, London, Hodder and Stoughton/The Open University (second edition) (Book One in this series).

SHIVA, V. (1991) *Ecology and the Politics of Survival*, London, Sage.

SIMMONS, I. (1996) 'The impact of human societies on their environments', Ch. 2 in Sarre, P. and Reddish, A. (eds).

SIMON, J. and KAHN, H. (1984) *The Resourceful Earth: a response to global 2000*, Oxford, Blackwell.

TAYLOR, A, (1996) 'World trade and its environmental impact', Ch. 6 in Sarre, P. and Blunden, J. (eds).

TOULMIN, S. (1990) *Cosmopolis: the hidden agenda of modernity*, New York, The Free Press.

WAGAR, W. (1992) *The Next Three Futures: paradigms of things to come*, Westport, CT, Greenwood Press.

# Answers to Activities

### Activity 3

My answer is built into the argument of Section 4. I see Brundtland as lying somewhere between the techno-liberal mainstream and eco-socialism. The exact position depends on how radically current orthodoxies are challenged by environmental problems and how much change will be accepted by the international community.

### Activity 4

My own view, which may be less critical than yours, is that most of the factors are reasonably sensibly treated. Ecological constraints have been given a major role through the effects of Malthusian checks, and seen as a major factor influencing the degree of change away from the techno-liberal mainstream. Technological change and economic growth are assumed to be probable, but as subject to change for political and environmental reasons, either through military conflict and even total breakdown at one end of the spectrum or through a focus on redistribution, welfare and voluntary simplicity at the other. Political options have been built in to the diagram, which embodies a rough left–right dimension, though one which recognises that regimes can change quickly and that there might be strong similarities between small-scale 'survivalist' and 'Buddhist' futures in many respects.

In my view, the least adequately treated factor is ideology, which has appeared in relation to politics and as a response to Malthusian problems, but without anticipating any substantial change as a result of environmentalist thought *per se*. Yet science produces new understandings of how the biosphere works as well as offering the basis for new production and weapons technologies. Increasingly, these understandings stress evolution, adaptability, diversity and creativity – virtues recognised by the anarchists but often stifled by states and markets alike. A greater regard for diversity, perhaps expressed in greater local power to reject unwanted development, could lead the whole debate about environmental futures into a new era, an era of debate about what people want rather than about what they are likely to have imposed upon them by political, economic or military power. Past trends in recreation and tourism suggest that as people grow more affluent they increasingly seek contact with wild nature and exotic lifestyles. Such desires suggest that we should aim for a future of diverse societies and substantial reserves of wild nature: the problem is how to achieve that degree of affluence without destroying the social and natural diversity that we and our descendants would value. The problem may well be soluble, but at present it seems unlikely to be pursued seriously unless a visionary environmentalist ideology appears which is more attractive than the consumerism that is currently in the ascendant through more of the world than ever before. Such an ideology would have to take a longer-term view of humanity's activities than any which have been proposed up to now.

## *Appendix: Acronyms*

| | |
|---|---|
| AASEI, II | Airborne Arctic Stratospheric Expeditions I and II |
| AOSIS | Alliance of Small Island States |
| ATBI | All Taxa Biodiversity Inventory |
| BAS | British Antarctic Survey |
| BGCI | Botanic Gardens Conservation International (NGO based in Kew, Surrey) |
| CAPE 21 | Campaign for Action to Protect the Earth (arising from Agenda 21) |
| CCD | Convention to Combat Desertification (United Nations) |
| CCGT | combined cycle gas turbine (power generation) |
| CFC | chlorofluorocarbon |
| CFP | Common Fisheries Policy (EU) |
| CGCM | coupled global circulation model |
| CGIAR | Consultative Group on International Agricultural Research |
| CIFOR | Centre for International Forestry Research (Bogor, Indonesia) |
| CITES | Convention on International Trade in Endangered Species of Wild Fauna and Flora |
| CoP | Conference of the Parties |
| CPR | chemically perturbed region |
| CSD | Commission on Sustainable Development |
| DoE | Department of the Environment |
| DPCSD | Department of Policy Coordination and Sustainable Development |
| DU | Dobson Unit (United Nations) |
| EBA | Endemic Bird Area |
| EEZ | Exclusive Economic Zone |
| EIA | Environmental Impact Assessment |
| EPA | Environmental Protection Agency (USA) |
| EU | European Union (formerly, EC or European Community) |
| FAO | Food and Agriculture Organisation (United Nations) (HQ in Rome, Italy) |
| FCCC | Framework Convention on Climate Change (United Nations) |
| GATT | General Agreement on Tariffs and Trades (now replaced by WTO) |
| GCM | global circulation model |
| GEF | Global Environment Facility (set up in 1990 and jointly operated by the World Bank, UNDP and UNEP) |
| GWP | global warming potential |
| HCFC | hydrochlorofluorocarbon |
| HFC | hydrofluorocarbon |
| IACSD | Inter-Agency Committee on Sustainable Development |
| IAEA | International Atomic Energy Agency |
| ICBP | International Council for Bird Protection (NGO based in Cambridge, UK) |
| ICDP | Integrated Conservation–Development Project |
| ICRAF | International Centre for Research in Agroforestry (Nairobi) |
| ICSU | International Council of Scientific Unions |
| ILO | International Labour Organisation/Office |
| IMF | International Monetary Fund |
| IMO | International Maritime Organisation |
| INBio | Instituto Nacional de Biodiversidad de Costa Rica |
| IPCC | Intergovernmental Panel on Climate Change |
| IPF | Intergovernmental Panel on Forests |

| | |
|---|---|
| IPGRI | International Plant Genetic Resources Institute (Rome) |
| IPM | integrated pest management |
| ISA | International Seabed Authority |
| ISIS | International Species Inventory System |
| IUCN | The World Conservation Union (formerly the International Union for the Conservation of Nature and Natural Resources – hence IUCN) (based in Gland, Switzerland) |
| IWC | International Whaling Commission |
| JUSCANZ | Japan, USA, Canada, Australia and New Zealand |
| LMO | living modified organism |
| MAB | Man and the Biosphere Programme (UNESCO) |
| NAS | National Academy of Sciences (USA) |
| NASA | National Aeronautics and Space Administration (USA) |
| NGO | non-governmental organisation |
| NMSC | non-melanoma skin cancer |
| ODP | ozone depletion potential |
| OECD | Organisation of Economic Co-operation and Development |
| OPEC | Organisation of Petroleum Exporting Countries |
| PCB | polychlorinated biphenyl |
| PCR | polymerase chain reaction technique |
| PSC | polar stratospheric cloud |
| QBO | quasi-biennial oscillation |
| RAP | Rapid Assessment Programme (created by the Washington-based NGO, Conservation International) |
| RDB | Red Data Book |
| RSPB | Royal Society for the Protection of Birds (NGO based in Sandy, Bedfordshire, UK) |
| SBSTTA | Subsidiary Body on Scientific, Technical and Technological Advice (under Convention on Biodiversity) |
| SCOPE | Scientific Committee on Problems of the Environment (of ICSU) |
| SORG | Stratospheric Ozone Review Group (UK) |
| SSC | Species Survival Commission (of IUCN) |
| TOMS | Total Ozone Mapping Spectrometer |
| TRAFFIC | Trade Records Analysis of Flora and Fauna in Commerce network (NGO based in Cambridge, UK) |
| ULCCs | ultra large crude carriers |
| UNCED | United Nations Conference on Environment and Development, or Earth Summit (Rio de Janeiro, Brazil, June 1992) |
| UNCHE | UN Conference on the Human Environment in Stockholm, 1972 |
| UNCLOS | United Nations Convention on the Law of the Seas (1982) |
| UNCOD | United Nations Conference on Desertification (Nairobi, 1977) |
| UNDP | United Nations Development Programme |
| UNEP | United Nations Environment Programme (HQ in Nairobi, Kenya) |
| UNESCO | United Nations Educational, Scientific and Cultural Organisation |
| UV | ultraviolet |
| VLCCs | very large crude carriers |
| WCED | World Commission on Environment and Development (chaired by Gro Harlem Brundtland; reported in 1983 in *Our Common Future*) |

| | |
|---|---|
| WCMC | World Conservation Monitoring Centre (Cambridge, UK) (managed jointly by IUCN, UNEP and WWF) |
| WHO | World Health Organisation |
| WMO | World Meteorological Organisation |
| WRI | World Resources Institute (based in Washington, DC) |
| WTO | World Trade Organisation |
| WWF | World Wide Fund for Nature (originally the World Wildlife Fund) (HQ in Gland, Switzerland; WWF-UK based in Godalming, Surrey) |

# Acknowledgements

Grateful acknowledgement is made to the following sources for permission to reproduce material in this book:

## Cover

*Front cover, clockwise from top right:* Roy Lawrance; Ron & Valerie Taylor/Ardea; © Greenpeace/Midgley; Charles Mason/Colorific!; © Jeremy Hartley/Oxfam; Simon Fraser/Science Photo Library; Paul M. Smith; Dr David Snashall; © François Gohier/Ardea; *centre:* © Ann Purcell/Colorific!; *back cover:* H.Girardet/The Environmental Picture Library.

## Colour plate section

*Plate 1:* Science Photo Library; *Plate 2:* Reproduced with permission of the Earth Observation Satellite Co. (EOSAT), Lanham, Maryland, USA; *Plates 3 and 4:* Reproduced with permission of The Hadley Centre, The Meteorological Office; *Plate 5:* NASA/Science Photo Library; *Plate 6:* NASA GSFC/Science Photo Library; *Plate 7:* Philippe Plailly/Science Photo Library; *Plate 8:* Primrose Peacock/Holt Studios; *Plate 9:* Brian Rogers/Biofotos; *Plate 10:* © Heather Angel; *Plate 11:* © Michael Depraz/Still Pictures; *Plate 12:* 'Coalbrookdale by Night' by Philip James de Loutherbourg, 1801, National Museum for Science and Technology; *Plate 13:* Simon Fraser/Science Photo Library; *Plate 14:* Paul M. Smith.

## Figures

*Figures 1.2, 1.3 and 1.4:* pp. 14, 15, 17 and 43 from S. Smith (1982) *Discovering the Sea*, Trewin Copplestone Books Ltd, London, a division of Longman Group Ltd, and Sceptre Books, Hodder & Stoughton Ltd; *Figures 1.5, 1.6, 1.7, 1.12, 1.20, 1.22, 1.24 and 1.25:* Figures 2.5, 2.13, 9.6, 9.7a & b, 9.8, 11.15 and 11.17 in P. S. Meadows and J. I. Campbell, *An Introduction to Marine Science*, Blackie & Son Ltd, 1988; *Figure 1.8:* Steele, J. H. (1989) 'The message from the oceans', *Oceanus*, Vol. 32, No. 2, Summer 1989, Woods Hole Oceanographic Institute; *Figure 1.9, 3.6, 3.9, 3.11, 3.20:* Houghton, J. T. (1994) *Global Warming: the complete briefing*, Lion Publishing plc; *Figure 1.11:* 'The main oil transport routes by sea in 1980', in 'Oil pollution control in the East African region', *UNEP Regional Seas Reports and Studies*, No. 10, 1982, © UNEP/IMO 1982; *Figure 1.14:* illustration by Allen Bechel in S. J. Holt, 'The food resources of the ocean', in *Scientific American*, September 1969; *Figure 1.17:* 'Exclusive economic zones', *Leaflet – Focus 68*, The Department of Energy and Marine Technology Directorate Ltd, © 1988 Crown Copyright. Reproduced with the permission of the Controller of Her Majesty's Stationery Office; *Figures 1.18 and 1.19:* C. Sanger (1986), pp. 57, 179, *Ordering the Oceans: the making of the Law of the Sea*, Zed Books and University of Toronto Press; *Figure 1.23:* adapted from Figures 12.4 and 12.5 in G. L. Small, *The Blue Whale*, Columbia University Press, New York, 1971; *Figure 1.26:* adapted from Figures 2.4 and 2.1 in 'Pattern of residual near-surface currents' from *The North Sea Forum Report*, The Council for Environmental Conservation (Co ENCO), 1987. Reproduced with permission of Ciba-Geigy Plc; *Figures 2.9, 2.16(a), 2.17 and 2.21:* Diagrams from SORG (*UK Stratospheric Ozone Review Group*), 1987, 1988, 1990. © Crown Copyright. Reproduced with the permission of the Controller of Her Majesty's Stationery Office; *Figure 2.10:* K. Warr, 'The path to ozone loss', p. 37, *New Scientist*, 27 October 1990, IPC Magazines Ltd, World Press Network 1990; *Figures 2.11:* M. B. McElroy and R. J. Salawitch, *Science*, p. 764, Vol. 243, 10 February 1989, p. 927, 21 November 1986, American Association for the Advancement of Science, © AAAS; *Figure 2.13:* R. P. Wayne, *Chemistry of Atmospheres*, Figure 4.19, p. 159, The Clarendon Press, 1985; *Figure 2.14, 2.16(b), 2.20, 2.22, 2.23, 2.26: Stratospheric Assessment of Ozone Depletion*, Report No. 37, WMO 1994; *Figure 2.18: Stratospheric Ozone 1996*, SORG (*UK Stratospheric Ozone Review Group*), © 1993 Crown Copyright. Reproduced with the permission of the Controller of Her Majesty's Stationery Office; *Figure 2.24:* Häder, Worrest, Kumar and Smith, 1995, 'Effects of increased solar ultraviolet radiation on aquatic ecosystems', *AMBIO: A Journal of the Human Environment*, Vol. 2, No. 3, 3 May 1995, Royal Swedish Academy of Sciences; *Figure 2.25:* Figure 1, p. 326 from Elwood *et al.*, 'Relationship of melanoma and other skin cancer mortality to latitude and ultraviolet radiation in the US and Canada', *International Journal of Epidemiology*, Vol. 3, 1974, by permission of Oxford University Press, 1974; *Figure 2.27:* Brown, L. R., Lenssen, N. and Kare, H. 1995, *Vital Signs 1995–1996: The trends that are shaping our future*, p. 63, Figure 1, Earthscan Publications Ltd; *Figure 3.1:* (top) *The Observer*, 26 November 1989, © The Observer, (*middle right and bottom)* Pearce, F. (1995) 'Hottest year heralds global warming' and 'Global warming "jury" delivers guilty verdict', *New Scientist*, 23/30 December 1995 and 9 December 1995, IPC Magazines, Schoon, *(middle left)* N. (1995) 'Global warming gets cold shoulder', *The Independent*, 27 March 1995; *Figures 3.3 (b) and (c):* R. G. Fleagle and J. A. Businger from *An Introduction to Atmospheric Physics*, © Academic Press, 1963; *Figures 3.8, 3.12 and 3.13:* J. Gribbin, *Hothouse Earth*, Figures 3.3, 5.2 and 5.1, Bantam Press, a division of Transworld Publishers Ltd, 1990/Bantam Books, a division of Bantam Doubleday Dell Publishing Group, Inc., 1990, © John and Mary Gribbin, 1990, and by permission of Murray Pollinger; *Figures 3.10, 3.16, 3.17, 3.18, 3.19, 3.24, 3.25 and 3.26:* Houghton, J. T. *et al.* (1995) *Climate Change 1994: Radiative Forcing of Climate Change*, Cambridge University Press, © 1995 Intergovernmental Panel on Climate Change; *Figure 3.15 and 3.22:* J. T. Houghton, J. T. Jenkins and J. Ephraums, *Climate Change: the IPCC scientific assessment*, Cambridge University Press, 1990. © WMO/UNEP/IPCC, 1990; *Figure 3.23 and 3.27:* adapted from SCOPE 29, *The Greenhouse Effect, Climatic Change, and Ecosystems*, edited by B. Bolin *et al.*, John Wiley & Sons, Chichester, UK, 1986; *Figure 4.1: Systematics Agenda 2000: Charting the Biosphere – Technical Report*, 1994, Systematics Agenda 2000: A Consortium of the American Society of Plant Taxonomists, the Society of Systematic Biologists, and the Willi Hennig Society, in cooperation with the Association of Systematics Collections; *Figure 4.2:* Wilson, E. O. (1992) *The Diversity of Life*, Penguin Books Ltd, by permission of E. O. Wilson and Penguin Books Ltd; *Figure 4.4:* Peters, R. L. and Darling, J. D. (1985) 'The greenhouse effect and nature reserves', *Bioscience*, Vol. 35, No. 11, © 1985 American Institute of Biological Sciences.

### Tables

*Tables 3.1 and 3.2:* Houghton, J. T. *et al.* (1995) *Climate Change 1994: Radiative Forcing of Climate Change*, Cambridge University Press, © 1995 Intergovernmental Panel on Climate Change; *Table 5.1: Partnerships in Practice*, © 1994 Crown Copyright. Reproduced with the permission of the Controller of Her Majesty's Stationery Office.

### Photographs and cartoons

*p. 21:* David Parker/Science Photo Library; *p. 23:* Popperfoto; *p. 29:* Topham Picture Library; *p. 32:* Simon Fraser/Science Photo Library; *p. 39:* Paddy/Clyde Sanger/Zed Press, London, and The University of Toronto Press, Toronto, 1986; *p. 49:* Graham Burns/Environmental Picture Library; *p. 51:* Mary Evans Picture Library; *p. 55:* Rapho/M. Fraudreau/Science Photo Library; *p. 61:* Peter Dunne/The Times; *p. 64: (top left)* Pete Fryer/ Environmental Picture Library, *(top right)* © Greenpeace, *(bottom)* David Parker/Science Photo Library; *p. 70: (left)* National Cavity Insulation Association Ltd, *(right)* Vanessa Miles/The Environmental Picture Library, *(centre)* © Eric Brissaud/Gamma/Frank Spooner Agency; *p. 92:* © Pete Addis; *pp. 94, 95:* © British Antarctic Survey; *p. 95: (left)* NASA/Science Photo Library, *(right)* © British Antarctic Survey; *p. 96:* NASA/Science Photo Library; *p. 100:* Frank Lane Picture Agency/ photo by Alfred Saunders; *p. 110:* James Stevenson/Science Photo Library; *p. 111:* Courtesy Professor Alan Teramura; *p. 138:* Mary Evans Picture Library; *p. 140:* © British Antarctic Survey; *p. 144:* Fred Espanak/Science Photo Library; *p. 145:* © Greenpeace/Midgley; *p. 147: (both)* Adam Hart-Davis/Science Photo Library; *p. 148:* Peter Menzel/Science Photo Library; *p. 151:* Popperfoto; *p. 170:* Barnaby's Picture Library/Oliver; *p. 183:* Mark Edwards/Still Pictures; *p. 205:* Copyright The Universal Press Syndicate/Intercontinental Features Ltd; *p. 211:* Courtesy World Conservation Monitoring Centre; *p. 214:* Sue Cunningham; *p. 215:* The Far Side by Gary Larson is reprinted by permission of Chronicle Features, San Francisco, CA. All rights reserved; *p. 216:* J. Allan Cash Ltd; *p. 218:* courtesy of the Royal Society for the Protection of Birds; *p. 222:* © John Cleare/Mountain Camera; *p. 225: (top)* by permission of Royal Botanic Gardens, Kew, *(bottom)* courtesy of Botanic Gardens Conservation International; *p. 226:* courtesy of The Zoological Society of London; *p. 227:* by permission of the World Wide Fund for Nature; *p. 230:* © Philippe Revelli/Still Pictures; *p. 240:* Furness Museum, Barrow; *p. 241 (both):* Mansell Collection; *p. 242: (left)* Crown Copyright/Trustees of the Victoria and Albert Museum, *(right)* By courtesy of the Board of Trustees of the Victoria and Albert Museum; *p. 243: (both)* Mary Evans Picture Library; *p. 248: (top)* Norwegian Embassy Press and Information Service, *(bottom)* IIDC Campaign Newsletter, No. 14, October 1988/89; *p. 262:* Associated Press; *p. 280:* Copyright The Universal Press Syndicate/Intercontinental Features Ltd; *p. 297:* Jeremy Hartley/Oxfam; *p. 314:* Reproduced by kind permission of the Master, Fellows, and Scholars of Trinity College, Cambridge; *p. 315, 321, 335 and 336:* The Ronald Grant Archive; *p. 318:* Ancient Art and Architecture Collection; *p. 323:* Mansell Collection; *p. 325:* University of London Library.

# *Index*

(Page numbers in *italics* refer to figures, tables and photographs.)

**The Open University U206 *Environment* Course Team**
Mary Bell, Staff Tutor, Cambridge
Roger Blackmore, Staff Tutor, London
Andrew Blowers, Professor of Planning
John Blunden, Reader in Geography
Stuart Brown, Professor of Philosophy
Michael Gillman, Lecturer in Biology
Petr Jehlicka, Lecturer in Environment
Pat Jess, Staff Tutor, Northern Ireland
Alistair Morgan, Lecturer in Educational Technology
Eleanor Morris, Producer, BBC
Suresh Nesaratnam, Lecturer in Environmental Engineering
Alan Reddish, Course Team Chair
Philip Sarre, Senior Lecturer in Geography
Varrie Scott, Course Manager
Jonathan Silvertown, Senior Lecturer in Biology
Sandra Smith, Lecturer in Earth Science
Kiki Warr, Lecturer in Chemistry

**Consultants**
Christine Blackmore, Lecturer in Environment and Development, The Open University
Ian Bowler, Senior Lecturer in Geography, University of Leicester
Colin Clubbe, Co-ordinator of Graduate Studies, Royal Botanic Gardens, Kew
David Elliott, Senior Lecturer in Design, The Open University
Allan Findlay, Professor of Geography, University of Dundee
Patricia Garside, Professor of Environmental Health and Housing, University of Salford
Owen Greene, Lecturer in Peace Studies, University of Bradford
David Grigg, Professor of Geography, University of Sheffield
Cliff Hague, Professor, School of Planning and Housing, Edinburgh College of Art/Heriot-Watt University
David Lowry, Research Fellow in Energy and Environment Research Unit, The Open University
Christopher Miller, Senior Lecturer in Health, University of Salford
David Olivier, Energy Consultant
Alan Prior, School of Planning and Housing, Edinburgh College of Art/Heriot-Watt University
Jeremy Raemaekers, School of Planning and Housing, Edinburgh College of Art/Heriot-Watt University
Marcus Rand, Energy and Environment Research Unit, The Open University
Shelagh Ross, Staff Tutor, The Open University, Cambridge
Ian Simmons, Professor of Geography, University of Durham
Annie Taylor, Faculty of Social Sciences, University of Southampton
Philip Woodhouse, Research Fellow, Faculty of Technology, The Open University
John Wright, Reader in Earth Science, Faculty of Science, The Open University

**External Assessor**
Professor Sir Frederick Holliday, CBE, FRSE, former Vice Chancellor and Warden, University of Durham

**Tutor Panel**
Liz Chidley, Part-time Tutor, The Open University, Cambridge
Judy Clark, Part-time Tutor, The Open University, East Grinstead
Margaret Cruikshank, Part-time Tutor, The Open University, Northern Ireland
Helen Jones, Part-time Tutor, The Open University, Manchester

**Course Production at The Open University**
Melanie Bayley, Editor
David Calderwood, Project Controller
Margaret Charters, Secretary
Lene Connolly, Print Buying Controller
Margaret Dickens, Print Buying Co-ordinator
Harry Dodd, Print Production Controller
Janis Gilbert, Graphic Artist
Rob Lyon, Graphic Designer
Michèle Marsh, Secretary
Janice Robertson, Editor
Paul Smith, Liaison Librarian
Doreen Warwick, Secretary
Kathy Wilson, Production Assistant, BBC